THE
ARCHITECTURE
OF HUMANISM
A Study in the History of Taste

百年经典建筑艺术理论英汉对照读物

人文主义建筑艺术
一项关于审美趣味演变历史的研究

[英国] 杰弗里·斯科特（Geoffrey Scott） 著

吴家琦 译

东南大学出版社
SOUTHEAST UNIVERSITY PRESS
·南京·

图书在版编目（CIP）数据

人文主义建筑艺术：一项关于审美趣味演变历史的研究：英汉对照 /（英）斯科特（Scott, G.）著；吴家琦译 .—南京：东南大学出版社，2015.4
（百年经典建筑艺术理论英汉对照读物）
ISBN 978 - 7 - 5641 - 5404 - 2

Ⅰ.①人… Ⅱ.①斯… ②吴… Ⅲ.①人道主义 - 建筑艺术 - 建筑理论 - 英、汉 Ⅳ.① TU-80

中国版本图书馆 CIP 数据核字（2014）第 303108 号

人文主义建筑艺术：一项关于审美趣味演变历史的研究

出版发行	东南大学出版社
出 版 人	江建中
社　　址	南京市四牌楼 2 号（邮编 210096）
印　　刷	兴化印刷有限责任公司
经　　销	全国各地新华书店
开　　本	700 mm × 1000 mm　1/16
印　　张	17
字　　数	430 千字
版　　次	2015 年 4 月第 1 版　2015 年 4 月第 1 次印刷
印　　数	1—3000 册
书　　号	ISBN 978-7-5641-5404-2
定　　价	49.00 元

* 东大版图书若有印装质量问题，请直接向营销部调换。电话：025-83791830。

艺术也和人生一样,最重要的课题是懂得正确地放弃和牺牲一些东西。

——杰弗里·斯科特

Preface

The scope of this book requires a word of explanation, since from a very simple purpose it has developed to a rather complicated issue. My intention had been to formulate the chief principles of classical design in architecture. I soon realised that in the present state of our thought no theory of art could be made convincing, or even clear, to any one not already persuaded of its truth. There may, at the present time, be a lack of architectural taste: there is, unfortunately, no lack of architectural opinion. Architecture, it is said, must be 'expressive of its purpose' or 'expressive of its true construction' or 'expressive of the materials it employs' or 'expressive of the national life' (whether noble or otherwise) or 'expressive of a noble life' (whether national or not); or expressive of the craftsman's temperament, or the owner's or the architect's, or, on the contrary, 'academic' and studiously indifferent to these factors. It must, we are told, be symmetrical, or it must be picturesque—that is, above all things, unsymmetrical. It must be 'traditional' and 'scholarly,' that is, resembling what has already been done by Greek, Roman, Mediaeval or Georgian architects, or it must be 'original' and 'spontaneous,' that is, it must be at pains to avoid this resemblance; or it must strike some happy compromise between these opposites; and so forth indefinitely.

If these axioms were frankly untrue, they would be easy to dismiss; if they were based on fully reasoned theories, they would be easy, at any rate, to discuss. They are neither. We have few 'fully reasoned' theories, and these, it will be seen, are flagrantly at variance with the facts to be explained. We subsist on a number of architectural habits, on scraps of tradition, on caprices and prejudices, and above all on this mass of more or less specious axioms, of half-truths, unrelated, uncriticised and often contradictory, by means of which there is no building so bad that it cannot with a little ingenuity be justified, or so good that it cannot plausibly be condemned.

Under these circumstances, discussion is almost impossible, and it is natural

前言

关于这本书所涉及的范围,我需要在这里先说几句话,解释一下。这本书是从当初一个很简单的想法开始,结果是一发不可收拾,演变成一个相当复杂的课题。我最早的想法就是打算把古典建筑的设计问题,归纳总结出几条概括性的原则,但是很快便发现,在目前的思想状态下,现有的任何一种艺术理论,都无法让那些对这些理论原则毫无思想准备的人信服,甚至能把我自己的理论说清楚恐怕都成问题,除非读者在此之前已经认同并采取了同样的立场。在今天,我们的社会和我们的民众,或许缺乏关于建筑艺术的审美趣味,但是不幸的是,关于建筑艺术问题,目前绝对不缺乏任何自以为是的武断看法。有人认为,建筑艺术必须要"表现出它的使用目的";有的说,建筑艺术必须要"表现出它的真实建造结构";有的说要必须"表现出所使用的建筑材料";还有的说"要表现出国家民族的生活(无论这种生活是否崇高)";也有人说"要表现出崇高的生活(无论这种崇高的生活与国家民族有无关系)";有人说建筑要表现出工匠艺人的秉性和气质,也有人认为需要表现的是投资赞助人或者建筑师的个性;当然也有人持相反的意见,认为建筑应该表现出建筑艺术的纯粹"学术性",不应该受到上面那些因素的干扰。有人觉得建筑的构图必须是对称的,也有人觉得建筑的构图应该是很符合绘画构图原则的那种造型,亦即不对称的构图。有人坚信建筑必须是具有"传统"建筑的特征,必须显现出深厚的"学养和功力",换句话说,就是建筑物必须看上去与过去在希腊时期、罗马时期、中世纪、甚至是英国乔治时期出现过的那些建筑有明显形似的地方;也有人则坚持认为,建筑必须是"具有原创性"的艺术创作,必须"具有水到渠成般的自然",也就是说设计人要绞尽脑汁地避免任何与前人相似的地方;又或者从刚刚说到的两个极端中间找到一个皆大欢喜的折中方案等等,不一而足。

如果说上面这些为很多人所接受的观念是完全错误的,那么,我们反驳起来也不会是什么难事;如果说这些言论都是经过严格推理得出的正确结论,那么我们在深入讨论它们的时候,也应该是很容易的。但是,这两种情况都不是。我们很少有机会看到"经过充分的理性分析、推导"所得出来的理论。看到的各种理论都不加掩饰地与需要解释的事实相背离。我们生活中所接触到的,都是建筑艺术上流传下来的一些习俗、零零碎碎的传统片断、各种小把戏和花样、先入为主的偏见,当然最主要的就是上面提到过的那些人云亦云所形成的公理。这些公理都是一些真假混杂的东西,也与我们所关心的问题没有什么直接关系,也没有经过缜密的思考和甄别,而且常常是彼此矛盾的。如果根据这些公理来对建筑艺术进行评判,那么,我们根本找不出一座建筑物,说它是那样的糟糕,以至那上面的任何一个巧妙细节都无法获得合理化的解释,同时我们也找不出一个建筑,说它是那样的优秀,以至那上面的细节让人似乎根本无从加以指责。

在这样一种情形下,正常的讨论无法进行下去;这样产生的批评也自然而然地就变为陈

that criticism should become dogmatic. Yet dogmatic criticism is barren, and the history of architecture, robbed of any standard of value, is barren also.

It appears to me that if we desire any clearness in this matter, we are driven from a *priori* aesthetics to the history of taste, and from the history of taste to the history of ideas. It is, I believe, from a failure to appreciate the true relation of taste to ideas, and the influence which each has exerted on the other, that our present confusion has resulted.

I have attempted, consequently, in the very narrow field with which this book is concerned, to trace the natural history of our opinions, to discover how far upon their own premisses they are true or false, and to explain why, when false, they have yet remained plausible, powerful, and, to many minds, convincing.

This is to travel far from the original question. Yet I believe the inquiry to be essential, and I have sought to keep it within the rigorous limit of a single argument. On these points the reader will decide.

So far as this study is concerned with the culture of the Italian Renaissance, I am indebted, as every student must always be indebted, primarily to Burckhardt. I have profited also by Wöfflin's *Renaissance und Barock*. To the friendship of Mr. Bernhard Berenson I owe a stimulus and encouragement which those who share it will alone appreciate. Mr. Francis Jekyll of the British Museum has kindly corrected my proofs.

5 VIA DELLE TERME,
FLORENCE,
February 14, 1914.

词滥调的教条。教条化的批评是一片不毛之地,没有什么意义,而建筑艺术的历史在缺乏价值标准的情况下,也就跟着变得一样贫瘠了。

在我看来,如果我们想在这个问题上得到任何比较清晰的概念,那么我们就必须让自己远离纯思辨美学中的那些假设和前提,把关注的重点转移到人们审美兴趣的发展过程上来,在审美兴趣的发展过程中找出审美理念的发展过程。我相信,正是由于我们不能正确地判断人们的审美兴趣与审美理念之间的关系,不能正确了解二者之间存在的相互影响,因此才会有我们今天的各种思想的混乱。

从这个想法出发,我努力把这本书讨论的主题限定在一个比较狭窄的范围里,追踪一下在欣赏建筑艺术时我们个人审美兴趣喜好的发展脉络,看看这些观点本身有哪些是正确的,哪些是错误的。如果是错误的,我就努力找出原因,说明为什么这些错误的观念仍然给人一种好像很有道理的样子,仍然很有影响力,而且为什么仍然会有不少人对它深信不疑。

这个目标距离我最初的简单问题已经相当远了,但是我觉得这种探究还是很有必要的,我也努力把这种探究严格地限定在一个简单的论述范围之内。读者可以在这些方面作出判断。

这本书所关注的主要内容是意大利文艺复兴时期的文化,如同这个领域里的每一位深受柏克哈特(Burckhardt)影响的学生一样,我也是深受其惠,同时我也从沃尔夫林(Wölfflin)教授的著作《文艺复兴与巴洛克》(*Renaissance und Barock*)中受益匪浅。与柏哈德·贝伦森先生(Mr. Bernhard Berenson)的友谊让我得到资助和鼓励。凡是得到过他的鼓励的人们都会赞同我的说法,对这些资助与鼓励仍然心存感激。大英博物馆的弗朗西斯·泽凯尔先生(Mr. Francis Jekyll)热情地校阅了我的手稿。

一九一四年二月十四日
于佛罗伦萨

Preface to the Second Edition

An epilogue at the end of the volume contains what I have wished to add in this edition.

There are a few changes in the text; but these do not affect the argument of the book.

G.S.

March 1924

第二版前言

在本书第二版的最后,我添加了一段后记,用来说明一下我希望借此机会所增添的一些内容。

在原书的行文上,我对个别的字句做了一些更动,但是这些更动对本书原来的论点与论据均没有任何影响。

杰弗里·斯科特
一九二四年三月

CONTENTS

	Preface	II
	Preface to the Second Edition	VI
INTRODUCTION	The Architecture of Humanism	002
ONE	Renaissance Architecture	016
TWO	The Romantic Fallacy	038
THREE	The Romantic Fallacy (*continued*): Naturalism and the Picturesque	066
FOUR	The Mechanical Fallacy	094
FIVE	The Ethical Fallacy	120
SIX	The Biological Fallacy	164
SEVEN	The Academic Tradition	186
EIGHT	Humanist Values	210
NINE	Conclusion	240
	Epilogue, 1924	248

目录

前言 III

第二版前言 VII

导言　人文主义的建筑艺术 003

第一章　文艺复兴时期的建筑艺术 017

第二章　浪漫主义艺术理论的谬论 039

第三章　浪漫主义艺术理论的谬论（续）：崇尚自然与追求绘画构图式的造型 067

第四章　建筑技术决定论的谬论 095

第五章　道德决定论的谬论 121

第六章　基于生物进化论的艺术理论之谬误 165

第七章　讲究学术理论的传统 187

第八章　人文主义的价值 211

第九章　结论 241

后记，1924年 249

译者的话 258

建筑物是一面能够反映出我们自己的镜子。

——杰弗里·斯科特

INTRODUCTION
The Architecture of Humanism

'**Well**-building hath three conditions: Commodity, Firmness, and Delight.' From this phrase of an English humanist[1] a theory of architecture might take its start. Architecture is a focus where three separate purposes have converged. They are blended in a single method; they are fulfilled in a single result; yet in their own nature they are distinguished from each other by a deep and permanent disparity. The criticism of architecture has been confused in its process; it has built up strangely diverse theories of the art, and the verdicts it has pronounced have been contradictory in the extreme. Of the causes which have contributed to its failure, this is the chief: that it has sought to force on architecture an unreal unity of aim. 'Commodity, firmness, and delight'; between these three values the criticism of architecture has insecurely wavered, not always distinguishing very clearly between them, seldom attempting any statement of the relation they bear to one another, never pursuing to their conclusion the consequences which they involve. It has leaned now this way and now that, and struck, between these incommensurable virtues, at different points, its arbitrary balance.

Architecture, the most complex of the arts, offers to its critics many paths of approach, and as many opportunities for avoiding their goal. At the outset of a fresh study in this field, it is well, at the risk of pedantry, to define where these paths lead.

Architecture requires 'firmness.' By this necessity it stands related to science, and to the standards of science. The mechanical bondage of construction has closely circumscribed its growth. Thrust and balance, pressure and its support, are at the root of the language which architecture employs. The inherent characters of marble, brick, wood and iron have moulded its forms, set limits to its achievement, and governed, in a measure, even its decorative detail. On every hand the study of architecture encounters physics, statics, and dynamics, suggesting, controlling, justifying its design. It is open to us, therefore, to look in buildings for the logical expression of material properties and material laws. Without these, architecture is impossible, its history unintelligible. And if, finding these everywhere paramount, we seek, in terms of material properties and material laws, not merely to account

---

[1] Sir Henry Wotton, *Elements of Architecture*. He is adapting Vitruvius, Bk. I. chap. iii.

导言
人文主义的建筑艺术

"**好**的建筑有三个条件：实用、坚固和令人愉悦。"从一位英国人文主义学者[1]的这句话出发，我们可以建立起一套建筑理论。建筑艺术是三个不同的目的汇集到一起之后所形成的焦点。三者以一种单一的方法混合在一起；通过一个单一的结果而同时实现。但是，这三个目的中的每一个，仍然保持有自己根深蒂固的特点，彼此各不相同，其对立的性质根本无法融和。建筑艺术理论就是在这个建筑物的生成过程中迷失了自己，形成了奇怪的截然不同的理论，且各种理论所得出的结论彼此矛盾，有的甚至是无法调和的对立。导致这些问题的原因是多方面的，但是其中最主要的一点就是：建筑理论试图把三个彼此冲突的目标很不现实地强行扭在一起，试图形成一个单一的统一目标。在"实用、坚固和美观"这三种不同价值之间，建筑理论很不自信地摇摆不定；在很多情况下，理论家并不能清醒地了解自己在其中的左右摇摆，也不能很明确地在自己的理论中对三者加以区别。当前的理论很少试图说明其中的某一个方面与另外两个方面的关系，更不会去深入探究每一个方面各自发挥作用时会导致的结果。这些理论一会儿偏向其中的一个要素，一会儿又偏向另一个，在这三种有着不同衡量标准的元素之间，很武断地选择其中某一种状态，从而取得暂时的平衡。

建筑艺术是所有艺术形式当中最为复杂的一种，它为建筑艺术理论的研究提供了众多的途径与方式，同时也存在着同样多的歧途与陷阱。在进入到这个领域开始我们全新的探索之前，还是让我们先看看这些不同途径都会把我们引导到什么地方吧。虽然这样做似乎有点迂腐及卖弄学问之嫌。

建筑要求"坚固"，这是不可或缺的。这项要求把建筑与科学联系起来，也让建筑必须要合乎科学的标准。建筑施工中涉及的机械、力学等具体操作手段，直接限制了建筑艺术的发展与成长。作用在建筑上的张力与对于张力的抵消，压力与承受压力的支撑物等，它们都是建筑艺术语言的根本来源。石头、砖瓦、木头、铁件的固有性质让使用这些材料的建筑艺术，逐渐发展出自己的形式，同时也限制了建筑艺术的发展范围。在某种程度上，这些建筑材料的使用方式也决定了建筑装饰艺术的细节。从各个方面来讲，建筑艺术的研究都会遇到物理学、静力学和动力学等问题，这些问题的研究给了建筑设计一些启示，给了一些必须遵循的规则，同时也证明了某种设计的合理性。因此，从建筑物本身寻找出建筑材料的性质与物理规律的逻辑表现形式就成为我们自己的任务。离开了建筑材料的物理性质与规律，建筑艺术就无从谈起，建筑历史也就成了无法理解的东西。了解到这些问题的重要意义，如果这

---

[1] 亨利·华顿爵士.《建筑艺术的基本要素》。他是根据维特鲁威的著作第一卷第三章的内容作出自己这个说法的。

for the history of architecture, but to assess its value, then architecture will be judged by the exactness and sincerity with which it expresses constructive facts, and conforms to constructive laws. That will be the scientific standard for architecture: a logical standard so far as architecture is related to science, and no further.

But architecture requires 'commodity.' It is not enough that it should possess its own internal coherence, its abstract logic of construction. It has come into existence to satisfy an external need. That, also, is a fact of its history. Architecture is subservient to the general uses of mankind. And, immediately, politics and society, religion and liturgy, the large movements of races and their common occupations, become factors in the study. These determine what shall be built, and, up to a point, in what way. The history of civilisation thus leaves in architecture its truest, because it's most unconscious record. If, then, it is legitimate to consider architecture as an expression of mechanical laws, it is legitimate, no less, to see in it an expression of human life. This furnishes a standard of value totally distinct from the scientific. Buildings may be judged by the success with which they supply the practical ends they are designed to meet. Or, by a natural extension, we may judge them by the value of those ends themselves; that is to say, by the external purposes which they reflect. These, indeed, are two very different questions. The last makes a moral reference which the first avoids, but both spring, and spring inevitably, from the link which architecture has with life—from that 'condition of well-building' which Wotton calls commodity.

And architecture requires 'delight.' For this reason, interwoven with practical ends and their mechanical solutions, we may trace in architecture a third and different factor—the disinterested desire for beauty. This desire does not, it is true, culminate here in a purely aesthetic *result*, for it has to deal with a concrete basis which is utilitarian. It is, none the less, a purely aesthetic *impulse*, an impulse distinct from all the others which architecture may simultaneously satisfy, an impulse by virtue of which architecture becomes art. It is a separate instinct. Sometimes it will borrow a suggestion from the laws of firmness or commodity; sometimes it will run counter to them, or be offended by the forms they would dictate. It has its own standard, and claims its own authority. It is possible, therefore, to ask how far, and how successfully, in any architectural style, this aesthetic impulse has been embodied; how far, that is to say, the instincts which, in the other arts, exert an obvious and unhampered activity, have succeeded in realising themselves also through this more complicated and more restricted instrument. And we can ask, still further, whether there may not be aesthetic instincts, for which this instrument, restricted as it is, may furnish the sole and peculiar expression. This is

INTRODUCTION

时我们按照建筑材料的性质以及物理规律来寻找它们对于建筑历史所起到的作用,同时也评估一下这些材料性质与物理规律的价值,那么,建筑艺术的评估与判断就有了它的精确性与真实性,这一点与建筑在建造过程中的方式方法是一致的,同时这种评估与判断也符合建造过程中的物理规律。这便是建筑艺术的科学标准:它是把建筑艺术与科学联系到一起的一种符合逻辑的标准。了解到这一点,对于我们来说已经足够了。

建筑同时也要求自己是很"实用"的。建筑物虽然具备了自身内在的一致性,满足了科学的建造逻辑规律,但这都还不够。建筑之所以会产生,那是因为它必须要满足某种外在的需求。这也是建筑历史中的一项重要事实。建筑是服务于人类的一般需要的。与此密切相关的,便是从社会生活需求中产生的政治与社会、宗教与礼仪、人类的大规模活动与日常的职业等等,这些都成为建筑艺术的研究内容。正是这些内容决定了人类将会建造什么样的建筑,也在一定程度上决定了这些建筑物应该如何建造。人类的文明历史在建筑上留下了最真实的记录,因为这些记录都会不假思索地留下最客观的痕迹。如果说我们可以认为建筑艺术能够表现机械力学方面的规律,那么我们就同样也可以认定,建筑艺术也可以表现人类的生活。这一点所要求的价值判断标准与前面我们提到的科学方面的标准完全不同。建筑可以根据当初在设计的时候所希望满足的实际要求来进行衡量,看看它是否达到这些目的。或者,由此引申一下,看看当初设定的这些目标都是什么,然后根据这些目标的价值来做一个判断;这都是在采用建筑自身之外的目的作为检验建筑艺术的尺度,来衡量这些建筑是否达到最初的要求。到目前为止,我们在这里所说过的两个问题,实际是两个根本不同的问题,后面一个涉及道德的问题而前面一个则没有,但是二者都来自同一个根源,那就是建筑服务于人的生活,而且只能是来自于这个根源,也就是沃顿爵士所说的"好的建筑必须具备的三个条件"中,属于"实用"的那一个条件。

建筑也要求"令人愉悦"。这一要求实际上是已经被交织在实用和坚固这两条里面的了,它让我们追踪到第三个要素,是与其他要素不同的另外一个要素,也就是抛开任何功利考虑而单纯追求美观的意愿。需要强调一点,这里追求美观绝对不会是一个纯粹的唯美主义的*结果*,因为建筑必要要与实用等很具体的功能需求发生互动。不可能脱离实际而成为纯思辨活动。尽管这个结果不是纯粹的唯美主义的产物,但是它的初始追求却是一种唯美主义的*冲动*,这种追求美的冲动,与建筑活动中其他那些也必须同时满足的各种要求是很不一样的。就是因为有了对于美的追求这样一种冲动的存在,建筑才成为艺术的。这是一种与其他技能完全不同的才能,也算是人类的一种本能。这种能力有时候会从与坚固有关的力学中获得启发,有时会从实用中得到灵感,但是有时候也会与它们发生矛盾,抑或它也会对由坚固和实用两项要素所决定的建筑形式产生不满。它有自己的标准,它也想施展出自己的权威。这种出自唯美主义美学的*初始冲动*在其他艺术领域里是不受任何阻碍限制的,可以在艺术创作中自由地发挥自己的作用,但是在建筑艺术中,无论是哪一种建筑风格,这种美学的追求会受到建筑物这种更为复杂媒介的各种制约,因此,我们有理由会问,在建筑艺术中,这种美学的初始冲动能够实现到什么程度?或者更引申一步,我们可以追问:是否根本就不需要那个美学的追求与能力,单凭建筑本身的制约就能从中产生出属于建筑自己独特且唯一的表现形式呢?从最严格的意义上来讲,这就要求把建筑当作是一种艺术来

to study architecture, in the strict sense, as an art.

Here, then, are three 'conditions of well-building,' and corresponding to them three modes of criticism, and three provinces of thought.

Now what, in fact, is the result? The material data of our study we certainly possess in abundance: the statistics of architecture, the history of existing works, their shape and size and authorship, have long been investigated with the highest scholarship. But when we ask to be given not history but criticism, when we seek to know what is the value of these works of art, viewed in themselves or by comparison with one another, and why they are to be considered worthy of this exact attention, and whether one is to be considered more deserving of it than another, and on what grounds, the answers we obtain may be ready and numerous, but they are certainly neither consistent nor clear.

The criticism of architecture has been of two kinds. The first of these remains essentially historical. It is content to describe the conditions under which the styles of the past arose. It accepts the confused and partly fortuitous phenomenon which architecture actually is, and estimates the phenomenon by a method as confused and fortuitous as itself. It passes in and out of the three provinces of thought, and relates its subject now to science, now to art, and now to life. It treats of these upon a single plane, judging one building by standards of constructive skill, another by standards of rhythm and proportion, and a third by standards of practical use or by the moral impulse of its builders. This medley of elements, diverse and incommensurated as they are, can furnish no general estimate or true comparison of style.

Doubtless, *as a matter of history*, architecture has not come into existence in obedience to any *a priori* aesthetic. It has grown up around the practical needs of the race, and in satisfying these it has been deflected, now by the obstinate claims of mechanical laws, now by a wayward search for beauty. But the problem of the architect and that of the critic are here essentially different. The work of the architect is synthetic. He must take into simultaneous account our three 'conditions of well-building,' and find some compromise which keeps a decent peace between their claims. The task of the critic, on the contrary, is one of analysis. He has to discover, define, and maintain the ideal standards of value in each province. Thus the three standards of architecture, united in practice, are separable, and must be separated, in thought. Criticism of the historical type fails to apply an ideal and consistent analysis, for the insufficient reason that the *practice* of architecture has, of necessity, been neither consistent nor ideal. Such criticism is not necessarily misleading. Its fault is more often that it leads nowhere. Its judgments may be

INTRODUCTION

加以研究。

到这里,我们看到了"一座好的建筑应当具备的三个条件"是什么,与这三个条件相对应的自然就产生了艺术理论的三种不同模式,这也说明围绕着建筑艺术,它的理论也具有三种不同的思想领域。

那么,建筑艺术的理论到今天具有什么样的结果呢?关于建筑艺术的研究材料和数据当然是十分丰富:建筑的基本统计资料,建筑作品的历史背景,建筑的造型和尺寸以及设计者等等,这些工作很久以来一直在不断地积累,而且都是最高水平的学术研究成果。但是,当我们询问关于这些建筑艺术内容的评价,而不是单纯地叙述建筑作品的历史事实的时候,我们想要了解的是这些作品的价值是什么,包括这些作品的自身价值,或者彼此之间的相对价值是什么,或者希望了解为什么某些建筑作品让人们觉得比其他的更有价值,更值得学习,以及为什么(或者根据什么)我们做出这样的判断。目前我们就能够得到现成的答案,而且还不止一个。但是,这些答案前后既不一致又不清楚。

长期以来大致有两种关于建筑艺术的理论。第一种理论从本质上讲,只是陈述历史事实。这种评论满足于叙述发生在一件建筑艺术作品周围的情况,说明过去的那些建筑艺术风格是在什么样的情况下才得以产生的。建筑成为一种混乱,同时又充满了各种偶然现象的一种艺术形式。在评论这种艺术形式时,这一混乱的特点则被全盘接收下来,让理论本身也同样地缺乏严密的逻辑性,同时又充满了很多偶然性。论述的过程是在三种不同的思想领域里穿来穿去,一会儿把论述的主题靠近科学的方面,一会儿又靠近艺术,再过一会儿又与生活密切地联系起来。这种论述的方法是把三种不同的因素摆放在同一个平面之上,有时用建造技艺来衡量建筑,有时又用韵律和比例关系来衡量,要不然就以使用方面的实用性或者建造这座建筑时所涉及的道德水平问题为标准衡量。这种方法因为各种元素及评判标准的混杂、缺乏一致性,使得它很难形成一种普遍适用的概括结论,也无法在不同风格之间进行有意义的比较。

有一个与建筑相关的*历史事实*是不容置疑的:建筑艺术从来都不是按照某*一种抽象的艺术前提条件*演变而成的。建筑艺术都是从满足人类的某种具体实际需求中产生出来的。在满足这些实际需求的时候,建筑艺术也会受到一定的局限,有些是来自工程技术方面的制约,有些则是受到某种艺术风格趣味对于建筑艺术的左右。但是我要强调的是,从根本上讲,建筑师与理论家所面对的问题有着根本的不同。建筑师的工作属于综合性质的,他必须把三种"好建筑的条件"同时加以考虑,努力提出一种解决办法,让三种不同的要求在同一个时刻得到适当的满足,在三者之间取得一种妥协与平衡。但是对于一位理论家来说,他的工作则恰恰相反,是属于分析性质的。理论家必须让三种不同的要求,在各自的势力范围内被挖掘出来、摊开来,让它们得到确认,并且在整个建筑艺术中贯彻这些原则。因此,建筑艺术的三个方面的三种不同原则在建筑实践中是结合成为一体的,但在理论评论中则是分离的,也必须在清晰的思维中保持住这种分离。那种按照历史演变过程来评判一个建筑的艺术则正是缺乏这种最理想的准则,缺乏始终一贯的分析,因为建筑实践本身就是这样缺乏某种理想化状态,也缺乏一贯性。这类建筑理论一般说来是不会把我们带入歧途的,但它的根本问题在于,它并不会给我们指出任何一个方向。这种评论的个别结论可

individually accurate, but it affords us no general view, for it adopts no fixed position. It is neither simple, nor comprehensive, nor consistent. It cannot, therefore, furnish a theory of style.

    The second type of criticism is more dangerous. For the sake of simplicity it lays down some 'law' of architectural taste. Good design in architecture, it will say, should 'express the uses the building is intended to serve'; 'it should faithfully state the facts of its construction,' or again it should 'reflect the life of a noble civilisation.' Then, having made these plausible assumptions, it drives its theory to a conclusion, dwells on the examples that support its case, and is willing, for the sake of consistency, to condemn all architecture in which the theory is not confirmed. Such general anathemas are flattering alike to the author and his reader. They greatly simplify the subject. They have a show of logic. But they fail to explain why the styles of architecture which they find it necessary to condemn have in fact been created and admired. Fashion consequently betrays these faultless arguments; for whatever has once genuinely pleased is likely to be again found pleasing; art and the enjoyment of art continue in the condemned paths undismayed; and criticism is left to discover a sanction for them, if it can, in some new theory, as simple, as logical, and as insufficient as the first.

    The true task of criticism is to understand such aesthetic pleasures as have in fact been felt, and then to draw whatever laws and conclusions it may from that understanding. But no amount of reasoning will create, or can annul, an aesthetic experience; for the aim of the arts has not been logic, but delight. The theory of architecture, then, requires logic; but it requires, not less, an independent sense of beauty. Nature, unfortunately, would seem to unite these qualities with extreme reluctance.

    Obviously, there is room for confusion. The 'condition of delight' in architecture—its value as an art—may conceivably be found to *consist in* its firmness, or in its commodity, or in both; or it may consist in something else different from, yet dependent upon these; or it may be independent of them altogether. In any case, these elements are, at first sight, distinct. There is no reason, *prima facie*, to suppose that there exists between them a pre-established harmony, and that in consequence a perfect principle of building can be laid down which should, in full measure, satisfy them all. And, in the absence of such a principle, it is quite arbitrary to pronounce dogmatically on the concessions which art should make to science or utility. Unless it can be proved that these apparently different values are in reality commensurable, there ought to be three separate schemes of criticism:

能都很准确,但是它却不能给我们提供一个宏观上的评估和判断,因为这种评论本身是从不断变换着的角度和视点来看问题的,缺乏固定的立场。这种理论虽不粗浅、简单,但也不全面深刻。它又常常变幻莫测,让人很难抓住。因此,这种理论基本上不能归纳总结出某种建筑风格的理论。

建筑理论中的第二种思想方法则是很危险的方法。为了让叙述问题变得简单起见,这种理论从一开始就设定了自己对于建筑艺术进行品鉴时将会采用的"某些规则"。这些理论家们会提出这样或那样的规定,如"好的建筑要表现出建筑想要满足的功能";"好的建筑应该真实地表现出建造技术中的实际情形";或者,好的建筑"应该反映出一种高贵文明的生活"。在规定了这些定义之后,这些理论家们便开始从这些假设前提出发,推导出自己理论中的某些结论。这种理论都是建立在那些与理论家的结论相一致的具体案例之上的。为了保持这种理论的一贯性,对于与自己的理论不一致的那些建筑则是大加挞伐。对于理论家和读者两方面来说,这种对于异己的挞伐会带给他们一种很受用的快乐感觉,问题也因此变得简单起来,更显示出逻辑感。但是,这种理论却没有说明那些被他们痛加斥责、挞伐的建筑风格当初为什么会被创造出来,为什么会受到那么多人的喜爱。时尚潮流的改变到后来便会出卖这种完美的理论,因为曾经被人喜爱过的东西迟早还会再次受到人们的喜爱。无论有人怎样地诅咒、谩骂某些艺术形式,丝毫都不会影响他人对它们的喜爱,丝毫不影响他人对于这些艺术的欣赏和享受。倒是这些理论家们,他们必须要绞尽脑汁地再次更新自己的理论,再次寻找惩罚这些艺术形式的新理论,最好是如同上一版的理论一样,那样简单且容易被人接受。最好还是那样具有强烈的逻辑性。当然新的理论还是同以前的理论一样,根本不能充分地说明问题的本质。

建筑理论的真正任务就是把人们在审美体验中真实感受到的那种愉悦心情上升到理论高度,并且从中归纳总结出一些规律和结论。但是,理论分析是不会产生出美感体验的,当然也不会消除这种体验。因为艺术的存在目的从来就不是逻辑性的东西,而是为了让人从艺术作品中获得愉悦的享受,所以,建筑艺术的理论既要求具有逻辑分析,同时也要求具有完全独立于逻辑分析的审美感受,建筑艺术理论对于后者的要求一点儿也不比对于前者的少。但是,造化似乎是最不愿意看到这二者结合到一起的情形发生。

显然,这里有很多容易让人产生疑惑的地方。建筑艺术中的"愉悦成分"是使得建筑成为艺术的价值所在,它可能是由建筑中的坚固部分衍生出来的,也可能是从建筑的实用性中衍生出来的,也可能二者兼而有之;也可能是从其他地方或者其他元素中衍生出来的,这些元素有可能与坚固和实用之间存在着某种关系;也可能与上面这一切均毫无关系。无论是哪一种情况,建筑中的那些美学元素给人的第一眼印象便是很容易被感受得到的。但是,我们绝不能就此假设在这些美学元素之间存在着一种先验的和谐规律,认为我们可以从中归纳出完美的建筑原则,而这些原则又是可以完全满足那种先验的和谐规律的。眼下,我们不可能得到这样完美的建筑原则,于是,有人便教条地宣布,在某种情况下艺术应该在科学技术和实用功能面前做出让步。除非我们能够证明这三个不同方面的价值观与价值判断之间的差别只是表面的,它们在本质上是有着某种等价关系的,否则的话,我们就应该按照三种不同的思路和标准来展开我们的建筑评论:第一种是根据建造技术来判断,第二种是根

the first based on construction, the second on convenience, the third on aesthetics. Each could be rational, complete, and, within its own province, valid. Thus by degrees might be obtained what at present is certainly lacking—the data for a theory of architecture which should not be contradicted at once by the history of taste.

The present study seeks to explain one chapter of that history. It deals with a limited period of architecture, from a single point of view.

The period is one which presents a certain obvious unity. It extends from the revival of classical forms at the hands of Brunelleschi, in the fifteenth century, to the rise of the Gothic movement, by which, four hundred years later, they were eclipsed. The old mediaevalism, and the new, mark the boundaries of our subject. At no point in the four centuries which intervened does any line of cleavage occur as distinct as those which sever the history of architecture at these two points. And between them there is no true halting-place. Thus the term 'Renaissance architecture,' which originally denoted no more than the earlier stages, has gradually and inevitably come to be extended to the work of all this period.

It is true that during these years many phases of architectural style, opposed in aim and contradictory in feeling, successively arose; but the language in which they disputed was one language, the dialects they employed were all akin; and at no moment can we say that what follows is not linked to what went before by common reference to a great tradition, by a general participation in a single complex of ideas. And incompatible as these several phases—the primitive, classic, baroque, academic, rococo—may at their climax appear to be, yet, for the most part, they grew from one another by gradual transitions. The margins which divide them are curiously difficult to define. They form, in fact, a complete chapter in architecture, to be read consecutively and as a whole. But at the two moments with which our study begins and ends, the sequence of architecture is radically cleft. The building of the Pazzi Chapel in Florence marks a clear break with the mediaeval past, and with it rises a tradition which was never fundamentally deserted, until in the nineteenth century traditionalism itself was cast aside.

It is in Italy, where Renaissance architecture was native, that we shall follow this tradition. The architecture of France in the seventeenth and eighteenth centuries and, in a lesser degree, that of the Georgian period in England, might furnish brilliant examples of the same manner of building. The Italian experiment enabled the architects of France, amid their more favourable environment, to create a succession of styles, in some ways more splendid, and certainly more exquisite and complete. Yet, if we wish to watch architectural energy where it is most concentrated, most vigorous, and most original it is to Italy that we must turn.

INTRODUCTION

据使用的便利性,第三种是根据美学标准。这三种途径中的每一种都是理性的、完整的、在各自的范围内也是有足够说服力的。因此,从这种工作方法中是有可能找到我们目前所需要的东西的,亦即组成建筑理论的基本资料与数据,至少在某种程度上应该如此。这样建立起来的建筑理论与我们所了解到的民众对于建筑的审美趣味应该是一致的,就是说这种理论与客观观察到的事实是不矛盾的。

我们这本书所关注的范围是建筑历史中的一个简短的章节而已。它只是从一个单一的视角来观察发生在一个非常有限的时间段里的建筑艺术。

我们所说的这个有限时间段里的建筑,它们具有一种很强烈的共同特征,看起来非常明显地一致。从十五世纪布鲁乃列斯基(Brunelleschi)第一次再现古典形式的那一刻开始,直到后来哥特建筑风格的再次流行为止,前后经历了四百年的时间。古老的中世纪建筑风格和后来假古董式的哥特复兴,标志着我们所关注的这个时间段的两端。这四百年期间,在建筑风格上没有出现过明显的断裂,没有出现过任何如同在它两端那样明显不同类别的建筑语言。正因为如此,虽然"文艺复兴式的建筑"一词最早出现的时候是特指布鲁乃列斯基那个时期的建筑,但是后来逐渐演变成现在的情况,用这个词汇来概括整个四百年的全部建筑作品。

在这四百年里,建筑的确经历了许多不同的阶段,不同的艺术风格接二连三地出现,它们的目标不同,带给人们的感受也是相互矛盾的,这一点是毫无疑问的。但是,这些建筑风格中所使用的建筑语言,从根本上讲只有一种,每个阶段所使用的方言都是同源的。可以这样说,出现稍晚一些的建筑风格没有一个不是像在它之前的那样,都是参考了过去同一个伟大传统,所有的人都在积极地阐述着同一个建筑观念。这些不同的阶段在各自达到高峰的时候,看上去的确彼此很不相同,比如古典建筑初始阶段的那种原始朴拙、经典时期的典雅、巴洛克的热情、学院派建筑、洛可可风格,等等。但是,在绝大多数情况下,这些不同风格都是从前面一个逐渐向下一个演变过渡的,它们之间的界线很难划得清。因此,我们可以说,这些风格组成了建筑历史中的一个完整的章节,这些内容应当作为一个整体来阅读、理解。但是,在这四百年之前和之后,建筑风格则非常明显地出现断裂。在佛罗伦萨出现的巴齐礼拜堂(Pazzi Chapel)清楚地标志着中世纪的建筑思想即将成为过去,随之而来的是一个崭新的时代,而这个新时代在本质上一直持续到十九世纪,从来没有中断过。到了十九世纪,一切以传统形式出现的建筑艺术都逐渐衰落,这种古典主义建筑风格也逃脱不了同样的命运,最后衰落而成为了历史。

当我们研究这段建筑艺术历史的时候,我们把目光主要集中在意大利,因为那里文艺复兴时期出现的建筑,实际上又是当地土生土长的民间传统建筑。十七世纪和十八世纪的法国建筑中,也产生过属于这一传统风格的很多辉煌精彩的作品。英国乔治时代的建筑虽然与法国的建筑比较起来,显得略微差一点,但也出现过一些很精彩的建筑作品。法国的建筑师们,在意大利人进行过的各种尝试基础之上,结合了自己有利的自然环境,创造出一个接一个的不同风格样式,其中不乏更为辉煌宏伟的作品,有些建筑甚至要比意大利的建筑更为精湛细腻,也比意大利的建筑显得更加完整。但是,如果我们要想看看最为集中、最为旺盛,同时又最具独创性的建筑艺术生命力,那么就必须要看意大利的建筑。因为我们最关心的

And in a study which is to deal rather with the principles than with the history of Renaissance architecture, it will be convenient thus to restrict its scope.

From what point of view should this architecture be judged so as best to reveal its unity and its intent? A general survey of the period will show grounds for deciding that, while a mechanical analysis or a social analysis may throw light on many aspects of Renaissance architecture, it is only an aesthetic analysis, and an aesthetic analysis in the strictest sense, which can render its history intelligible, or our enjoyment of it complete. If the essence, and not the accidents merely, of this architectural tradition is to be recognised, and some estimate of it obtained that does not wholly misconstrue its idea, this ground of analysis must be consistently maintained. The architecture of the Renaissance, we shall see reason to conclude, *may* be studied as a result of practical needs shaped by structural principle; it *must* be studied as an aesthetic impulse, controlled by aesthetic laws, and only by an aesthetic criticism to be finally justified or condemned. It must, in fact, be studied as an art.

Here, however, is the true core of the difficulty. The science, and the history, of architecture are studies of which the method is in no dispute. But for the art of architecture, in this strict sense, no agreement exists. The reason has few problems so difficult as those which it has many times resolved. Too many definitions of architectural beauty have proved their case, enjoyed their vogue, provoked their opposition, and left upon the vocabulary of art their legacy of prejudice, ridicule, and confusion. The attempt to reason honestly or to see clearly in architecture has not been very frequent or conspicuous; but, even where it exists, the terms it must employ are hardened with misuse, and the vision it invokes is distorted by all the preconceptions which beset a jaded argument. Not only do we inherit the wreckage of past controversies, but those controversies themselves are clouded with the dust of more heroic combats, and loud with the battle-cries of poetry and morals, philosophy, politics, and science. For it is unluckily the fact that thought about the arts has been for the most part no more than an incident in, or a consequence of, the changes which men's minds have undergone with regard to these more stimulating and insistent interests. Hardly ever, save in matters of mere technique, has architecture been studied sincerely for itself. Thus the simplest estimates of architecture are formed through a distorting atmosphere of unclear thought. Axioms, holding true in provinces other than that of art, and arising historically in these, have successively been extended by a series of false analogies into the province of architecture; and these axioms, unanalysed and mutually inconsistent, confuse our actual experience at the source.

To trace the full measure of that confusion, and if possible to correct it, is

INTRODUCTION

是文艺复兴时期建筑艺术的一般原则，而不是它的具体演变历史，因此，为了研究的方便起见，我们把研究范围加以限制。

从哪一个角度来研究这个时期的建筑艺术才最抓住它的全貌和意图呢？对这个时期作过一个全面的考察以后，我们就会有充分的理由相信，虽然从建造技术方面和社会需求方面的分析，的确是可以帮助我们理解文艺复兴时期的建筑艺术，但是，只有通过对这些建筑进行的美学方面的研究，而且是从纯粹美学的角度来研究，我们才能对这个时期的建筑艺术演变过程有一个充分的理解，才能对这个时期建筑艺术有一个全面的欣赏。如果我们希望把握住文艺复兴时期建筑的本质，而不是某些个具体偶然的实例；如果我们希望那些被我们得出的结论能够反映这一时期艺术的主流思想，而不至于发生误导，那么，我们就必须严格地把握住分析研究的基本点，并始终坚持在这个基本点之上。文艺复兴时期的建筑艺术虽然可以通过对社会需求是如何通过当时的结构技术原则来满足的这一线索来进行研究，但是，我们通过这本书的讨论，可以了解到我们有理由得出这样一个结论，即文艺复兴时期的建筑，必须从它对于美学的追求、接受美学中的定律的角度加以研究，才能够揭示这个时期建筑艺术的本质，只有通过美学的判断才能正确地把握住哪些东西是值得肯定的，哪些东西是必须加以谴责的。换句话说，必须把这个时期的建筑当作一门艺术来研究才行。

然而说到这里，我们遇到了一个真正的核心困难。关于建筑科学与历史的研究方法，没有人会提出任何质疑。但是把建筑当作一门艺术来研究，严格来说，大家还没有形成一种普遍接受的共识。理性还从来没有遇到过这样困难的问题，已经就这些问题给出过多次答案。关于什么是建筑美已经有过太多的说法，它们都曾经证明过自己的价值，都曾经流行于一时，也都曾经引起过争议，在当前的艺术语汇中留下了自己的遗产，或许让人产生偏见，或许让人嘲弄，或许给人带来迷惑。诚实地进行理性分析的尝试和能够清醒地看问题的努力，长期以来在建筑艺术领域内并不多见，也没有显著的案例。即使有人这样努力了，在分析中所使用的术语大概因为以往的滥用而变得生硬，从这些术语中所能联想到的形象也都因为这些陈腔滥调而被扭曲变了形。眼下，我们不但要全盘承受过去遗留下来的残破结果，同时也继续被那些造成这些残破景象的战火弥漫着的硝烟，那些诗意与道德、哲学政治、科学，等等，在英勇战斗中的呐喊声影响着。一个不幸的事实是，关于艺术的任何思考，在绝大多数情况下都是很偶然地产生的，可能是在大脑不断地接受各种信息与本来固有的持久兴趣突然产生改变的那一瞬间产生的。除了与建筑技巧相关的研究之外，建筑艺术就艺术本身的研究很少能够得到很诚挚的关注。因此，哪怕是最简单地概括一下建筑艺术都要在扭曲的氛围中受到各种混乱不清的思想影响。目前关于建筑艺术的各种见解，实际上都是对一些与建筑无关的其他领域中的个别正确事实加以无限的引申与类比，而这些见解既没有经过理性分析，也没有经过证明在引申到建筑领域里是否依然正确，结果从一开始便对我们的审美认识带来了思想混乱。

找出产生这种混乱的原因，有可能的话，对此加以校正，这就成了我们现在这本书的第

therefore the first object of this book. We enter a limbo of dead but still haunting controversies, of old and ghostly dogmatisms, which most effectively darken the counsel of critics because their presence is often least perceived. It is time that these spectres were laid, or else, by whatever necessary libations of exacter thinking, brought honestly to life.

The path will then be clear to attempt, with less certainty of misconception, a statement of the aesthetic values on which Renaissance architecture is based.

To follow, in concrete detail, this *Architecture of Humanism*, to see how the principles here sketched out are confirmed by the practice of the Italian builders, and to trace their gradual discovery, may be the task of another volume.

INTRODUCTION

一个目标。我们现在正走进一片阴森的地带,过去的亡灵仍然缠绕在那里,那里有鬼怪一样的教条主义总是在神不知鬼不觉地影响着我们的思想,而这些教条的存在总是不大容易被人们所察觉。现在是让我们借助一切必要的严密理性分析手段作为祭奠仪式,把这些蛊惑人心的亡魂埋葬,或者,真诚地让其他的灵魂再次复活的时候了。

这样一来,对于作为文艺复兴时期建筑艺术基础的美学价值做一个总结就可以开始尝试着展开了,道路上的障碍也将得到清除,至于概括总结当中是否会出现错误的理解和错误的概念则现在还说不好。

至于在《人文主义建筑艺术》这本书之后,通过具体的实例说明了我们总结出的这些原则在意大利建筑艺术家的具体实践中是如何得到运用的,记录下这些艺术家们后续不断的新发现,那就需要出版另外一本书来完成这个任务了。

ONE
Renaissance Architecture

The architecture of Europe, in the centuries during which our civilisation was under the sway of classical prestige, passed in a continuous succession through phases of extraordinary diversity, brevity and force. Of architecture in Italy was this most particularly true. The forms of Brunelleschi, masterful as they appeared when, by a daring reversion of style, he liberated Italian building from the alien traditions of the north, seem, in two generations, to be but the hesitating precursors of Bramante's more definitive art. Bramante's formula is scarcely asserted, the poise and balance of classic proportion is scarcely struck, before their fine adjustments are swept away upon the torrent that springs from Michelangelo. In the ferment of creation, of which Italy from this time forth is the scene, the greatest names count, relatively, for little. Palladio, destined to provide the canon of English classic building, and to become, for us, the prime interpreter of the antique, here makes but a momentary stand among the contending creeds. His search for form, though impassioned, was too reactionary, his conclusions too academic and too set, for an age when creative vigour was still, beyond measure, turbulent. With that turbulence no art that was not rapid and pictorial in its appeal could now keep pace. The time was past when an architecture of such calculated restraint as Sammichele had foreshadowed could capture long attention; and the art of Peruzzi, rich though it was with never-exhausted possibilities, seems to have perished unexplored, because, so to say, its *tempo* was too slow, its interest too unobtrusive. Vignola, stronger perhaps than these, is before long forgotten in Bernini. Architecture becomes a debatable ground between the ideals of structure and decoration, and from their fertile conflict new inventions are ever forthcoming to please a rapidly-tiring taste. Fashions die; but the Renaissance itself, more irresistible than any force which it produced, begets its own momentum, and passes on, with almost the negligent fecundity of nature, self-destructive and self-renewing.

We are confronted with a period of architecture at once daring and pedantic, and a succession of masters the orthodoxy of whose professions is often equalled only by the licence of their practice. In spite of its liberty of thought, in spite of its

第一章
文艺复兴时期的建筑艺术

当欧洲建筑艺术的风潮再次转向,让古希腊、古罗马那一类的古典主义建筑再次成为主导的艺术风格之后,在接下去的几个世纪里,这种建筑风格经历了很多个不断变化的阶段。在这些不同阶段里的建筑,它们个性的迥然不同、持续时间的短暂、作品所显示的巨大力量等等,都是极其明显的。尤其是把这样的一种概括性的描述,应用于这个时期里意大利的建筑艺术上面的话,就更能显现出这个概括的正确性。布鲁乃列斯基所创造出的杰出建筑形式,把意大利的建筑艺术从异族北方人崇尚的建筑形式中解放出来了。这一事实也预示着在两代人之后,将会出现伯拉蒙特(Bramante)的精湛与完美。然而,当伯拉蒙特的建筑艺术还没来得及得到明确,他那种庄重稳健的古典比例关系还没来得及完全牢固地确立起来的情况下,米开朗基罗(Michelangelo)的那些富有动感的挣扎、扭曲、变形,便出来取而代之了。在这种接连不断的迅速更迭创新中,显赫的名声根本算不了什么,创造出新东西才是最重要的。从这时起,意大利一直处在不断的探索中。比如说,了不起的帕拉第奥(Palladio),命运让他的理论成为后来英国古典主义建筑的标准,让他成为我们心目中关于古希腊、古罗马时期建筑艺术的最重要的阐释者。但是在当时众多艺术流派的竞争当中,帕拉第奥仅仅是昙花一现般地留下那么一点点痕迹而已。帕拉第奥对于建筑形式的追求尽管充满了热情,但是他的形式本身则是过于保守的,过于学究式的严谨,而且是过于僵化,条条框框很多而没有圆通的空间,这与当时艺术界甚至全社会的气氛不相符合。当时的社会,创造力仍然是极其旺盛的,那种活跃、跳动着的创造力是语言无法描述的。在这种追求创造力的快速起伏跳跃中,那些速度跟不上、造型没有吸引力的作品则注定是要被淘汰的。像萨米彻利(Sammichele)的那种严谨的建筑在过去可能会得到相当的重视,或许会被人久远地继承下去,但是在这个时期里却不再风光;伯鲁齐(Peruzzi)的艺术虽然不乏丰富的细节与故事,变化多样,但是现在看来,似乎在没有机会引起周围的人对它表现出足够兴趣的时候便消失了。原因无他,就是它们的节奏太慢了,它们的效果太含蓄,不够抢眼。韦尼奥拉(Vignola)的作品比上面两位要好一点,但是随着伯尔尼尼(Bernini)的出现而被人很快地遗忘。建筑艺术成了建筑结构和建筑装饰两股势力较量的地方,结构和装饰各有各的理想,在二者的不断冲突中,建筑艺术中的新表现手段不断地涌现,以满足人们的兴趣与爱好。时尚的东西无疑会死亡,但是文艺复兴这场文化运动本身,要比这个运动中创造出来的任何作品与流派都更有活力,具有自己的动力和惯性,它会冲破任何障碍而不停地向前行进,具有一种如同大自然所具有的那种新陈代谢的能力,可以自我灭亡,又自我更新。

我们面前的这个文艺复兴时代,它的建筑艺术既雄心勃勃地大胆追求创新,又讲究博学多闻的研究传统;我们的面前还有一批接一批的艺术大师们,他们在自己在艺术领域里,通过展现自己的巨大成就,努力让自己相应地成为随心所欲的艺术家。尽管这个时期在思想

keen individualism, the Renaissance is yet an age of authority; and Rome, but pagan Rome this time, is once more the arbiter. Every architect confesses allegiance to the antique; none would dispute the inspiration of Vitruvius. For many the dictates of the Augustan critic have the validity of a papal deliverance upon a point of faith. Yet their efforts to give expression to this seemingly identical enthusiasm are contradictory in the extreme. Never were the phases of a single art more diverse. For to consistency the Renaissance, with all its theories, was vitally indifferent. Its energy is at every moment so intense that the forms, not of architecture alone, but of every material object of common use, are pressed into simultaneous and sympathetic expression; yet it is guided on no sure or general course. Its greater schemes too often bear evidence to this lack of continuity, this want of subordination to inherited principle. Upon the problem of St. Peter's were engaged the minds of Bramante, Michelangelo, Raphael, Peruzzi, Sangallo, Fontana, Maderna and Bernini. So much originality could not, without peril, be focussed at a single point; and those of Bramante's successors who were fortunate enough to carry their schemes into execution, obscured, if they did not ignore, the large idea which he had bequeathed to them. The history of St. Peter's is typical of the period. Shaped by a desire as powerful as it is undefined, its inventive impulse remains unexhausted, and style succeeds to style in the effort to satisfy the workings of an imagination too swift and restless to abide the fulfilment of its own creations. In this the Renaissance stands alone. The mediaeval Gothic had indeed been equally rapid, and equally oblivious of its past, so rapid and so oblivious that few of its principal buildings were completed in the style in which they were begun. Nevertheless it pursued one undeviating course of constructive evolution. Beside this scientific zeal the achievement of the Italian builders might appear, at first sight, to be as confused in aim as it was fertile in invention. Contrast it with the cumulative labour, the intensive concentration, by which the idea of Greek architecture, ever reiterated, was sharpened to its perfection, and the Renaissance in Italy seems but a pageant of great suggestions. Set it beside the antique styles of the East, compare it with the monumental immobility which for eighteen centuries was maintained in the architectural tradition of Egypt, and it might pass for an energy disquieted and frivolous. Yet, at every instant in the brief sequence of its forms, it is powerful and it is convinced; and from the control of its influence Europe has attempted to free itself in vain.

上是自由的,每个人都在追求着个人的价值,但是文艺复兴时期作为整体来说,却是在表现着某种权威。罗马再次成为仲裁者,但是此时的罗马已经不是过去基督教全面控制的罗马,而是异教徒文化的罗马。在这个时期,让每一位建筑师心甘情愿地贡献出自己忠诚与才华的不再是宗教信仰,而是古代罗马的艺术文明;这毫无疑问是受到了维特鲁威(Vitruvius)著作的影响。在很多人的眼中,这位古罗马奥古斯都时期的理论家,他的话具有如同教皇宣讲宗教信仰的布道词般的权威。然而,每一位建筑师在把这些话转换成自己具体的建筑设计时,则是把同一种热情转变成极其不一样的各种建筑语言,他们彼此之间的不同,简直可以用水火不相容来形容。在艺术历史上,从来没有像这样地在同一个时期,同一个艺术领域里,作品却有着如此的不同。文艺复兴时期的全部艺术理论,从本质上讲,根本不在意彼此是否矛盾。这个时期里的每一个瞬间,艺术家的热情总是那样的饱满,他们的创作总是带有当时一致的特点与表现手段。这不仅表现在建筑上,而且表现在一切日用器皿、物件的设计上。但是,这种表现手段的下一步走向与趋势是什么,则是无法预料的,随时都有可能改变方向。从大的方面来看,文艺复兴时期的艺术发展常常显露出这种明显的缺乏延续性,缺乏公众对于某一个原则的认同与坚持,缺乏为了一个共同目标自己甘当配角的认识。如在圣彼得大教堂的建造过程中,众多精英集中于一处,伯拉蒙特、米开朗基罗、拉菲尔(Raphael)、伯鲁齐、桑伽罗(Sangallo)、封丹纳(Fontana)、马德尔纳(Maderna)、伯尔尼尼,这些杰出大师们都曾在不同时间参与其中。这么多富于创新思想的艺术家集中到一件作品上,不用想就会猜到,他们不可能围绕着同一个主题展开自己的创作。伯拉蒙特的后继者们如果不忽略伯拉蒙特留给他们的大的想法与构思,那么有机会参与并实施自己方案的那些人必将会变得无声无息,一切的荣耀都将归功于伯拉蒙特。圣彼得大教堂的建造过程是这个时期非常典型的案例。这个建筑就是要表现出一种超越他人的力量,但同时又没有人知道这种力量到底是什么东西。大家前仆后继地不断探索创新,为了满足处于不断改变中的想象力对于这种力量的理解,建筑的样式一个接一个地出现。样式的变化在不断地出现,但是,文艺复兴的精神却依然故我。中世纪时期的哥特建筑也是一种成长迅速的艺术风格,也是同样在成长过程中常常忘记自己从前的样子,一座大教堂在完成的时候,很多情况下都是与当初开始建造时的风格有所不同。但是,在哥特教堂上面,所有参与的人所追求的却都是沿着同一个奋斗目标,都是为了建造技术的完善而在努力勤奋工作着。 回头来看意大利的建造者们所取得的成就,除了在科学技术上的热情与成就以外,他们在建筑艺术上那些不断创新的努力背后,我们看到的好像是一幅迷茫的面孔,他们不知道自己的目标在哪里。与此形成强烈对比的是古希腊时期的建筑艺术。在古代希腊,人们投入巨大的精力,非常专注地把一代又一代的注意力集中在一个目标,那就是反复不断地推敲建筑各个细节的比例关系,让这些细部的比例关系日臻完善,达到十全十美的境地。但是在文艺复兴时期的意大利,人们所炫耀的好像只是一些新奇的想法而已。把文艺复兴时期的意大利建筑与古代埃及的建筑摆放在一起,我们便会看到,古埃及延续了十八个世纪的建筑依然保持着当初的那种沉稳、庞大等传统艺术风格的特征,而文艺复兴时期的意大利则是浮躁与轻佻。但是在整个文艺复兴时期的过程里,它的每一个时刻的建筑形式都是充满了旺盛的生命力的,当时的人相信本来就应该是这样的。它所产生出来的巨大影响力,以至于让后来的整个欧洲根本无法摆脱它的魔咒。

We shall seek without success, among conditons external to art, for causes adequate to an effect so varied, so violent, and so far-reaching. The revolutions which architecture underwent in Italy, from the fifteenth to the eighteenth century, corresponded to no racial movements; they were unaccompanied by social changes equally sudden, or equally complete; they were undictated, for the most part, by any exterior necessity; they were unheralded by any new or subversive discovery whether in the science of construction or in the materials at its command. All these, and other such conditions, did indeed contribute to the architectural result. Sometimes they set their limits to what was accomplished, sometimes they provided its opportunity. But none of them separately, nor all in conjunction, will sufficiently explain the essential character of the whole movement, or of each successive step, nor afford any clue to the sequence of its stages. They are like the accidents of a landscape which might shape the course of a wandering stream. But the architecture of Italy is a river in the flood. Race, politics, the changes of society, geological facts, mechanical laws, do not exhaust the factors of the case. Taste—the disinterested enthusiasm for architectural form—is something which these cannot give and do not necessarily control. Nevertheless it is by reference to these external factors that the architectural forms of the Renaissance are persistently explained.

Let us see how far such explanations can carry us. It is probably true that a 'Renaissance' of architecture in Italy was, on racial grounds, inevitable. Already in the twelfth century there had been a false dawn of classic style. Indeed, it seems evident that mediaeval art could exercise but a temporary dominion among peoples who, however little of the authentic Roman strain they might legitimately boast, yet by the origin of their culture stood planted in Roman civilisation. Classic forms in Italy were indigenous and bound to reappear. And this fact is important. It enables us to dismiss that unintelligent view of Renaissance architecture, once fashionable, and still occasionally put forward, which regards it as a pedantic affectation, or perverse return to a manner of building that was alien and extinct. But it is a fact which in no way helps us to understand the precise form of classic culture which the Renaissance assumed. It does not explain the character, number, and variety of its phases. And it tells nothing of classic culture in itself. Racial considerations are here too general and too vague.

The field of politics might seem more fruitful. The growth of the new style is undoubtedly associated, at Florence, Milan, Naples and other city states, with the rise to power of the Italian 'tyrants,' themselves another echo of antiquity, and another characteristic expression of the Renaissance, with its cult for individuality and power. Cosimo the Great, whom Michelozzo followed into exile at Venice,

尽管我确信我们在艺术领域以外的周围地带是找不到解释文艺复兴现象的直接原因的,但是还是让我们到周边来看看吧,看是否能找到帮助我们理解这种多变、充满力量、影响深远的文艺复兴时期的艺术相对应的一些线索。意大利的建筑艺术在从十五世纪到十八世纪这段时间里所经历过的演变过程,没有发生过任何与民族有关的历史事件;也没有发生过与文化艺术领域里出现的这种突然、彻底的变革相对应的社会变革;绝大部分的意大利,在这个时期里没有外来族裔或者文化的统治;这时的意大利也没有在科学或者建造技术方面出现过什么颠覆性的重大发现和发明。这一切,当然连同其他的条件一起,影响了当时的建筑艺术的结果。有的时候,这些因素成为建筑艺术发展的阻碍,有的时候却也为建筑艺术的发展提供了机会。但是,无论是把所有的这些因素集中起来综合地看,还是单独分开来看,它们都不足以说明为什么文艺复兴时期的建筑艺术会成为当时那种状况,都不足以说明为什么在整个过程中会出现那么多不同的一个接一个的变化,也无法解释为什么在一个阶段过后会出现下一个情况。这些因素就好比是大自然中的那些偶然存在着的地形地貌,因为它们的存在,影响了本来自由流淌着的溪水往后的流淌路径。但是,这时意大利的建筑是一条洪水正在泛滥的河流。种族、政治、社会变革、地理条件、技术上的进步等等因素,无法阻止泛滥的洪流。人们的喜好与兴致,让他们在建筑形式上投入了无限的热情,并且是没有任何功利的动机在驱使他们这样做。人们的这种喜好与兴致是我们在前面提到过的那些自然与社会因素所不能提供的,也是无法控制的。但是在后人的各种理论中,试图从那些外部因素入手,来对于文艺复兴时期的建筑艺术形式的产生与发展进行解释,这样的努力却是从来就没有间断过的。

让我们来看看这种解释能把我们带到哪里,看看这种解释到底能够走多远吧。我们也许可以说,文艺复兴的建筑出现在意大利,从种族的角度来看的确是有些道理的,好像也无可避免。在十二世纪的时候,意大利就已经出现过一次古典风格复兴的假象。实际情况表明,中世纪文化艺术在当时的意大利虽然暂时占主导地位,但是那里的文化从任何一个局部来看,仍然是深深地根植于罗马文明之中。而中世纪的这些人并不能百分之一百地代表正统的罗马人的文化。古罗马时期出现过的那些建筑是这里土生土长的,它们当然还会重新出现。这一点十分重要。它能够帮助我们摒弃一种错误的观念,这种错误的观念认为,文艺复兴时期的建筑是卖弄学问的结果,不过是学究们搞出来的一种让早已灭绝了的旧的建筑形式死灰复燃的建筑形式,这些旧东西已经不再属于我们今天。这种观念实际上是很愚蠢的,然而在过去曾经很是流行过一阵子,现在偶尔也还会冒出来。虽然从文化和种族的角度来看,这个问题可以帮助我们澄清一些疑问,但是,这种解释仍然没有说明为什么文艺复兴时期的建筑会演变成我们所看到的那个样子,也不能解释为什么整个文艺复兴时期会有那么多的不同阶段,会有那么多的不同特征与不同风格。它也没有解释古罗马的文化到底是什么。从种族角度来看这个问题,其结果就是过于笼统,也过于模糊不清。

从政治的角度来看,或许它能比从种族的角度更能说明一些问题。在佛罗伦萨、米兰、那不勒斯等城邦共和国,这种新的建筑风格的确与这些意大利城邦里的"暴君"掌权有着同步的关系。这些暴君的出现的现象本身也反映了古代的一种传统,也是文艺复兴时期的一个具有代表性的特征,即对于个人的崇拜与对权力的崇拜。美第奇家族的祖先科西莫一世(Cosimo the Great)、美第奇家族的劳伦佐(Lorenzo)、意大利南方的阿尔方索(Alphonso)和

Lorenzo, the protector of Giuliano da Sangallo, Alphonso in the South, in the North the Sforzas—these, and others like them, were certainly influential patrons. But it would be difficult to maintain that they left a deep imprint of themselves, or their government, upon the character of the art. Gismondo Malatesta, tyrant of Rimini, the rough soldier who caused a Gothic church to be converted into the equivalent of a pagan temple dedicated to his mistress, and flanked it with the entombed bones of Greek philosophers and grammarians, may well impress us with his individuality; but, as between him and Alberti, his architect, himself of noble family and one of the greatest humanists of his time, there can be little doubt where the paramount imagination lay. The influence of patronage on art is easily mis-stated. Art may be brought to the service of the state and its rulers; but the most that rulers can do towards determining the *essence* of an art is to impose upon it a distinctively courtly character, and the coherence which comes of a strongly centralised organisation. We should, for instance, misconstrue the inmost nature of Augustan art, or of the art of Louis XIV., if we were to ignore this factor. But nothing similar is true of the Renaissance city-state. Here the conditions were merely such as to give free play to an architecture which, intrinsically, in its character as an art, remained independent of them. The sole centralising influence, in any imaginative sense, was that of the Church, and even this was not felt as such till after the art had acquired its own natural momentum in the free, secular life of Florence.

It must be recognised, however, that the existence, in the sixteenth century papacy, of a soil perfectly suited to receive the roots of the restored art was in itself a piece of rare good fortune. The return to the antique, however tentative and, so to say, provincial, at the first, was in essence and by implication a return to the 'grand style' —to an imperial, and, in the literal sense a 'catholic' architecture. For the assertion and development of such a style the papacy was the ideal instrument: the papacy with its imperial court, its boast of ancient continuities, its claim to universal dominion, its pagan inheritance, and its pomp. All such qualities were favourable to the vigour of a partly retrospective enthusiasm, fascinated by the broken ruins in which ancient Rome had embodied splendours so similar to these. And this was not all. For, in proportion as the classic movement was no empty revival, in proportion as it represented a rising to the surface of the preferences, still vital and potent, of an ancient and indigenous culture, which claimed a future as confidently as it possessed the past, just in that measure it required a field in which to realise its own creative resources, its own untried originality. It could not have found itself in any rigid discipline or imposed continuity such as that which, later, in the France of Louis XIV., gave to architecture a formal and restricted, aim. It needed the patronage of

意大利北方的斯福尔扎（Sforzas）家族，以及很多如同他们一样有权有势的人，他们的确通过赞助某些大艺术家而对于艺术产生过相当大的影响。米开洛佐（Michelozzo）追随科西莫流亡到过威尼斯，劳伦佐资助过朱利亚诺·达·桑伽罗（Giuliano da Sangallo）的艺术活动，等等。但是，这些事实都很难说明这些风云人物或者他们控制下的政府完全左右了当时艺术作品的最终结果。比如说，在里米尼城邦（Rimini），它的当权者是暴君吉斯莫多·马拉特斯塔（Gismondo Malatesta），他是一位军旅出身的粗暴家伙，把一座哥特时期的天主教教堂改造成了一个异教的教堂，供自己的女友使用。在改造过的教堂里，他放置了两具古希腊贤哲的尸骨，一具是古希腊哲学家的，另外一具是语言学家的。他的这种古怪另类的想法的确让我们印象深刻。但是，他有一位御用的建筑师，叫阿尔伯蒂（Alberti）。阿尔伯蒂本人也是贵族出身，自己又是文艺复兴时期最伟大的人文主义学者之一。在这种情况下，不用说我都会知道，那些建筑的造型与样式最终到底是出自谁的手笔了。出资兴建某些建筑物的赞助者所能够发挥的作用，通常会很容易被夸大，或者误传。艺术的确会被统治者以及他的政权所利用，但是这些统治者或者他的政权，最多只能决定艺术必须要表现出庄严、隆重等本质特征，在必须要表现他的集中权力和意志等方面发表自己意见，并提出自己具体的要求，其余的一切还是由建筑师来替他决定。如果不了解这一点，那么我们就会误解古罗马奥古斯都时代的建筑艺术，误解法国路易十四时代的建筑艺术。但是在文艺复兴时期的城邦共和国里，情况则根本不同。在文艺复兴时期，建筑艺术作为一种艺术，它的创作完全留给了艺术家去自由发挥，从艺术角度来看，完全不受赞助者的影响。唯一出现过对艺术家创作有些限制的地方，可以想象得出，只是来自于教会。即便是这样，那也是在世俗的社会生活左右了佛罗伦萨，自由的艺术获得了足够惯性之后，教会才会出面对艺术家在宗教建筑上的创作活动加以适当的限制。

有一点很重要，我们必须要注意到。在十六世纪，在教皇统治一切的情况下，这种复兴古代艺术的种子能够找到这么一块土壤，并且能够生根发芽成长，不能不说是一个千载难逢的机缘和运气。回归古代的荣耀，这种想法在开始的时候或许是很不成熟的，或许只是很小一部分人的想法，但在本质上却暗示着回归一种"伟大的风格样式"，它暗示着过去的帝国辉煌，从字面上理解，也就是恢复到真正的"普天大同"的建筑风格［普天大同（Catholic）在字面上与天主教（Catholic）的词形是完全一样的］。为了确立这种建筑风格的地位并充分地发展这样一种建筑风格，教皇是最理想的工具：教皇拥有一个类似于古代皇帝君王的宫廷、教皇试图确立自己自古以来的正宗地位、教皇宣扬的对于宇宙的主宰、教皇所具有的类似古代异教徒的那种权力承传方式、教皇对于浮华排场的热衷，等等。所有这些，对于一件本质上是回归过去的荣耀这个理想来说，当然是有利的，教廷当然愿意借助古罗马残迹中所象征的辉煌来装点他们自己所向往的那些东西。事实上，事情还不仅仅如此简单。因为这场复兴古典文化的运动绝对不是简单的模仿外在形式而已。人们直接看到的，是它表现在外部的现象，看上去是在努力恢复过去古代的文化，但是在本质上，这种文化又是当地的本土文化，依然具有旺盛的生命力、依然茁壮。这种文化对于未来的信心，如同它过去拥有过的辉煌一样，是不容置疑的。它所需要的只是一个可以实践的机会，只要有了这个机会，那么这种无尽的文化源泉和无穷的创造力就会再次得到喷发和释放。当时，这种文化的复兴还没有给自己设定一系列的条条框框，或者把自己限定在继承某一条固定的线索上。这与后来

a large idea, but it required also space and scope, that it might attempt every mode of self-realisation yet stand committed to none. This space, and this patronage, the papacy was fitted to provide. The rivalry of successive popes, their diverse origins and sympathies, their common passion to leave behind them an enduring monument of their power; above all, their detached office, controlling the different states of Italy and forcing each of them to bring its own artistic temperament within the spell of Rome, gave architecture, in perfect combination, the focus and the liberty, the varied impulse and the renewed vitality necessary for making a great imaginative experiment under the influence of the antique.

  The papacy, then, may be considered to have predetermined in some degree the formation of Renaissance style. Yet we must not exaggerate its contribution. By its imperial quality it will appear to have furnished the large idea to which the new classic architecture might stand in service. But we must not overlook the extent to which the papacy was itself indebted, for that quality, to the artists of the Renaissance. It is a common fallacy to account for artistic expression by external conditions for whose very being that expression is in some cases responsible, and which, but for that expression, would never, perhaps, have been supposed to exist. In the present case, no doubt, this point could not be pressed very far. Yet St. Peter's and the Vatican, and the great monuments of restored Rome, are witnesses no less to the power of architecture to create and define the imaginative value of the Renaissance papacy, than to the encouragement and inspiration which the papacy contributed to art. Moreover, the character of the papacy in this period was largely formed by the character of its popes; and such men as Pius II., Leo X., and Julius II., were fit patrons of Renaissance architecture, partly for the reason that they were cultivated enthusiasts, awake to the ideals of an art which, quite independently of themselves, had given evidence of its nature, and which was already, in the eyes of all men, an energy so vigorous and splendid, that the popes could conceive no securer means of adding to their fame than by inviting its support.

  So, too, with the more particular religious and social movements by which the phases of Renaissance architecture have sometimes been explained. When the Counter-Reformation made its bid for popularity, it erected on every hand churches in the baroque manner frankly calculated to delight the senses and kindle common enthusiasms. Never, perhaps, has architecture been more successfully or more deliberately made the tool of policy than by this brilliant effort which transformed the face of Italy; nor has the psychological insight of the Jesuits been manifested with greater sureness than when it thus enlisted in the service of religion the most

在路易十四时代的法国不一样,法国的古典建筑被设定了许多形式上的规则,限制了它的发展空间。在文艺复兴时期的意大利,建筑只需要赞助者提出一个大的构想、建筑规模和建筑内容。建筑可以是任何能够实现的东西,没有限制,人们也没有打定主意一定要实现某一个具体的想法。对于这种性质的建筑,在具有这种能力与资源的众多赞助者当中,教皇是最理想的人选。每一位新教皇总是从心里都要和他之前的教皇比较一番。教皇与教皇的出身、背景各不相同,喜好自然不同,但是他们却都有一个共同的心愿,那就是要在他们身后留下一些永久的纪念碑,来纪念他们曾经拥有的权力与荣耀。再有一点,分布在意大利各地的代表着教皇发号施令的代理人,控制着意大利的每一个角落,这些人收罗了各地的艺术精华与风格,听凭教皇的差遣。这样,教皇资助的这些建筑就有了不同的艺术追求与不间断的生命力,在追求古典建筑这个大目标下面,展开了各式各样的实践,不同的艺术焦点加上自由发挥的空间,让这些建筑形成了一个完美的组合。

如此说来,教皇在某种程度上或许预先决定了文艺复兴时期艺术风格的基本走向,我们的确是可以这样认为的。但是,我们绝不能夸大教皇在其中的作用。由于教廷具有某种古代帝国的结构形式,而新出现的古典建筑很可能为之所用,所以,教廷让这样的建筑艺术获得了某种宏大构想看起来是很自然的。然而同时我们也不能忽略,在一定程度上,教皇的这种宏大构思也是受到文艺复兴时期艺术家的启发才形成的。关于艺术表现问题,有一个常见的谬误就是通过外在的条件来解释艺术表现形式。但是实际上,这种所谓的外在条件恰恰是由于这种表现手段的出现才得以存在的,如果没有这种表现形式,那个所谓的外在条件可能根本就没有机会出现。在当下,这个结论的普遍适用性无疑是不能随意被夸大的。但是在圣彼得大教堂、在梵蒂冈、甚至在用纪念性建筑重新塑造的罗马,这些地方都见证了建筑艺术的力量是如何创造、确立文艺复兴时期的各位教皇在人们心目中的地位。这些建筑艺术带给教皇的荣耀绝不亚于教皇带给艺术的鼓励与启发。再有一点,这个时期的每一位教皇执政时期的特点完全取决于当时的教皇本人,例如,庇护二世(Pius II)、利奥十世(Leo X)、尤里乌斯二世(Julius II)都是文艺复兴时期建筑艺术的理想赞助人,一方面是因为他们本人都是有文化修养的热心人士,对于艺术有着自己的理解,也有自己的理想。另一方面,这一时期的艺术已经独立于教廷的控制,形成了自己的特色。在所有人的眼中,文艺复兴时期的建筑艺术已经是富有活力和成就辉煌的,它的地位已经很崇高了,教皇为了增加自己的身价,也极力把建筑艺术拉拢到自己的身边来。除了采用这样的笼络手段之外,教皇恐怕也没有别的更好的办法让建筑艺术为自己服务。

某些个别的宗教活动和社会变革运动,也曾经被拿来解释文艺复兴时期建筑艺术中的某些个别阶段。当天主教全面反对宗教改革,千方百计、绞尽脑汁地打算吸引民众的时候,教会便利用各种机会,在教堂建筑中大肆运用巴洛克艺术风格,目的就是要让人们的视觉感官得到愉悦的享受,燃起人们的热情。历史上好像从来没有过这么成功且又目的性明确的做法。这让建筑艺术成为政治斗争的工具,而且这个精彩的做法同时也改变了意大利大街小巷的面貌。历史上也从来没有像耶稣会(Jesuits)那样,在把握人们的心理方面如此地准确、有效,把人类最富有戏剧效果的设计拿来为宗教服务。但是,我要再一次强调的是,这次宗教运动的成功,完全是因为它充分认识到这些被拿来应用的建筑艺术的价值,对巴洛克艺

theatrical instincts of mankind. But, once more, the very success of the movement was occasioned by the fact, so well appreciated by the Jesuits, that the taste for such an architecture was already there. The readiness of the *seicento* Italians to respond to an architectural appeal, their delight in such qualities as these baroque churches embodied, are pre-existent facts. The achievement of the Jesuits lay in converting these preferences of a still pagan humanity to Catholic uses, aggressively answering the ascetic remonstrance of the Reformation by a still further concession to mundane senses. The artistic significance of the style which the Jesuits employed, remains something wholly independent of the uses to which they put it. To explain the first by the second is to misconstrue the whole matter. To condemn the first on account of the second, as has repeatedly been done, is nothing less than childish.

Somewhat similar objections will apply when the architectural history of Italy is interpreted as the outcome of social changes. The 'increase of wealth,' the 'rise of great families,' the 'luxurious habits of a more settled society' — those useful satellites of architectural history—helped, no doubt, to create the demand which architecture satisfied. But the significant point is precisely that it was to artistic uses that this wealth, this power, and these opportunities, were devoted, and to artistic uses of a particular kind. Rich and flourishing societies have not seldom grown up, and are growing up in our own time, without any corresponding result. Prosperity is a condition of great achievements; it is not their cause. It does not even stand in any fixed relation to their progress. It provides power, but does not, artistically, control its use. The economic conditions which, in Italy, assisted the architecture of the Renaissance to assume such prominence, did not vary with the marked and swift alterations of its style. The style had an orbit, and an impetus, of its own. In Italy nothing is commoner than to find an architectural display wholly disproportionate, and even unrelated, to the social purpose it ostensibly fulfils, and to the importance or prosperity of the individuals or communities responsible for its existence. Princely gates, more imposing than those of a great mansion, lift up their heads in the loneliest places of the Campagna, but nothing glorious goes in. They lead, and have always led, to unpeopled pastures or humble farmsteads. The baroque spirit delighted in this gay inconsequence. It appreciated grandeur for its own sake, aesthetically; and it had a sense of paradox. In Tuscany, on the other hand, though Cosimo had to rebuke the too lordly schemes of Brunelleschi, and though the Strozzi Palace frowns in unfinished grandeur, the noblest occasions are often met by an exquisite humility of architecture. Yet, chastened as it was to its extreme refinement, this modest style of Tuscany must sometimes have formed the frame to very mediaeval manners. A great critic, Professor Wöfflin, reviewing the numerous changes in style which marked the entrance of the Baroque, is content to

术的喜爱是在宗教需要它之前就已经存在了的，只是耶稣会充分认识到了它的价值，对它表现出欣赏并加以采用。十七世纪的意大利人（seicento Italians）对于建筑艺术的喜爱，对于诸如巴洛克教堂建筑所代表的那种艺术品质的喜爱都是早已存在的。耶稣会的贡献，在于他们把那些当时仍然属于异教徒的人文主义的东西带到了天主教会里面来了，让它们为天主教会服务，而且是针锋相对地采用比世俗更加注重世俗享乐的东西，来回应宗教改革派所提倡的禁止物欲的主张。这种出现在耶稣会建筑中的艺术，其本身所代表的意义完全独立于它在教堂里面的具体的使用目的。用后者解释前者必将会误解整个问题的关键，而根据后者来谴责前者则根本就是幼稚的儿戏，但是这种儿戏般的理论观念却是一而再、再而三地反复出现。

意大利的建筑历史有的时候被解释成是社会变革的产物，对于这种观点，我们也坚持与上面类似的反对态度。"财富的增加"、"大家族的兴起"、"一个稳定社会对于奢华的追求"等现象，的确创造了某些需求，而这些需求也的确为建筑艺术提供了展现自己的机会。这一点是没有什么疑问的。但是这些因素都是建筑艺术发展的周边次要因素。关键点恰恰在于，这些财富也好，权势也好，机遇也好，一切都在追求艺术，而且是在追求某种特定的艺术风格。财富与繁荣的社会经常会出现，我们这个时代也不乏财富与社会繁荣，但是在艺术风格方面与之并没有什么相应的直接结果。财富是取得伟大艺术成就的一个条件，但绝对不是它的动力和原因。财富与艺术的进步甚至没有任何固定关系。财富能够带来权力，但是无法用财富左右艺术。意大利的经济发展状况帮助文艺复兴时期的建筑艺术取得了崇高的地位，但是同样的经济背景状况也见证了这种建筑艺术的明显改变与剧烈变化。艺术风格有自己的轨迹，也有自己的内在动力。在意大利，一个建筑的外观与该建筑所扮演的社会功能可能根本不相匹配，甚至是毫无关系的，一个建筑的外观与在背后资助兴建这个建筑的个人或者团体的社会地位、经济实力也是毫无直接关系的。这一现象在意大利十分普遍。意大利南部的城市坎帕尼亚（Campagna）的一个偏僻角落，孤零零地矗立着一座大门，大门有着皇家的气势，比很多豪宅府邸的大门还要气派，骄傲地站在那里，但是那里只有这一座大门，除了这座大门之外，其他什么也没有。从大门走进去只有无人的田野，或者是简陋的农庄。这种情形在过去是这样，现在还是这样，从来就是这样。在这种没有后续动作的大门口，我们感受到了属于巴洛克艺术的那种欢乐精神。这座门只是从美学的角度赞美巴洛克的大气，同时对表面浮华的做法又有一点反讽的味道。相反，在托斯卡纳（Tuscany），科西莫曾经为了布鲁乃列斯基设计方案过于庄严隆重而大声吼叫过，斯特罗齐宫（Strozzi Palace）也因未完成的宏伟设计而无奈地皱着眉头，这些贵族赞助的最能代表自己身份的建筑实际上总是采用精湛又谦卑的建筑。经过反复的推敲，托斯卡纳地区的这些含蓄又低调的建筑，在整体上变得非常精彩、细腻，形成了一种类似于中世纪的那些精致的处理手法。一位伟大的艺术理论家，沃尔夫林教授（Professor Wölfflin）研究了出现在巴洛克艺术风格之前一系列艺术风格的演变，他因此提出一个概念，叫做"时代的精神"，说导致那些艺术演变的原因是时代的精神在改变。十九世纪出现的那些许多让人半懂不懂的艺术理论，非常喜欢使用"时代精神"这个概念，好像"时代精神"是一种真实存在的社

refer them to a change in 'the Spirit of the time.' Nineteenth century mythology is favourable to the phrase; and 'the Spirit of the time' is often spoken of as a social power. But 'the Spirit of the time' does not exist independently of the activities which manifest it. It is the atmosphere which results from their combined operation; or it is the influence of the earlier and more spontaneous of these activities as felt by those which come more tardily or more reluctantly into play. Now, among those activities, art and architecture were in Italy ever to the forefront, as spontaneous and vital a preoccupation as existed in the national life. It is hardly philosophical, among a number of parallel manifestations of energy, to explain the stronger by the weaker; yet that is what an appeal to 'the Spirit of the time,' if it means anything, here implies. When, therefore, we have interpreted a change in architecture by a change in 'the Spirit of the time' we have in this case demonstrated a mere tautology.

Nor shall we fare much better in the attempt to find the key to Renaissance architecture in constructive science. There have been occasions when the discovery of a new structural principle, or the use of a new material, has started architectural design upon a path which it has followed, as it were of necessity, unable to desist from its course until the full possibilities of the innovation had been explored. Each step is determined by a scientific logic; and beauty lingers in the art by a fortunate habit, or comes, in some new form, by accident to light. Such, in some sense, was the case with the mediaeval Gothic; and so it might be with some future architecture of steel. But such was not the case with the architecture of the Italian Renaissance. No constructive innovation explains the course which it pursued. The dome of Brunelleschi, unquestionably, by its audacity and grandeur, the effective starting-point of the Renaissance, was indeed a great triumph of engineering skill; but it involved no fundamental principle which was not already displayed in the dome of Pisa or the Baptistery of Florence. On the contrary, although the construction of the Renaissance was often vast in extent and courageous in conception, it was at the same time simpler and less scientific than that of the centuries immediately preceding, and it was based for the most part upon the simplest traditional Roman forms. In proportion, moreover, as the use of stucco became prevalent, the construction which it concealed became an object of indifference.

The one constructional practice which distinguishes the Renaissance does but confirm the insignificant interest which construction, as such, possessed for the men of this period. That practice is the constant and undisguised use of the tie-rod to strengthen and secure arches and vaults which of themselves were insufficient to withstand the outward thrusts. This was an expedient by no means unknown to the Gothic builders. But what in mediaeval construction had been an exceptional remedy, was accepted by the Renaissance builders as an obvious and legitimate

会力量似的。但是,所谓的"时代精神"这个概念,它是根本不可能独立于体现这个精神的那些具体现象而单独存在的。它是一种氛围,是各种现象综合到一起的结果;或者说是迟钝者与后来者对于前辈那种不留痕迹的自然举措所表现出的一种感受。在文艺复兴时期的意大利出现的所有活动中,艺术和建筑活动是最受世人瞩目的,是走在最前面的,是国民生活中最不可或缺的活动,也是国民最主动参与的活动。在面对这样能量巨大的各种表现形式当中,用那些次要的现象来说明其中最主要的活动,很难说是明智之举。如果说"时代精神"这个概念能够告诉我们任何一点东西的话,那就是它在试图利用一个时代里的某些次要的东西来说明当时最主要的东西,我们大概只能得出这样一个结论。所以说,当我们试图用"时代精神"的变化来解释建筑艺术之变化的时候,实际上我们只是在犯逻辑上的循环论证的毛病。

如果试图从建造技术方面来理解文艺复兴时期的建筑艺术,寻找一把开启奥秘的钥匙,那么,这种尝试的结果也肯定不会好到哪里。历史上的确有过一些情形,某种新的结构体系的发明,或者某种新材料的使用,的确带来了某些新的建筑设计手法,并引起后续的不断更新。出于新的要求,为了挖掘出这种新技术的全部潜能,建筑设计也跟着不断探索,直到这种新技术、新材料被完全掌握为止。这个过程的每一步都有科学的依据;美感的出现完全是碰运气,或者是以一种新的形式呈现出一个事先并没有想到的效果。从某种意义上说,哥特建筑应该属于这一类的结果;未来的钢铁建筑或许也可能产生这种意想不到的美感。但是在文艺复兴时期的意大利建筑中,绝对不可能有这种情况。建造技术方面的创新不能够解释这个时期意大利建筑到底在追求什么。标志着文艺复兴起点的布鲁乃列斯基的巨大穹顶,从宏大尺度与胆魄上,无疑是工程技术上的伟大成就;但是,这项工程并没有什么在原理上的创新,大穹顶上面所用到的那些计算、设计原理,与早先在比萨的穹顶,或者佛罗伦萨洗礼堂上面所采用过的原理没有什么差别。不但没有差别,而且刚好恰恰相反。文艺复兴时期的建筑尽管在规模上、在气魄方面都很惊人,但是与刚刚过去的上一个世纪里的建筑相比,在科学方面要简单得多。文艺复兴时期的建筑技术绝大多数是基于最简单的传统的古罗马建筑形式。不但如此,随着在建筑外表面上抹灰的做法大量使用,那些被抹灰盖住的真正建造技术则变得无所谓了。

文艺复兴时期倒是有过一种技术,在当时的工程中非常普遍地使用。但是它的使用只能证明这个时期的人们对于建造技术的漠视。这里所说的这项技术就是在拱券、穹顶根部位置常常采用的铁筋、铁链、拉杆来加强和保护这些拱券、穹顶,因为它们自身不足以抵抗向外的水平推力。这种权宜之计式的技术,对于中世纪的人们来说,绝对不会是什么秘密,只是他们不接受这种手段罢了。但是就是这种中世纪的建筑师不大喜欢采用的技术手段,到了文艺复兴时期却成为最常见、最有效地解决问题的方式。这方法中没有任何新东西,它的不断出现根本不能说明有什么新科技原理在发挥作用,只能说明人们关于艺术美感的新观

resource. There was nothing novel in the expedient. Its frequent recurrence signifies not the adoption of a new constructive principle, but the adoption of a new artistic point of view. The suggestive point about its use is that the element on which, in real fact, the stability of the construction depended was ignored, frankly and courageously, in the aesthetic design. The eye was expected to disregard it as completely as it disregards the prop which in ancient sculpture supports a prancing horse. That is to say, between the aesthetic purpose of the work, and the means by which, in actual construction, it could be realised, a sharp distinction was now admitted. How far such a distinction between construction and design is legitimate for architecture is open to dispute. The question, which is a difficult one, must be examined more closely in a later chapter. Here we may notice it merely as a confirmation of our statement, that it was not from any new constructive interest that the impulse of the Renaissance style was derived, or its progress defined. On the contrary, it is frequently objected that the decorative use of the Orders so conspicuous in Renaissance architecture did *not* express structure, that it was contrary to construction, and, for that reason, vicious.

Lastly, architectural design was not dictated, except to a slight degree, by the materials employed. This physical explanation of style is much favoured by modern critics, but it is singularly inapplicable to the period we are considering. Italy is rich in every kind of building material, and the architect could suit his needs. No doubt the great blocks of stone which could be quarried at Fiesole assisted the builders of the Pitti Palace, as it had assisted the Etruscans before them. Probably the inspiration lay rather in the Etruscan tradition than in the material itself. Still, had the Florentine builders rested content with the Etruscan masonry, it might be said, without essential untruth, that their materials determined their style. But the Florentines brought to perfection not only the most massive of Italian styles, but also the lightest. Their most remarkable achievement was a sudden power of quiet delicacy and grace. Conversely, when the baroque architects of Rome desired a monumental and Cyclopean effect, they obtained it without the Florentines' advantages. Again, the smooth *pietra serena* of Tuscany may lend itself to fine carving; but the passion of the Florentines for exquisite detail is no less marked in their painting, where no such factors operated, than in their architecture. Clearly, therefore, it sprang in both cases from an independent and native preference of taste. And, conversely, once more, the rough travertine of Rome did not yield up its 'natural' effect, its breadth of scale and roundness of feeling, until the baroque imagination, trained in painting to seek for soft transitions and broad shadow, began to require those qualities in architecture. Till then, travertine had been used, against its nature, in the Florentine tradition of sharp detail. In the Renaissance

念出现了。这个现象说明：技术上依赖它抵抗外向推力、保持建筑结构稳定的这个拉杆或者铁链，在艺术美学上、在视觉上则完全被忽略，而且是毫不掩饰地对它视而不见，仿佛它根本就不存在一样。建筑形式上根本没有考虑它的位置，观赏者的眼睛是根本不希望停留在这个地方上的。就好像在古代雕刻中出现的那些跑动中的马匹需要借助后面的支撑构件才能获得稳定一样，但是这些支撑构件在艺术构图中完全被排除在视觉效果之外。这证明了在这个时期，一件建筑作品中，它的美学设计造型与具体建造技术产生了分离，而且这种分离是为这一时期的人们所接受的。至于造型设计与建造技术之间的分离保持在什么程度范围以内才能为人们所接受，这还是一个没有定论的问题。这个问题很难一下子回答清楚，我们将在本书后面的章节里再作进一步的讨论。在这里，我们只是借此证明一个观点：文艺复兴时期的建筑风格起源绝对不是来自建造技术方面的创新与进步。不仅不是来自技术方面的创新和进步，而且恰恰相反，文艺复兴时期建筑中常常出现的装饰性使用古典柱式结构根本不是在表现建筑的真实结构，而是违背建筑技术原则的。对于文艺复兴时期建筑艺术的批评也常常基于这一点理由，也正因为如此，这种虚假的做法才是堕落的。

最后想说明一点是，建筑设计除了在极小的程度上受到建筑材料的一点影响以外，基本上与之没有什么关系。现代的建筑理论家十分热衷于用建筑材料等的客观物质因素来解释艺术造型风格，但是这个方法在讨论文艺复兴时期建筑的时候则根本行不通。意大利的建筑材料十分丰富，建筑师可以找到自己喜欢使用的各种东西。不用说，从菲耶索莱（Fiesole）采石场开采出来的巨大石块，对于建造彼提宫殿（Pitti Palace）的建筑师、石匠来说，的确帮助很大，这些石材在远古时代对于伊特拉斯坎（Etruscans）文明的形成也帮助非常大。但是，对于建造彼提宫殿的人们来说，他们所受到的启发更多的是来自伊特拉斯坎的文化，而不是那些巨大的石块。或许有人会说，如果我们看到佛罗伦萨的石匠们从来就对伊特拉斯坎的砖石材料的砌筑艺术觉得非常满意，那么就应该说是当地的石材决定了当地的建筑艺术风格。这个说法乍看起来也不是完全没有道理的。但是，佛罗伦萨人所精通的不仅仅只有粗大的石材的砌筑艺术，而且也精通精雕细刻的石雕技艺，也精通建造最为轻巧的砖石结构。他们最为著名的成就，正是他们在用精湛的技艺表现出安详与优雅中的一种爆发力。而事实上，当巴洛克时期的罗马建筑师想要获得庞大的纪念性尺度和巨大毛石石块砌筑效果的时候，他们并不需要继承佛罗伦萨人已经掌握的那些成就与技术。托斯卡纳地区出产的光滑灰色大理石为当地的雕刻家们雕刻出很精细的细部提供了条件；但是佛罗伦萨人在建筑中表现出来的那种精雕细刻的精湛技艺，同样也可以通过绘画的方式加以表达，而且不会受石材的影响。这两种情况说明人们都根据各自的喜好来选择自己的表现方式。再举个反例来进一步说明，罗马出产的石灰岩孔洞很多，但是，罗马人从来都不会去大肆炫耀这种多孔岩石的"自然"效果，如石板的体量规模，圆形的感觉，直到巴洛克时期，才有人受到作画手法的影响，强调柔性的过渡与宽大的光影的效果，开始把这种材料的特点加以发挥。在此之前，罗马人使用这种多孔的石灰岩的时候，总是违反它的自然属性，用它来模仿佛罗伦萨地区的那种材质很细腻的石材。在文艺复兴时期，想象力与创造力是第一位的，凡是艺术

the imagination came first; and where it existed it never failed to find materials for its expression. No doubt one material was better than another, and an architect accustomed, as were the Italians, to his tools, would take the best he could; but the men of the Renaissance were notoriously, and perhaps viciously, indifferent to the matter. If they conceived a design which called for a material difficult to obtain, they made no scruple about imitating it. Their marbles and their stones are often of painted stucco. When the blocks of masonry with which they built were not in scale with the projected scheme, the real joints were concealed and false ones were introduced.[1] And these practices were by no means confined, as is sometimes suggested, to the later and supposedly decadent phases of the art. Material, then, was utterly subservient to style.

Enough has now perhaps been said to suggest that Renaissance architecture in Italy pursued its course and assumed its various forms rather from an aesthetic, and, so to say, internal impulsion than under the dictates of any external agencies. The architecture of the Renaissance is pre-eminently an architecture of Taste. The men of the Renaissance evolved a certain architectural style, because they liked to be surrounded by forms of a certain kind. These forms, as such, they preferred, irrespective of their relation to the mechanical means by which they were produced, irrespective of the materials out of which they were constructed, irrespective sometimes even of the actual purposes they were to serve. They had an immediate preference for certain combinations of mass and void, of light and shade, and, compared with this, all other motives in the formation of their distinctive style were insignificant. For these other motives, being accidental, exerted no consistent pressure, and, consequently, were absorbed or thrust aside by the steady influence of a conscious taste for form. As an architecture of taste, then, we must let it rest, where our historians are so unwilling to leave it, or where, leaving it, they think it necessary to condemn: as though there were something degraded in liking certain forms for their own sake and valuing architecture primarily as the means by which they may be obtained.

What is the cause of this prejudice? What is the reason of the persistent attempt to force upon architectural art such external standards, and to explain it by such external influences? Clearly, it is this. Taste is supposed to be a matter so various, so capricious, so inconsequent, and so obscure that it is considered hopeless to argue about it in its own terms. Either, it is thought, we must resign ourselves to chaos,

---

[1] *e.g.* in the Strozzi Palace many apparently vast blocks of stone are made up of shorter ones with concealed vertical joints. In the Cancelleria, conversely, long stones are made to appear shorter than they are, by 'joints,' which are in reality only channels on the surface. In both cases the purpose is to maintain 'scale'; the unit of design, that is to say, is not material but aesthetic.

家能够想象出来的东西,他们总能找到合适的材料来表现自己的这种想象力。毫无疑问,建筑师对于自己的工具、做法最为熟悉,他们一定会去挑选自己最喜爱的材料。某一种石头对于他们来说一定比另外一种要好些。不单单是建筑师会这样做,每一位意大利人都会如此。但是,文艺复兴时期的人们最为世人所熟知的一点,就是他们对于建筑材料没有兴趣,任何材料对他们来说都一样的。如果他们想象中的理想设计需要某种材料,而这种材料又一时找不到的话,那么他们就会毫不犹豫地用其他材料仿造一个。罗马人有很多看上去是大理石、花岗岩的地方,实际上都是在抹灰的墙面上画出来的效果而已。当砌筑墙体的石材在尺度上与想象中所要求达到的设计不一致的时候,他们就把真正的接缝掩盖掉,然后把自己想要的虚假分缝强调出来。[1]这种做法绝对不是像某些理论家所说的那样,是艺术到了后期逐渐走向堕落时才出现的产物,实际上这个时期的人们一直就在不停地这样做。建筑材料归根到底是服务于建筑风格的。

意大利文艺复兴时期建筑所追求的形式以及建筑形式的演变过程都是出于美学方面的考虑,是出自艺术家主观内在的动力,而不是什么来自外界的影响所导致的结果。关于这一点我们已经说得足够多了。文艺复兴时期的建筑形式绝对是那个时期的人们审美喜好的结果。就是因为他们喜欢看见自己身边出现某种建筑形式,这些形式才因此得以出现。这些为当时人们所喜爱的建筑形式与它们的建造方式没有任何关系,与建造过程当中所使用的建筑材料也没有关系,有时甚至与这些建筑的实际用途也没有关系。这个时期的人们就是喜欢某种虚实对比、光影明暗等组合效果,相比之下,其他的所有因素在这种艺术形式的产生过程中都显得没有意义。因为其他的那些影响因素都是在具体的情况下出现的偶然因素,不可能产生持久的影响,而人们的审美喜好则是有意识地在不断发挥作用,所有那些偶然因素迟早会被人们的喜好所吸收或者排斥,最后的结果中依然是人们的喜好占据上风。既然是人们喜爱的样式,那么我们就应该接受这个事实。而历史学家们却偏不这样,他们总要找别的理由来解释它,最后实在不能解释的,就一定要找些理由来谴责批评它:好像单纯地为了满足人们喜爱而创造出来的艺术形式就是大逆不道,而帮助获得这种形式的建筑艺术就必定是腐败堕落的。

这种偏见的根源在哪里呢?为什么总是有人不断地试图用建筑艺术之外的因素或理由来解释建筑艺术本身呢?其实原因很简单。人们普遍地认为,每个人的审美喜好是多种多样的,而且变化无常,无法预测它的变化规律,同时个人的审美观念又是一些模模糊糊根本说不清楚的东西,因此无法加以严格的讨论。面对这个局面,我们要么就听凭这种混乱状

---

[1] 举例来说,在斯特洛齐宫殿(Strozzi Palace)上,有很多看上去是很大块的石头,其实是一些很小的石块拼凑起来的,真实的竖向接缝尽可能地被掩盖住。但是在坎切雷利亚(Cancelleria),又称文书院宫,这里的做法刚好相反。长条的石块看上去是几块石头垒起来的。真实情况是,那些看起来像接缝一样的地方,实际上是在石头表面的沟槽。在这两个案例中,艺术家的目的是相同的,就是要保证整体的尺度感。也就是说,决定设计的基本单位不是材料,而是审美习惯。

or we must exclude taste from our discussion, or we must reduce taste to terms of something more constant and reliable. Only by so reducing it can we control it, or hope to understand it. The tendency, in fact, springs from the impatience of the intellect in the presence of a factor which seems to disown its authority, and to be guided, if it is guided at all, by instincts of which the intellect can give no immediate account. It is an unconscious attempt to drill art into the ready-made categories which we have found useful in quite other fields, and to explain the unfamiliar by the familiar. It is the application to art of the methods of science, which sometimes are less concerned with the ultimate truth about its facts than with bringing them within the range of a given intellectual formula. But it is unscientific to persist in the application when it is clear that the formula does not fit.

We have dealt in this chapter with a point of historical fact. It is historically true that the distinctive control in Renaissance architecture lay not in construction or materials or politics, but, chiefly and typically, in the taste for form. It follows that it is reasonable to analyse the Italian styles primarily in terms of taste: to ask, how far do they fulfil that third 'condition of well-building' which Wotton names 'delight.'

But it is one thing to state how Renaissance architecture arose; it is quite another to estimate its value. For it may be rejoined that good taste in architecture consists in approving what is truthfully built—expressive alike of the methods and materials of its construction on the one hand, and, on the other, of the ends it has to serve; and that if the taste of the Renaissance was indifferent to these points it was bad taste, and the architecture which embodied it bad architecture. Thus, the very factors which, on the point of *history*, we have relegated to a secondary place, might still, on the point of *aesthetics*, resume their authority.

This view of architecture has many adherents. It finds confirmation—so at least it is claimed—in the greater styles of the past, in the practice of the Greek and Gothic builders. To ignore this rejoinder would be to fall into the common error of dogmatic criticism, and to neglect a large part of actual artistic experience. But it is a view of architecture which the Renaissance builders, at least, were far from holding. It is at variance with buildings which were enjoyed, and enjoyed enthusiastically, by a people devoted, and presumably sensitive, to art.

Confronted by those rival dogmatisms, how can we proceed? The natural course would be to examine the buildings themselves and take the evidence of our own sensations. Are they beautiful, or not? But on our sensations, after all, we can place no immediate reliance. For our sensations will be determined partly by our opinions and, still more, by what we look out for, attend to, and expect to find. All these preoccupations may modify our judgment at every turn, and

态继续左右我们的言行，要么就在理论中把审美喜好排除在外，用一种不变、可靠的概念来加以说明。把人们的审美喜好压缩到一个很小的范围里，希望借助于此举可以从此控制住它，并且最好能够理解它。这种做法实际上是反映了我们思维模式的缺欠，在面对众多其本身不具有说服力、也不具权威性的杂乱因素的时候，我们的思绪便本能地表现出一种急躁情绪，其结果也就造成在这一刹那，我们的直觉便不假思索地乘虚而入。这是我们自己无意识的尝试——强行地把艺术硬塞进到现成的那些在别的地方被证明很管用的框框里，然后，便可以用我们熟悉的东西来解释那些我们不熟悉的。这是把科学的方法搬到艺术领域里来了，但是科学的方法有时候有极大的局限性，只是在某种特别的情况下，它的公式才成立，但绝不是在任何情况下都成立的。当我们明明知道某一个公式实际上并不成立的时候，仍然强行使用这个公式，这样的做法显然是极不科学的。

　　在这一章里，我们重点强调了历史事实带给我们的一个启示：决定文艺复兴时期建筑艺术特征的因素不是建造技术、不是建筑材料、也不是社会政治，而是当时人们对于某种建筑艺术形式的审美喜好，这不仅是最主要的原因，也最能说明当时的实际情况。正因为如此，我们在研究文艺复兴时期建筑艺术的时候，很自然地应当从那个时期人们的审美喜好开始入手。我们也就会很自然问自己：在沃顿爵士给"优秀的建筑"所开列的三条件中的"令人愉悦"这一条面前，文艺复兴时期的人们到底做到怎样一种程度呢？

　　能够把文艺复兴时期的建筑艺术是如何兴起的过程说清楚是一回事儿，对它进行恰当的评估则完全是另外一件事。在评价建筑艺术的时候，我们可能又重新遇到反对我们的意见，即建筑艺术中好的品位应该是包含了对真实建造起来的东西予以肯定，一方面肯定建造技术的方法与建造材料得到真实地反映，另一方面，肯定它的使用功能得到应得的重视。如果文艺复兴时期的建筑爱好完全无视这几个方面的差别，那么当时人们的爱好就不是什么好的爱好，他们的品位也就不是什么好品位，而把这些审美品位具体化的建筑作品也就不是好建筑。因此，根据历史事实被我们推后至次要位置的那些因素，现在也许因为美学判断的理由会重新恢复了它们固有的权威性。

　　坚持用这一观念看待建筑艺术的人有很多。过去历史上的那些伟大的艺术风格，例如古希腊建筑、中世纪的哥特建筑，它们的创造者都证明了以上理论的真实性，至少有些人是这样认为的。如果对这种反驳意见视而不见，我们就落入了教条主义的最常见错误，也与绝大多数的艺术实践经验不相符合。但是，这种观念起码是文艺复兴时期的建筑师与建造者很少认同的观念。而我们发自内心真正喜爱、报以极大热情进行欣赏的文艺复兴时期的建筑无疑是与这种观念相左的，而打心底里喜爱文艺复兴时期建筑艺术的人都是对艺术十分敏感的人，他们全心全意地喜爱着这样的建筑艺术。

　　在这两个针锋相对的观念面前，我们该如何做出我们自己的抉择呢？最明显的答案是先认真研究这些建筑物本身，再结合我们自己在这些建筑面前的真实感受来做一个判断。这些建筑具有美感吗？还是没有？但是，话又说回来了，我们自己的感受其实并不可靠，因为这些感受无疑受到我们自己主观观念的左右，不仅如此，我们观察的对象在很大程度上已经受到我们主观的影响，因为我们一直在努力寻找的东西、我们感兴趣的东西、我们希望从中发现的东西让我们产生了某种期待，这一切早已不再是在客观地欣赏与感受了。这些先

interpose between us and the clear features of the art an invisible but obscuring veil. Before we put faith in our sensations, before we accept the verdict of others, it is necessary to examine, more closely than has yet been done, the influences by which contemporary opinion, in matters of architecture, is unwittingly surrounded and controlled.

入为主的成见实际上在每一步都影响了我们的判断,在我们自己与艺术作品之间插进了一道肉眼看不见的帷幔。在我们对自己的感受进行正确的判断之前,在我们接受别人的结论之前,还是让我们先来仔细地看看那些当前流行的各式各样的观念,而且是比以往更加严肃认真地看看这些观念,看它们是如何神不知鬼不觉地包围在我们周边,如何在不被察觉的情况下左右着我们的思想。

#
The Romantic Fallacy

The Renaissance produced no theory of architecture. It produced treatises on architecture: Fra Giocondo, Alberti, Palladio, Serlio, and many others, not only built, but wrote. But the style they built in was too alive to admit of analysis, too popular to require defence. They give us rules, but not principles. They had no need of theory, for they addressed themselves to taste. Periods of vigorous production, absorbed in the practical and the particular, do not encourage universal thought.

The death of the Renaissance tradition should have enabled men, for the first time, to take a general view of its history, and to define its principles, if not with scientific exactness, at least without provinciality or bias. Of the causes which precluded them from so doing, the first was the prolonged ascendency of the Romantic Movement.

The Romantic Movement created in all the arts a deep unrest, prompting men to new experiments; and, following on the experiments, there came a great enlargement of critical theory, seeking to justify and to explain. So it was with the theory of architecture. How far, in this change of thought, has it been strengthened and enriched; how far encumbered and confused? A clear view of Renaissance architecture requires an answer to this question.

Although, in every department of thought, there are principles peculiar to it, necessary to its understanding, and with reference to which it should properly be approached, yet all the elements of human culture are linked in so close and natural a federation, that when one among them becomes predominant, the others are affected to an instantaneous sympathy, and the standards appropriate to the one are transferred, with however little suitability, to all.

Such, towards the close of the eighteenth century, was the case of the Romantic Movement, which, from being an enlargement of the poetic sensibility, came, in the course of its development, to modify the dogmas and control the practice of politics and of architecture. By the stress which it laid on qualities that belong appropriately to literature, and find place in architecture, if at all, then only in a secondary degree,

第二章
浪漫主义艺术理论的谬论

我们不能说文艺复兴时期产生过真正意义上的建筑艺术理论。产生过的只是一些关于工程做法的规则：方济各会的教士乔冈多（Fra Giocondo）、阿尔伯蒂（Alberti）、帕拉迪奥（Palladio）、塞利奥（Serlio）及其他人，这些人不仅仅建造自己的房子，而且也执笔著述。但是这些人在建造时所采用的建筑风格还在成长当中，在当时还无法对其进行分析；同时，这些风格也深深受到人们的喜爱，自然又不需要为之辩护。因此，在他们所有的论述中，这些理论家们只是给出建筑艺术中具体做法上的一些规则，没有什么大的思想原则方面的主张。他们不需要理论的支持，因为他们所遵从的是发自内心的喜好。在一个大家都乐此不疲地忙于具体建造的时代，一个兴趣与爱好都非常独特的时代，它是不可能产生出什么"放之四海而皆准"的带有普遍性的艺术思想的。

文艺复兴时期消亡之后，人们应该有机会好好反省这一段历史了吧？来看看它的全貌，找出其中的一些规律，即使不那么严格与科学，也总不至于犯下狭隘偏见的错误。但是，这种总结却一直就没有出现过。在这么长的时间里没能出现对于文艺复兴时期的检讨，原因有多方面的，但是最主要的原因就是长期以来流行的浪漫主义文化运动。

浪漫主义运动给各种艺术形式带来深刻的影响，在艺术领域引发了各种骚动与探索，它鼓励人们尝试新的艺术表现手段；伴随着这些探索与尝试，艺术批判理论的作用便膨胀起来，它们出来论证与批判各种创作，试图解释这些新作品。建筑艺术理论也以这样的方式大量地出现。在这种思想的转变中，有多少建筑理论是得到了强化并且丰富了其中的内容呢？又有多少反而是被弄得踉踉跄跄，甚至把自己本来的思想也搞乱了呢？要想弄清楚文艺复兴时期的建筑艺术，上面这类问题必须首先得到回答。

在人们思想的每一个方面都有属于自己的原则，离开了这些原则就无法理解它，有了这些原则，也就找到理解问题的途径与参照系统。但是同时，人类文化中所有的元素是紧密地联系在一起的，也很自然地形成了一个庞大的联合体，其结果就是，联合体中的某一个元素成为主导之后，其他元素的会立刻受到影响，自动地与主导元素互动。而当对应于主导元素的准则得到改变的时候，无论这种准则对于其他元素来讲是否适用，这些其余的元素也都跟着一起向新准则看齐。

到了十八世纪快要结束的时候，上面说的这种情形正是浪漫主义运动的真实写照。浪漫主义运动的出发点本来是诗歌领域里的一种情感，却在演变的过程中被不断地扩大，最后渗透到政治与建筑艺术里面来了，成为当时的一种主导倾向，改变了当时的艺术理论。这种浪漫主义艺术理论借助于它在文学领域里所发挥的作用来对建筑艺术施加影响，只是在程度上略微减弱，但它还是把建筑艺术的真正意义扭曲了，让人迷失方向。即使到了今天，浪

it so falsified the real significance of the art that, even at the present time, when the Romantic Movement is less conspicuous in the creation of architecture, the fallacies we shall trace to it are still abundantly present in its criticism.

Romanticism may be said to consist in a high development of poetic sensibility towards the remote, as such. It idealises the distant, both of time and place; it identifies beauty with strangeness. In the curious and the extreme, which are disdained by a classical taste, and in the obscure detail which that taste is too abstract to include, it finds fresh sources of inspiration. It is most often retrospective, turning away from the present, however valuable, as being familiar. It is always idealistic, casting on the screen of an imaginary past the projection of its unfulfilled desires. Its most typical form is the cult of the extinct. In its essence, romanticism is not favourable to plastic form. It is too much concerned with the vague and the remembered to find its natural expression in the wholly concrete. Romanticism is not plastic; neither is it practical, nor philosophical, nor scientific. Romanticism is poetical. From literature it derives its inspiration; here is its strength; and here it can best express its meaning. In other fields—as in music— it has indeed attained to unimagined beauties; but always within certain limits and upon fixed conditions. For here, on a borrowed ground, if it fail to observe the laws which music, or architecture, or life, as concrete arts, may impose, then even that element of value with Romanticism introduced, becoming mute and ineffective, is sacrificed in the failure of the whole.

It would be a mistake to imagine that Romanticism was in any way a new force at the time when, with the French Revolution, its various manifestations came into such startling prominence as to require attention and receive a name. Any movement strong enough to become conspicuously dominant must long previously, it is safe to suppose have been latently operative. And, in architecture, although the Romantic Movement of the nineteenth century dealt the final death-blow to the tradition of the Renaissance, yet that tradition, it must not be forgotten, was itself a romantic movement. The cult of mediaevalism, stimulated by the revival of ballad literature and by antiquarian novelists, is not more romanticist than the idealisation of antiquity, four centuries earlier, stimulated by the revival of classic poetry and the enthusiastic antiquarianism of Paduan scholars. Nor, for that matter, is it more romanticist than the neo-Greek architectural movement of the Hellenising emperors in antiquity itself. Why, then, it is natural to ask, should a motive which in the second and fifteenth centuries proved a source of strength, be regarded, in the nineteenth, as a disastrous weakness?

漫主义不再明显地左右着建筑艺术创作的时候,它所产生的各种谬论仍然大量地出现在建筑理论当中。

有人认为,浪漫主义如果沿着自己的理想方向发展,它会具有一种向往遥远的过去或者未来的性格,具有诗歌般的情感,事实的确就是这样的。它把一切遥远的东西都理想化,这里所说的遥远不但是时间上的遥远,也是空间距离上的遥不可及;它崇尚奇特、怪异为美。在古典艺术审美意识中那些被千方百计予以回避、排斥的追求新奇与采用极端手段,这时成了创作的源泉,古典审美意识所反对的那些朦胧细节在浪漫主义时期成了灵感的来源。浪漫主义艺术思潮在大多数情况下,总是不停地缅怀过去,无论当前的生活多么有价值,都因为它们是当前的东西,都因为大家对它们太过熟悉而被抛弃。浪漫主义总是在试图表现出它把一切都要理想化的倾向。浪漫主义理论家把自己没能实现的梦想转化成一种虚幻的过去,再把这个假想出来的所谓过去投影到屏幕上给大家看。他们最典型的手法就是对那些早已绝迹的东西表现出自己的无限崇拜。从根本上来讲,浪漫主义对于任何需要有具体可塑性造型的艺术来讲,几乎都是行不通的。它过于钟情于那些虚无缥缈的东西,过于钟情于记忆中的东西,而这些东西却无法用真实具体的手段很自然地表现出来。浪漫主义不是一种造型的艺术——它也不具有可以操作的实用性;它不是思辨哲学给人以启发;它也不是科学讲究事实。浪漫主义是一种诗歌意境,它的灵感来自于文学,文学领域才是它的用武之地,是它的强项,也只有在文学领域,它才有可能最好地表现出自己的意义。当它进入到其他领域里,比如说音乐,它也的确产生过某些意想不到的效果与美感,但是,这种美感的获得是要有很多条件限制的。在新的领域里,诸如音乐、建筑艺术,甚至生活本身,这些都是很具体的艺术领域,它们都有自己的固有规律和原则,如果违反了这些固有的原则与规律,那么,浪漫主义就自然会变得毫无作用,也没有发出自己声音的机会,伴随浪漫主义而来的新的价值观也就随着浪漫主义的失败而消亡。

浪漫主义刚好与法国大革命出现在同一个时期,浪漫主义在很多方面同时出现在世人面前,形成一种巨大的影响力而得到人们的注意,而且它的影响力之大,使得我们必须要给它起一个独立的名字才能在以后的讨论中清楚地表述。但是,我们因此就认为浪漫主义是一种全新的力量,那就错了。任何能够成为社会上一种具有广泛影响的力量,在获得社会的注意之前,一定是已经酝酿得很久了,一定是在还没有人注意到它的情况下一直在暗处发挥着作用的。十九世纪兴起的建筑艺术里的浪漫主义,可以说是带给文艺复兴运动的最后致命一击的力量。但是,请大家不要忘记,文艺复兴运动本身其实就是一种非常浪漫的文化运动。后来由于中世纪传奇文学叙事体故事得到流行,因为以考古为主题的故事小说的兴起,中世纪文化从而得到了社会上极大的崇拜。但是,我们还不能说后来出现的这种对于中世纪文化的崇拜就一定要比在它四个世纪之前出现的对于古希腊、古罗马文化的崇拜更多一些浪漫色彩。四个世纪以前对于古典文化的热衷也是因为对于古代诗歌的热爱,以及帕多瓦(Paduan)的学者们对于古典时期文化的热爱。按照这个思路再往前看,古希腊后期的泛希腊化时代的帝王君主们也兴起过复兴古希腊建筑风格的运动。十九世纪浪漫主义时期对于中世纪的崇拜也不会比海伦时代的这些做法更加富有浪漫色彩。那么我们不禁要问:在二世纪和十五世纪里的那些富有生命力的做法,为什么到了十九世纪就会被认为是具有灾

The reason is simple. Architecture is the art of organising a mob of craftsmen. This, the original meaning of the word, expresses an essential fact. You can pass, in poetry, at a leap from Pope to Blake, for the sleepiest printer can set up the most original remarks. But the conceptions of an architect must be worked out by other hands and other minds than his own. Consequently, the changes of style in architecture must keep pace with the technical progress of the crafts. And if, at the bidding of a romantic fashion, an abrupt change of style be attempted, then the technique and organisation required by the new ideal must not be more exacting than those employed by the existent art. For neither technique nor organisation can be called into being suddenly and at will.

For this reason the romantic return to mediaevalism failed, where the classic revival succeeded. The latter was concerned to restore the space, proportion and coherence of abstract design; and these the individual architect like Brunelleschi might hope to provide. He asked less, not more, of his craftsmen. The resources of skill acquired during centuries of Gothic practice were, technically, more than sufficient for the new tasks. Imaginatively, his sculptors were already imbued with the new classicism. And if, when they carved a frieze, they still betrayed some Gothic fancy, it mattered little: the point lay less in what they carved than in where they carved it, and this he could determine and control. But in the nineteenth century these conditions were reversed. To recreate the mediaeval vision was incongruous with men's life. The new ideal required a variety of skilled crafts that were irrecoverably lost, and the architect, with nothing but his scholarship, set out to restore a style that had never been scholarly.

The purpose of romanticism should have been the fusion of a poetical interest with the forms and principles of an existing art. Had the Romantic Movement complied, even in some degree, with the essential conditions, a genuine architectural style might have been created, formed, as it were, out of the materials of that which it superseded. In some directions, while the good sense of the eighteenth century still controlled the situation, this was indeed accomplished. For the first signs of the change had been innocent enough. In the middle of the eighteenth century, that romantic attitude, which later was to culminate in a wholly false aesthetic, can already be recognised in a certain restlessness and satiety with native and traditional forms, and in a tendency to take interest in remote kinds of art. One of the earliest indications of this spirit is the taste, prevalent at that time in French society, and imitated to a less degree in England and in Italy, for the art of China, which Eastern

难性的缺陷了呢?

其实理由很简单。建筑艺术是把一群暴徒一样的工匠们纠集在一起,大家共同来完成的一件艺术作品。工匠这个词非常能够说明建筑创作活动过程中的一些本质。在诗歌领域,人们可以很容易地从蒲柏(Pope)诗歌风格转变成布莱克(Blake)诗歌的风格,因为最无精打采的印刷工人也可以很出色地完成这份印刷工作,诗歌创作的任务早已经在诗人那里就完成了。但是在建筑领域,建筑师的构想则必须借助于其他人的帮助和参与才能完成。正因为这个缘故,建筑艺术中的风格转变必须有相应的匠人技艺与建造技术作为后盾。在浪漫主义艺术风格兴起的时候,如果一种全新的建筑风格突然地被提出来,那么这种风格的改变,它对于工匠技艺的要求,对于施工组织的要求,都不能超越当时的具体实际情况,因为技艺和施工组织工作不是在人们想要的时候就会自动出现的,它们是不以人们的意志为转移的。

正因为如此,浪漫主义为回归中世纪的艺术所作出的种种努力也就失败了,而文艺复兴时期的古典主义复兴则成功了。因为后者追求的是还原空间、比例关系以及各种抽象东西彼此之间的和谐,布鲁乃列斯基以及其他建筑师所追求的正是这些东西。他们对于工匠没有提出更高的要求,而是比从前那个时代对于工匠的要求还要低。而在那之前几个世纪里,熟悉哥特建筑风格的工匠们在技艺上早已超出文艺复兴时期的需求,因此很容易满足新风格的要求。在想象力方面,当时的工匠已经掌握了古典艺术所需要的技艺。在工匠建造古典建筑的时候,某些雕刻或许会需要这些工匠掺入一些哥特风格的细节,但是这些细节对于古典主义风格的建筑师来说,问题都不是很大,他们对于雕刻出现在什么地方的关注更胜过对于雕刻内容的关注,而雕刻出现的位置则正是建筑师能够决定并进行有效控制的地方。但是到了十九世纪的时候,这些过程刚好翻转过来。人们已经不再熟悉中世纪的生活,对于中世纪根本没有任何感觉。而新的艺术潮流所要求的那些特殊技艺早已经散失殆尽,根本无法再寻找回来。建筑师也只是从书本中了解了一些皮毛,但是还是不知道该怎样制作出中世纪风格的东西。建筑师依赖的是文献记载和考古研究,但是这种东西从来就没有成为学术而流传下来。

实际上,浪漫主义艺术运动所寻求的目标,从一开始就希望能把诗歌里的那种浪漫情感融合进当时已经存在的具体艺术形式与艺术原则当中去。如果浪漫主义的追求在一定程度上遵循了那些艺术中的基本条件和原则,哪怕只遵守了一部分,而不是试图去取代它们,那么,在那些已有的艺术手段的基础上,或许某种全新的艺术形式真的就因此被创造出来了。在某些地方,十八世纪里好的东西能够得到延续,并且仍左右局部情况,这些地方的确创造出了我们说的这类新作品。从古典主义向浪漫主义转化的最初阶段,也是非常自然、无意识地发生了的。到了十八世纪中期,浪漫主义的艺术已经开始表现出一些焦躁不安的症状,不再满足于原来传统的艺术形式,开始向往遥不可及的东西,但是,还不至于向后来那样干脆采用装腔作势的虚假手段。这种艺术趋势最初的症状就是当时欧洲人开始表现出对于来自中国的那些精巧玲珑东西的极大兴趣。崇尚来自中国的手工艺品在当时的法国上流社会非常流行,随之而来的便是英国人和意大利人又开始模仿法国人,但是不像法国那样疯狂。中国的东西借助于前往远东做贸易的商人以及耶稣会的传教士,才为欧

commerce and the missionary efforts of the Jesuits had made known.[1] In this case no condition of concrete art was offended. For one of the phases of Renaissance art, which will fall in due course to be examined, was the translation into architectural language of our pleasure in rapid, joyous, and even humorous physical movements. In France, this phase was embodied in the art of Louis xv. It was contemporary with the climax of that interest in the Chinese which, we have said, was an early instance of the romantic spirit. Now, in its predilection for gay and tortuous forms, as also in its love of finish, the art of China (as the French understood it) was perfectly congruous with their own. It required no organisation which contemporary art was not able to supply; and the zeal for it came at a time when architecture was so vigorous that it readily assimilated such elements of the new material as suited its requirements, and produced, in the *Chinoiseries* of the eighteenth century, a charming invention, which, while it gratified the romantic instinct of the age, added, at the same time, to its appropriate decorative resources.

  The successive stages of the Gothic taste exhibit very clearly the character of romanticism, and the point at which it overweighs the sense of form. Up to the middle of the eighteenth century the mediaeval style merely spelt discomfort, desolation, and gloom.[2] Noble owners, so far as their purse allowed, converted their Gothic inheritances, as best they could, to the Georgian taste, or rebuilt them outright. Then enters the spirit of history, the romance of the distant and the past, with archaeology at its heels. The connoisseurs, about 1740, are full of zeal for the stylistic distinctions between the Egyptian, the Gothic, and the Arabesque, and charmingly vague about their limits. Their studies are pursued without calling in question the superior fitness of the classical tradition. Nevertheless, the orthodoxies of archaeology now hold sway. They are submitted to not without reluctance. Gray, in 1754, writes of Lord Brooke, at Warwick Castle: 'He has sash'd the great Appartment... and being since told that square sash-windows were not Gothic, he has put certain whim-wams within side the glass, which, appearing through, are made to look like fret-work. Then he has scooped out a little Burrough in the massy walls of the place for his little self and his children, which is hung with chintzes in the exact manner of Berkley Square or Argyle Buildings. What in short can a lord

---

[1] The Chinese Trading Company of Colbert was founded in 1660; the Compagnie des Indes in 1664. From 1698 to 1703 the *Amphitrite* cruised in Chinese water. *Vide* J. Guérin, *Les Chinoiseries au XVIIIme Siècle*.

[2] There were not wanting those who maintained this opinion throughout the whole period of the romantic movement. In 1831, when it was at its height, even the stately and tempered mediaevalism of Knole still inspires the Duchesse de Dino with the utmost melancholy: 'Cette vieille fée (the housekeeper) montre fort bien l'antique et lugubre demeure de Knowles, dont la tristesse est incomparable.' —Duchesse de Dino, *Chronique*.

洲人所知晓[1]。到这个时候为止，新兴的浪漫主义还没有对任何一门艺术产生过任何特别的影响，因为文艺复兴时期中的巴洛克阶段就是把我们对于快速动感、愉悦、甚至幽默感转化成建筑语言。关于这一阶段我们会在下面合适的时机深入地加以探讨。在法国，这个时期刚好与路易十五时代同步，法国对于中国的艺术表现出极其浓厚的兴趣，这一点我们在前面说过，它是浪漫主义时代的滥觞。人们这时的兴趣是追求享乐，喜欢怪异的形式，在装修材料方面也是极尽奢华。恰好他们喜欢的这些东西，在中国艺术品（按照法国人所理解的中国艺术）中都能够找到。而且不需要什么特殊的技艺，现有的技艺足够应付这些需求。欧洲人对它展现出高涨的热情恰好是在建筑艺术充满旺盛活力的时刻，所以，当时很容易地模仿出十八世纪*中国风*（Chinoiseries）建筑风格所需要的各种材料与工艺，创造出一系列的精巧、讨人喜欢的东西，这些东西一方面迎合了当时的浪漫主义艺术潮流，同时也丰富了建筑装饰手段。

接下去风起云涌的哥特艺术风格，它的发展经历了一系列的不同阶段。这些发展阶段清楚地表现出浪漫主义的追求与特点，而且达到了为了浪漫效果而不顾建筑形式规律的程度。直到十八世纪中叶，中世纪的建筑带给人们的一般印象从来都是不舒服、凄凉、昏暗[2]。有钱有势的人、贵族，只要口袋里有了一定的积蓄，就都会忙不迭地纷纷张罗，把从祖上继承下来的哥特风格的建筑重新加以改造，变成乔治时期流行的古典建筑风格，有的甚至干脆推倒重建。后来，大家对历史开始感兴趣，对于过去与遥远的地方开始憧憬，考古学也随即兴起。1740年前后的艺术鉴赏家和爱好者们极其狂热地追捧古埃及风格、哥特风格、阿拉伯风格等争奇斗艳的艺术形式，觉得它们各种潜在的应用场合具有无限的发展空间。古典主义风格比所有这些风格更能适应不同场合的需求，这一点在针对那些风格的研究中并没有被质疑。不管怎么说，在当时，考古学占据了上风，这些艺术爱好者们也就身不由己地跟着风潮走，虽不十分情愿，也没有办法。1754年，葛瑞（Gray）曾经在文章中说到布鲁克勋爵（Lord Brooke）建造华威克城堡（Warwick Castle）的故事。他说："勋爵在给城堡里的居住空间安装窗扇……因为听人说，正方形的窗扇不是哥特风格，他就在玻璃的内侧衬上了一些装饰，从外面看上去有点花草一样的花纹。然后他在厚厚的城堡墙体里挖出几个小洞穴给他自己以及孩子们栖身，这些小洞穴都挂着窗帘，就和伯克利广场周边的住宅里挂的窗帘一样。换句话说，一位勋爵在一座庞大、坚固的古老城堡里能怎么样做呢？他只能找一个地方躲起来，就如同一个老鼠一样，慌忙中钻进它看见的第一个洞

---

[1] 库尔伯特的中国贸易公司创立于1660年；印度公司（Compagnie des Indes）始于1664年。从1698年开始到1703年为止，阿芙特里特号商船行驶在中国海域。详情参见 J. Guérin 编撰的《十八世纪中国丛书》（*Les Chinoiseries au XVIIIme Siècle*）。
[2] 在浪漫主义盛行的整个过程中，持这种观点的人为数众多。即便是在浪漫主义达到高峰的1831年，迪诺公爵夫人在看到气派又时髦的Knole哥特风格的豪宅的时候，忍不住说道："这个城堡的主人看起来的确是非常地好古，也让人沉闷不堪，他们的不幸真是让人无法忍受。"见《迪诺公爵夫人年谱》。

do nowadays that is lost in a great, old, solitary castle but skulk about, and get into the first hole he finds, as a rat would do in like case?'[1] But the vital taste of the time could not rest satisfied with archaeology. The Gothic forms were a romantic material, rich with the charm of history. Could they be fused with the living style? Batty Langley thought they could, and by no other mind more readily than his own. 'Ancient architecture, restored and improved by a great variety of grand and useful designs, entirely new, in the Gothick mode'; 'Gothic Architecture, improved by rules and proportions.' These were the titles Langley successively affixed to the first two editions of his work. They show two alternative ways of regarding the same question—the Gothic, steadied and sobered by 'proportion'; the ancient architecture made various with Gothic fancies. Here was no question of a mediaeval revival, as the next century understood it, but a true attempt at fusion. But then the two elements to be fused were utterly incongruous. If this was not clear before, Batty Langley's designs must have made it obvious to all who were not blinded by historical enthusiasm. And, on the whole, the right inference was drawn. 'Gothic Umbrellos to terminate a view'; Gothic pavilions for 'the intersection of ways in a Wood or Wildernesse,' were well enough. Here they might be admitted as curiosities—as literary reminders of the romantic past, or shrines to the poetry of nature with which the mediaeval style was conceived to be related. Above all, they might act as a foil to the classical elements themselves, and do a dual service by stimulating the sense of history while they set off the immaculate consistency of the time. The Gothic suggestions might even penetrate the house. They might, without discordancy, provide the traceries of a book-case or enrich the mouldings of a Chippendale table. Here and there, in the light spirit of fashionable caprice, they might furnish the decoration of a room, just as, elsewhere, an Eastern scheme might dominate. But to go further and Gothicise the main design, seemed at the first an obvious fault of taste. 'I delight,' writes Gray to Wharton, 'to hear you talk of giving your house some Gothic ornaments already. If you project anything, I hope it will be entirely within doors: and don't let me (when I come gaping into Coleman Street) be directed to the "gentleman's at the ten pinnacles" or "with the Church Porch at his door."' [2] And when, at Strawberry Hill, Horace Walpole allowed a quaint imitation of mediaevalism to furnish his whole design, the concession, startling and even absurd as it seemed to his contemporaries, was made in a spirit of amused pedantry and conscious eccentricity, or, at most, of archaeological patronage; nor could the amateurs of that time have credited the idea that the trefoils

---

[1] *Letters of Thomas Gray*, edited by D. C. Tovey, vol. I. No. cxiv.
[2] *Letters of Thomas Gray*, vol. I. No. cxiv.

穴。"[1]当时全社会的主要兴趣并不满足于考古发现与研究。哥特风格的艺术形式是无尽的浪漫话题，充满了丰富的历史故事。这些形式能否与当时的实际生活结合到一起呢？巴迪·朗格雷（Batty Langley）认为是可以的，而且是完全不用借助他人的观点来支持自己的见解。《通过公式化的哥特风格建筑对古老的建筑进行改进和整建》《哥特建筑比例关系与改进原则》是朗格雷出版的两本书的名称。这两本书说明了两种不同的方法来从事哥特风格的建筑设计，一个是注重稳重、严肃的比例关系，一个是用哥特的手法来改造古老建筑。我们看到，这时的哥特风格建筑还没有过渡到下一个世纪里出现的追求中世纪全面狂热复兴的程度，完全是试图在现有的基础上融合进一些哥特艺术的特征。但是这两个被融合进哥特建筑元素的地方则是完全地不合理。如果在此之前，大家还没有注意到这一点，那么，朗格雷的设计可以让大家很清楚地看到这些，除非是被复古的狂热冲昏了头脑。而且根据这些设计，他还归纳总结出一些结论。"哥特风格的亭子作为视觉通道的终点对景"，"哥特风格的亭子应该用在野趣与树林中的十字路口处"，仅这两个例子就足够了。这些建筑小品或许是因为主人出于好奇才引进来，作为一种具体的实物来提醒主人充满浪漫的过去，或是对大自然诗情画意的祭拜。中世纪的艺术风格在人们的观念中是与大自然密切相关的。总之，这些东西都是在古典主义的建筑上增加一些点缀和装饰而已，一方面带给人们某种历史感，另一方面又符合当时的潮流。哥特风格的艺术甚至渗透到建筑里面，如在书架上加上一点哥特的花纹，在桌子上加上一些哥特的线脚。这里一点、那里一点，甚至某一个房间也都装饰上哥特风格的花纹式样，就好像以前在某个房间全部装饰成伊斯兰风格那样，也不能说它们不协调。但是从这里把整个建筑物按哥特风格进行改造就开始显露这种做法的不当之处了。葛瑞给沃顿（Wharton）写信说："我听你说打算把你的房子按照哥特风格增加一些装饰。如果你真的要添加什么东西，我希望你把它们全部控制在室内。我不希望当我出现在科尔曼（Coleman）大街问路的时候，人家告诉我'就那个上面带有十个尖塔的男厕所'，或者'门口像教堂门廊'的那座房子。"[2]当霍瑞斯·瓦尔泊（Horace Walpole）在草莓山（Strawberry Hill）大兴土木的时候，他让建筑设计模仿了中世纪的风格，这在当时人们的眼中简直是令人吃惊、不可思议的。整个建筑是以玩票的心态对待中世纪的文化研究，是故意作出与众不同的姿态，最多是赞助了一次考古活动。当时对这个项目关注过的人们，他们也没有把瓦尔泊这个好玩的建筑当一回事儿，没有意识到这个建筑上面的那些哥特装饰和尖塔很快会成气候，而且最后终结了那些人自己还在实践着的东西，改变了人们对待艺术的态度。这种情况多少带有一点讽刺意味，这种讽刺也发生在法国，而且是因为法国贵族所鼓吹的"自然"平等的理论与启蒙思想（这些理论和思想本身就是浪漫主义的具体表现）而使得结果更加悲惨。平等的思想注定让这些堂皇又高尚的热情，连同那些哲学与启蒙思想一道，统统地被消灭。

---

[1] 见《托马斯·格雷书信集》，D. C. 托维整理，第一册，第 114 封
[2] 见《托马斯·格雷书信集》，D. C. 托维整理，第一册，第 114 封

and pinnacles of Walpole's toy heralded a movement which would before long exterminate alike the practice and the understanding of their art. The irony of this situation has an exact and tragic counterpart in the favour accorded at that epoch by the more philosophic and enlightened of the French aristocracy to those theories of 'natural' equality (themselves another expression of romanticism) which were destined to drive these noble patrons, their philosophy and their enlightenment, entirely out of existence.

  Side by side with this sense of Gothic as an amusing exotic—an attitude which was thoroughly in the Renaissance spirit and characteristic, above all, of the eighteenth century—there grew up a more serious perception of its imaginative value. When Goethe visits Strasburg Cathedral it is no longer, for him, the work of 'ignorant and monkish barbarians,' but the expression of a sublime ideal: and Goethe's mind foreshadows that of the coming century. At the same time he has no quarrel with the existing standards; a complete reaction against these is as yet unimaginable. But a change of attitude shows itself both with regard to Gothic and also to the living style. These now came more and more to be regarded *symbolically*, as standing for certain ideas. And in particular the habit arose of regarding Greek and Gothic art as contrasted, parallel and alternative modes of feeling. But the good taste of the period, although already permeated with Romanticism, recognised this distinction between them: the Gothic must remain an external object of admiration; the Greek feeling could be fused with the existing art, the Greek forms grafted on to, or extricated from, the living tradition. Just as it had required no impossible change to impart a Chinese turn to the gay Renaissance style of Louis XV., so, with equal facility, the romantic idealisation of Greece could be expressed by emphasising the elements of severity in the essentially Renaissance style of Louis XVI. But a species of literary symbolism becomes increasingly evident in the attempt. *The interest is shifted, more and more, from the art itself to the ideals of civilisation.* The Greek modes of the period are deliberately meant to 'suggest' its political or other doctrines; and the intrusion of Egyptian detail which followed Napoleon's African expedition is an instance of the same allusive tendency. Thus, though an apparent continuity is still maintained, a radical change has taken place. A romantic classicism of sentiment and reflection has overlaid and stifled the creative classicism which sprang up in the *quattrocento* and till now had held control. In imparting to the Renaissance tradition this literary flavour, in adopting this unprecedentedly imitative manner, the vigour of the Renaissance style was finally and fatally impaired. In obedience to the cult of 'ideal' severity it cut down too scrupulously all evidence of life; and when, with the passing of the old order of society, vanished also the high level of workmanship and exquisite

这种做法无疑是把哥特文化当作一种可以满足猎奇的对象,这种态度与文艺复兴时期艺术发展到了十八世纪时所表现出的精神与特征是一致的。与此同时,还有一种严肃对待哥特艺术形象的做法。当歌德(Goethe)造访斯特拉斯堡大教堂(Strasburg Cathedral)的时候,对于歌德本人来讲,这座教堂已经不再是"无知僧侣野蛮人"的东西了,而是一件表现出崇高理想的艺术作品,歌德的认识预示了接下来一个世纪里的主要思想观念。但是同时他也不会去责怪当时流行的行事风格与原则;我们无法想象,那个时候的人们会全盘否定自己当时普遍存在着的通行做法。但是,对于哥特的艺术与人们的生活方式,这时已经出现了与以往不同的态度。哥特艺术与生活方式越来越多地成为某种象征性的东西,它们只是代表了某种理念的符号。尤其是这种态度的转变,让我们产生了古希腊艺术与哥特风格的艺术是两种具有对比性质、互相平行、地位相当的艺术的观念,认为它们代表了两种不同的思想感受。但是,当时的主流审美观念即便是受到浪漫主义艺术思潮的影响,仍然认识到二者之间的根本差别:哥特风格的艺术必须与生活保持距离以便从外部进行观赏,而古希腊的艺术可以与其他艺术融合到一起的,古希腊的艺术形式可以嫁接到活着的传统里,也可以与传统脱离。正如在路易十五的时代,没有经过任何特别的艰苦努力,便在欢乐的文艺复兴风格中融进了中国风的东西一样,在路易十六时代也轻易地在以文艺复兴为主流的风格中融进了经过浪漫主义洗礼的古希腊艺术形式。但是,文学艺术里面的象征手法在艺术领域里越来越明显了。*大家的兴趣已经越来越多地从具体的艺术作品本身转向于代表各种文明的理想*。在这个时代,当人们提到希腊的东西,更多的是指它所代表的政治理想或者其他的思想;随着拿破仑在非洲的扩张,人们对于埃及的建筑兴趣也同样用来象征某种政治理想。因此,从表面看上去一切似乎还是依然如故,但是,思想意识的改变已经悄然地发生了。包含了感伤与思考成分的古典主义艺术胜过了具有独创性的古典主义艺术,而这种独创性是自十五世纪以来一直占主导地位的艺术力量。这种文学艺术的东西渗进文艺复兴的艺术传统里面以后,这种从前没有过的以模仿过去为时尚的做法,终于带给文艺复兴运动致命的创伤。为了追求严肃的理想,一切与生活相关的痕迹都被仔细认真地剔除干净;随着老一代人的陆续过世,从前的那些精湛的技艺也跟着他们永远地消失了,经过漫长时间积累形成的社会价值观念也被抛弃了。古典主义的风格真正地变成了废墟,粗制滥造的设计刚好有粗制滥造的技艺相配合。从布鲁乃列斯基开始的古典建筑到此时则完全变成了理性的机械制作,到了帝国风格的末期,古典主义建筑的源泉经历了这么长时间后似乎已经枯竭;拿破仑的建筑师采用古典主义手段为他建造了不少纪念碑,也为文艺复兴艺术撰写了墓志铭。

ordering of ideas which that society had exacted, then the ruin of the classical style was consummated, and poverty of execution completed what poverty of design had begun. The antique, which Brunelleschi invoked, was now realised with full self-consciousness; in the last stages of the Empire style the resources of classic architecture seem at length to be exhausted; in that style the architects of Napoleon built the monument, and wrote the epitaph, of Renaissance art.

But the romantic impulse, when it had thus dealt the death-blow to the living Renaissance tradition, still had its course to run. The attitude of mind of which the Empire style was the classical expression had yet to manifest itself in other forms less fit. Its final and definitive achievement was, of course, the general revival of Gothic. Towards this end the literary and sentimental currents of the time combined more and more powerfully to impel it, and as the nineteenth century progressed and the old standards became forgotten, romantic enthusiasm in architecture was concentrated upon this alone. Beckford, at Fonthill, finding in the Georgian mansion he inherited no adequate stimulus to the raptures of imagination, instructed his architect Wyatt to design 'an ornamental building which should have the appearance of a convent, be partly in ruins and yet contain some weatherproof apartments.'[1] The scheme at length developed into vast proportions. Impressive galleries of flimsy Gothic delighted their master with vague suggestions of the Hall of Eblis, and a tower, three hundred feet in height rose above them to recall the orgies of the wicked Caliph. Five hundred workmen laboured here incessantly, by day, and with torches in the night. But the wind blew upon it, and the wretched structure fell incontinently to the ground. The ideal of a monastic palace 'partly ruined' was ironically achieved. And the author of *Vathek*, contemplating in the torchlight his now crumpled, but once cloud-capped, pinnacles, may stand for the romantic failure of his time—for the failure of the poetic fancy, unassisted, to achieve material style.

It forms no part of our scheme to dwell upon the phases of the mediaeval revival. They exhibit the romantic spirit in a cruder, a less interesting, and a less instructive manner than the Greek movement which we have been criticising. Technique, organisation, vigour, understanding—everything, in fact, save learning and enthusiasm, were wanting to it. It illustrates, as abundantly as one could wish, the effect upon architecture of an exclusively literary attitude of mind; and as few today would do otherwise than lament its achievements, we may take leave of them.

[1] Vide *The Life and Letters of William Beckford*, by L. Melville. Beckford rebuilt his tower, but it again fell to earth. His life (1760—1844) bridges the interval between Walpole and Ruskin, and is an admirable example of the romantic spirit at its height. *Vathek* and Fonthill exhibit its power and its weakness.

到了这时,虽然浪漫主义的思潮把原本活生生的文艺复兴艺术扼杀了,但是,浪漫主义自身还有一大段路要走。帝国风格所表现的仍然是属于古典主义的东西,它只能通过其他不完全合适的途径对此加以表达。当然,它最后找到的表达方式就是哥特风格,这也是它的一个成就。终于,这个时代主流中的文学式感伤情绪越来越成为主要的推动艺术前进的力量,进入十九世纪以后,过去的技艺也已经大多被人遗忘,浪漫主义就只好把精力集中在哥特风格上了。在福特希尔(Fonthill)地区有一位名叫贝克福德(Beckford)的著名小说家,他觉得自己从父亲那里继承下来的乔治时期的古典主义风格建筑不能激发自己的想象力,要求自己的建筑师怀特(Wyatt)对此进行改建。"建筑的外观要看上去像一个修女住的修道院,一部分看起来像是废墟,但是居住部分当然要防雨防寒"。[1]这个项目经过一段时间后,演变成一个庞大的工程。哥特风格的大厅、走廊着实让主人很开心,让他觉得自己是住进了伊斯兰文化传说中的魔宫(Hall of Eblis),外面三百多英尺高的高塔也让他觉得自己成了寻欢作乐的哈利发。五百名工人在此不停地工作,不分白天黑夜。白天自不待言,黑夜则是灯火通明。但是,有点讽刺的是,施工期间遇到一次强风,大风把这座建筑中薄弱的部分吹倒在地。主人幻想中的"修道院废墟"就这样戏剧性地被实现了。这位《哈利发·瓦提克》(Vathek)一书的作者,举着火把,观看自己那曾经高耸入云的高塔,眼下已经成了一片废墟,这情景似乎象征了他自己这个时代浪漫主义的失败。仅有诗情画意般的幻想与憧憬是不能成为具体的建筑风格的。

这些还根本不是我们想要说明的中世纪艺术风格的复兴。它们还是处在浪漫主义发展的最初阶段,还很粗糙,不足以引发人们的兴致,也不能够像我们在前面提到过的那些希腊文化那样,带给人们启迪。在技术、组织施工上,在气魄上,在对于哥特文化的理解等几乎所有的方面都差强人意。只有在对哥特文化的狂热与对哥特文化研究学习两方面仍然保持着旺盛的热情。大量的证据表明,这恰恰是文字方面的成就被无限放大到建筑艺术上的结果。在今天,我们大家能做的只有对他们的努力表示一下惋惜而已,在这里我们就不继续这个话题了。

---

[1] 参见《威廉·贝克福德传记与书信选》,L.Melville 编撰。贝克福德重建了自己的高塔,不幸高塔又一次倒塌。贝克福德(1760—1844)生活的年代介于沃波尔(Walpole)和拉斯金(Rushkin)二人之间。在浪漫主义达到高峰的时候,他是一位非常有影响的人物。他的著作《哈利发·瓦提克》和他在福特希尔建造的城堡,都很能够说明他的野心与能力,同时也说明了他的弱点。

But among the consequences of that ill-timed experiment we have to emphasise this. The Romantic Movement, in destroying the existing architectural tradition, destroyed simultaneously the interest which was felt in its principles, and replaced it by a misunderstood mediaevalism out of which no principles of value could ever be recovered. The catastrophe for style was equally a catastrophe for thought. To this, without doubt, no small part of the existing confusion in architectural criticism may be traced. We laugh at Fonthill and Abbotsford and Strawberry Hill: Georgian architecture once again enjoys its vogue. Yet the Romantic Tendency, expelled from architecture, still lingers in its criticism. The Gothic revival is past, while the romantic prejudices that engendered it remain. And these it is important to define.

The first fallacy of Romanticism, then, and the gravest, is to regard architecture as *symbolic*. Literature is powerful to invest with fascination any period of history on which its art is imaginatively expended. Under the influence, directly or indirectly, of literature the whole past of the race is coloured for us in attractive or repellent tones. Of some periods inevitably we think with delight; of others with distaste. A new historical perspective, a new literary fashion, may at any time alter the feeling we entertain. Yet the concrete arts which these different periods produced remain always the same, still capable of addressing the same appeal to the physical senses. If, then, we are to attend impartially to that permanent appeal, we must discount these 'literary' preconceptions. But everything which recalls a period of the past may recall, by association, the emotions with which that period is, at the time, poetically regarded. And to these emotions, orginally engendered by literature, romanticism makes the other arts subservient. The element in our consciousness which ought to be discounted, it makes paramount. Its interest in the arts is that, like poetry, they should bring the mind within the charmed circle of imaginative *ideas*. But these ideas really belong to the literary imagination whence they sprang, and one result of applying them to architecture, where they are not inherent, is that all permanence and objectivity of judgment is lost. Thus, for example, the Gothic building from being the 'expression' of 'ignorant and monkish barbarians,' came to 'suggest' the idealised Goth— 'firm in his faith and noble in his aspirations' — who inspired the enthusiasm of Coleridge; and the forms of an architecture which later came to be admired as the lucid expression of constructive mathematics were about this time commonly praised as the architectural image of primeval forests. Some minds find in the work of the mediaeval builders the record of a rude and unresting energy; others value it as the evidence of a dreaming piety. Now, it is an 'expression of infinity made imaginable'; next, the embodiment of 'inspired' democracy. It is clear that there is no limit to this kind of writing, and we have only to follow the romantic criticism through its diverse phases to feel convinced of its

在前面刚刚说到的这个生不逢时的实践,它实际上也产生了一系列的后果,其中一点我们必须要在这里加以强调。浪漫主义运动在摧毁现存的建筑传统的同时,也摧毁了这场艺术运动中的一些基本原则,当人们对于浪漫主义的原则失去了兴趣之后,取而代之的便是被曲解的所谓中世纪复古主义,这个做法没有任何原则可言。这种艺术风格上的灾难同样也是一场思想上的大灾难。无疑,目前很多建筑艺术理论混乱的根源正在于此。我们嘲笑福特希尔、阿巴斯福德、草莓山等地区的那些故事:乔治时期的古典主义建筑又再次成为时髦的建筑风格了。浪漫主义在建筑实践中虽然受到了抵制,但是浪漫主义却仍然时常地侵入到建筑理论中来。哥特风格的复兴已经成为过去,但是催生哥特风格复兴的浪漫主义思想根源依然存在。因此,明确这些思想根源就变得非常重要了。

浪漫主义的第一个谬论,也是危害最大的一个谬论,就是它把建筑艺术当作一种*象征手段*。在发掘历史上某一个阶段的典故方面,文学艺术是一个强有力的工具,它可以帮助我们扩展对于那个时期艺术活动的理解与想象。在文学艺术的直接或间接影响下,某一时期与某个民族一些过去的文化活动就被渲染成能够吸引我们注意力的东西,或者令我们产生排斥心理并且厌恶的东西。在所有不同的历史时期,一定有某些是令我们欣赏,让我们感到愉悦的,另外也有一些让我们厌恶。这些都是很自然的。新的历史观、新的文学艺术手段会随时改变我们的感受,但是过去那些时代里出现过的具体艺术作品却不会因此而有所变更。这些艺术作品对于人们感官的作用从来都是一样的,过去是怎样的,今后还会是那样的。如果我们想排除各种文字的干扰,能够感受到某种艺术作品中固有的永恒美感,那么我们就必须排除那些文字带给我们先入为主的偏见。然而,能够让我们联想起过去历史的任何一件具体艺术品,很自然地也会让我们联想到那个时期留给人们的总体感受,免不了联想起那个时期所具有的某种诗意。这种诗意就是从文学艺术中产生的,而浪漫主义就是利用这种诗意,把其他的艺术手段变成自己手里的不同工具。在我们的建筑艺术领域,那些本来应该从我们的意识中排除的东西,却因为浪漫主义的干扰而突然变得高大起来。浪漫主义的目标就是要在人们欣赏艺术的时候,把幻想中的*理想*强行带进来,但是这些理想之类的东西完全是属于文学艺术领域里纯属想象的东西。当这种属于文学艺术中想象的东西发挥作用,在被带进建筑艺术中之后又成为主导,那么,这种本来不属于建筑艺术的东西就把建筑艺术中那些永恒、客观的判断标准给排挤掉了。比如,本来人们所客观认识的哥特建筑是"无知僧侣野蛮人"的具体体现,被浪漫主义熏陶以后,便摇身一变,成了"代表着坚定的信仰与高贵的期望",成了理想化中的哥特人的化身。哥特人就启发了英国大诗人塞缪尔·泰勒·柯勒律治(Coleridge)的艺术创作。哥特风格被后人誉为用建筑艺术表现出数学精美的建筑风格,在这个时期仍然是被看作原始建筑的一种表现。有些人认为,中世纪的建筑工匠们在自己的作品中留下粗野、充满激情的痕迹,而另外一些人则认为中世纪的工匠代表着理性与虔诚。有的说"中世纪的艺术让无限这个概念变得可以理解了",但是也有人会说"它代表了具有理想的民主社会",等等。有一点可以肯定是,这类文字将数不胜数,我们只要沿着浪漫主义艺术理论的演变过程走一走,就会发现那里面的每一个阶段都有数不尽的缺乏客观标准的空洞无物的辞藻。几百年来北方的民族所具有的性格,无论是实际存在的,还是意念中想象出来了,在这时都被发现了,而且是从十二世纪与十三世纪的那些大教堂建筑上面被发

total lack of any objective significance. Any characteristic, real or imagined, of a mixed set of northern races, during a period of several hundred years, is discovered at will in these cathedrals of the twelfth and thirteenth centuries, although it is more than doubtful how far such characteristics are capable of being embodied in architecture, or, if embodied, how far we, with our modern habits of thought, can extract them unfalsified, or, if extracted, how far they are relevant to the quality of the work. The whole process is purely literary, its charm is in the literary value of the idea itself, or in the act and process of association. Moreover, since literary exercises invite effects of contrast, the architecture of the Renaissance comes to be treated, like the villain in the melodrama, as a mere foil to the mediaeval myth. And because Renaissance life happened to yield no stimulus to the nineteenth century imagination, the architecture which ministered to the uses of that life became *ipso facto* commonplace. A combination of plastic forms has a sensuous value apart from anything we may *know* about them. Romanticism allows what it knows, or conceives itself to know, about the circumstances among which the forms were produced, to divert it from giving unbiassed attention to the purely aesthetic character, the sensuous value, of the concrete arts. If it is a question of architecture, the architectural design is taken as standing for the period which invented and is associated with it, and as suggesting, conventionally, the general imaginative state, the complex feelings of approval or disapproval which, the idea of that period happens to evoke. Architecture, in fact, becomes primarily symbolic. It ceases to be an immediate and direct source of enjoyment, and becomes a mediate and indirect one.

Under the romantic influence, then, the interest in architecture is symbolic, and taste becomes capricious. But that is not all. It becomes also unduly stylistic, and unduly antiquarian. For in proportion an architectural form is symbolically conventional, its precise character becomes far less important than its general so-called 'style'; just as in a handwriting the precise forms are less important than the meanings to which they refer, and exist only to call up the latter. Romanticism conceives styles as a stereotyped language. Nineteenth century criticism is full of this prepossession: its concern is with styles 'Christian' and 'un-Christian'; one 'style' is suitable to museums and banks and cemeteries; another to colleges and churches; and this not from any architectural requirements of the case, but from a notion of the *idea* supposed to be suggested by a square battlement, a Doric pillar, or a pointed arch.[1] And such criticism is far more occupied with the importance of

---

[1] Nor is this prepossession extinct. When, recently, the most eminent of English architects projected a basilica for the Hampstead Garden Suburb, the Bishop of London swept the admirable scheme aside, declaring he 'must have a spire point to God.' We trust his lordship is finding some solace at Golder's Green for the signal injury done him by Sir Christopher Wren.

TWO

现的。我们真的不知道有多少这样的人类性格是可以被转写到教堂建筑上面去的。即使有些的确是被转写到建筑上去了,但是我们现代人,根据现代的思维方式又能从中破解出多少也是一个大问号,其中有没有误解也不好说。而那些被破解出来的又与建筑艺术的品质有多少关系,这其中根本就没有对等的关系。整个的分析过程完全是文字游戏,属于文学性质的。它的魅力就在于它的文学艺术思想以及它通过文字所产生出的联想过程。不仅如此,文学必须通过比较才能对它进行分析,结果,文学的比较方法被引进建筑理论中之后,文艺复兴时期的建筑就自然地被拿来当作中世纪神话的陪衬,就好像是情节剧中的坏蛋角色。刚好由于十九世纪的人,对于文艺复兴时期的生活不感兴趣,代表那个时期生活的建筑艺术也就自然地成为牺牲品。造型艺术中的形体组合具有一种能够引起我们感官兴趣的品质,它不受我们理智的左右。浪漫主义理论利用自己掌握的资料和对于当时情形的了解,把建筑作品的形成过程按照自己的想法有选择地说给旁人听。他们不能客观地按照美学规律把作品中包含的感情、把具体的艺术作品,仅仅当作具体的艺术作品看待,把建筑作品的产生过程客观地描述给大家。在讨论建筑艺术的时候,浪漫主义理论就会把建筑设计当作是代表那个作品诞生的那个时代的一种象征,便把想象中与这个时代相关的各种联想统统强加在这个建筑作品之上,把大家对那个时代的一般肯定或者否定的态度也应用到这个具体的建筑作品上面。建筑作品果真便成了一种象征符号。它不再是具体生动的东西,不再是可以让我们从中感受到乐趣的具体物质。建筑成了冥思苦想的对象,成为间接的东西。

在浪漫主义思潮的影响下,建筑的意义仅剩下它的象征性,兴趣与爱好则变得反复无常。这还不算完。在这种情况下,建筑成了讲究时髦的幌子,也成了一种对于古董的嗜好。如果建筑形式没有什么新奇的地方,那么这个建筑的具体特征也就比它的建筑所采用的风格还要低一等。就好比书法,具体的作品要比它所传达的意义低一个等级,因为作品仅仅为了传达意思而已。浪漫主义把建筑风格当作了具有某些固定含义的形式语言。十九世纪里的建筑艺术理论到处是这类东西:它们认定某种建筑风格是"基督教"的风格,而别的则不是;某种风格适合于博物馆建筑;某些则适合做银行或者墓地建筑;有些建筑风格适合于学院或者教会建筑,等等。这些僵化的思维方式不是出于建筑的使用要求,而是根据建筑上面出现的墙头箭垛、多立克柱头,或者尖拱等形式自然联想到的。[1]这种理论与建筑评论更多地关注具体的建筑上面是否出现这些节点,而不在乎这种节点出现在作品里是否美观,不在乎这种节点与周围的其他部分组合到一起是否美观。浪漫主义形成了一种关于建筑风格的

---

[1] 这类成见远没有被消除。最近在汉普德花园郊区(Hampstead Garden Suburb),由英国最有声望的建筑师设计的一座教堂建筑,建筑形式是古代的巴西里卡,伦敦大主教不容分说就否定了这个受到广泛赞扬的设计,宣称:"我只要能和上帝沟通对话的尖塔形式。"我们确信这位主教大人一定是在利用这个机会,报复当年克里斯多夫·伦恩(Sir Christopher Wren)设计的圆形穹顶带给他的重大伤害。

having, or not having, these features in general, than with the importance of having them individually beautiful, or beautifully combined. It sets up a false conception of style and attaches exaggerated value to it. For it looks to the conventional marks of historical styles for the sake of their symbolic value, instead of recognising style in general for its own value.

And there ensues a further error. Every period of romanticism, ancient or modern, has, it is safe to say, been a period of marked antiquarianism. The glamour of the past, and the romantic veneration for it, are very naturally extended to the minutiae in which the past so often is preserved, and are bound to lend encouragement to their study. Nor is this study in itself other than beneficial. But the fault of the antiquarian spirit, in architectural thought, is precisely that it attaches an undue importance to detail as opposed to those more general values of Mass, Space, Line, and Coherence with which architecture properly deals, and which it will be the later purpose of this study to analyse and describe. For the present it is enough to emphasise the fact that between Renaissance architecture and the antiquarian criticism of the Romantic fallacy there is a fundamental opposition: and that opposition lies in their attitude to detail. For antiquarian criticism regards detail as the supreme consideration and Renaissance architecture regards it as a secondary and subservient consideration. And not only do they give it a different degree of importance, but, still more, they give it an importance of a wholly different kind. For in Renaissance architecture the purpose of detail, as we shall see, is primarily to give effect to the values of Mass, Space, Line, and Coherence in the whole design; and, secondarily, upon a smaller scale, to exhibit these qualities in itself. But for the romantic or antiquarian criticism it is required to be 'scholarly,' that is, to correspond exactly to some detail previously used in the period poetically approved. In this way, although it would seem highly unscholarly not to discover the aesthetic function of detail in general before dogmatising upon its use in particular cases, the antiquarian criticism of architecture has usurped the prestige of scholarship. And thus the romantic attitude which begins in poetry ends in pedantry, and the true spirit of architecture eludes it altogether. In the warfare of romantic controversy, Renaissance forms were defiantly multiplied, and sneeringly abused, as though the merit of the style consisted in the detached and unvalued elements common to the Piazzetta of Venice and the clubs of Pall Mall. Like the dishonoured fragments that mark the site of a forgotten temple, detail, mutilated by ignorant misuse—detail, and the conventional insignia of the styles—was all that remained of the broken edifice of a humanist tradition. And, as the merit of Renaissance architecture consists less in the variety than in the disposition of its forms, it became at last, as its enemies accused it of always having been, the lifeless iteration of a stereotyped material.

错误观念,并把风格的价值与作用无限制地夸大。它并没有从每一个具体的建筑风格本身入手,探讨它们各自真正的价值。

　　这样的做法紧接着导致了另一个错误的发生。我们可以很有把握地说,浪漫主义盛行的每个时期,无论是古代的,还是现代的,一定是一个以收集古董为主要特征的时期。过去的荣耀以及对过去的崇拜很容易地被引申到过去很多本来是微不足道的东西,而这些微不足道的东西又很容易地流传至今,因此便带动了对这些细枝末节的研究。这类的研究,除了研究自身以外没有任何别的益处。这种崇尚古董的风气在建筑界流行产生了一个弊端,那就是把某些本来没有什么重要性的细节的作用进行夸张性地放大,从而忽略了建筑艺术中需要面对的重大问题,诸如体量、空间、线条、整体的协调一致等。关于这些主要方面的问题,我们会在本书以后的章节中深入探讨,在目前,我们只强调一个事实:在文艺复兴时期的建筑艺术实践与浪漫主义鼓吹老古董的谬论之间存在着一个本质的差别,那就是在对待建筑细节的态度上,二者有着本质上的不同。崇尚古董的理论把细节看得无比重要,而文艺复兴时期的建筑艺术则把细节看作是第二位的,是服从于大的构思的。二者的不同不仅在于它们各自给予细节不同程度的重要性,而且是不同性质的重要性。对于文艺复兴时期的建筑来说,细节的目的在于烘托建筑的体量、空间、线条以及整体的协调一致,细节本身是第二位的,是在小尺度内欣赏的内容。对于崇尚古董文化的浪漫主义理论来说,细部必须表现出"学问",必须要与过去出现过的细节完全一致才行,否则就失去了原来的诗意。在了解到建筑细节的美学作用之前,便得出结论,坚持要人们教条地在实际案例中忠实地应用这些细节,这种做法本身尽管看起来很不符合做学问的基本原则,但是浪漫主义的建筑理论还是以很有学问的姿态出现在我们面前。浪漫主义的艺术观念以诗歌艺术作为开头,以满腹经纶的学究气作为结果。建筑艺术的真正精神在这种情况下便只好退避三舍,或者逃之夭夭了。在与浪漫主义较量的过程中,文艺复兴时期的建筑形式被肆意地歪曲,在讥笑声中被丑化,好像这个风格的全部优点仅在于它的某些孤立的、无价值的建筑细部。而这些细部或许看起来类似于威尼斯圣马可广场上那些建筑的某些细节,或者类似伦敦蓓尔美尔大街上的俱乐部建筑上的细节。残破的建筑遗迹只是表明早已不复存在的神庙过去曾经站立过的位置,支离破碎的细节是人文主义高楼大厦仅存的证明。但是,由于文艺复兴时期的建筑更多的是在于它的整体构图与组合,并不在于某些细部的花样和完美,文艺复兴时期的建筑艺术最终被说成只是复制过去的某些细节却又不得要领地粗制滥造,这正是它的敌人一直不断地对它进行攻击的地方。

The first pitfall, therefore, into which architectural criticism fell was that prepared for it by the Romantic Movement. The understanding of Renaissance architecture suffered from this, and still suffers, both by neglect, and by misinterpretation. It was inevitable that Romantic criticism should neglect the Renaissance style. Its antiquarian enthusiasts found in it no free scope, because the field was already well explored, the subject well formulated: they were revolted, moreover, by the unconventional use which the Renaissance artists often made of classical design; and, attracted to the mediaeval by its wealth of unexplored detail, they followed all the more willingly the summons of the romantic impulse which, by an accident of culture, had now set towards the middle ages. Its poetic enthusiasts, equally, were repelled from the Renaissance tradition because it was insufficiently remote, insufficiently invested with the glamour of the unknown; because it could be made symbolic of no popular ideas, and because it could not, like the Greek or the Gothic, be fitted at once into a ready-made, poetical connection. And thus, insensibly, the Renaissance style, since symbolic it had to be, became symbolic of ideas that were unpopular. The conditions in which it had grown up seemed relatively prosaic. Prosaic, therefore, and dull the Renaissance forms must necessarily be found.[1]

Such were the consequences of the prepossession which translates material forms into terms of 'literary' ideas. Yet it must not be said that literary ideas have no 'legitimate' place in architectural experience. Every experience of art contains, or may contain, two elements, the one direct, the other indirect. The direct element includes our sensuous experience and simple perceptions of form: the immediate apprehension of the work of art in its visible or audible material, with whatever values may, by the laws of our nature, be inherently connected with that. Secondly,

---

[1] Cf. Mr. Lethaby in a recent work: 'It must, I think, be admitted by those who have in part understood the great primary styles, Greek or Gothic, that the Renaissance is a style of boredom ... Gothic art witnesses to a nation in training hunters, craftsmen, athletes; the Renaissance is the art of scholars, courtiers ...' Such a statement, in a history which is content to dismiss the whole period in eight pages (or rather less than is devoted to the architecture of Babylon), may justify us in saying that, at the hands of our romantically-minded critics, the Renaissance suffers from neglect, and that it suffers from misinterpretation. For Mr. Lethaby further complains of its buildings that they are 'architects' architecture'; architecture, that is to say, not convertible, presumably, into terms of poetry or historical romance, but requiring a knowledge of architectural principles for its appreciation. Renaissance architecture, in fact, is here read off in terms of Renaissance society, and those who enjoy it as an art are stigmatised as 'architects.' When a critic, perhaps as learned and as eminent as any now writing on the subject of architecture in England, can offer us these censures, even m a popular work, as though they were accepted commonplaces, it is not easy to hope that the Romantic Fallacy is becoming extinct.—W. R. Lethaby, *Architecture*, 1912, pp.232-233.

因此,建筑艺术理论落入的第一个陷阱正是由浪漫主义艺术思潮专门为建筑艺术设下的。对于文艺复兴时期建筑艺术的正确理解因此而受到了错误的影响,而且这种影响仍然继续在误导着我们。对文艺复兴时期的艺术产生错误影响的手法要么是忽视它,要么是误解它。浪漫主义艺术理论不可避免地会忽略文艺复兴时期建筑风格。热衷古董的那些人,在文艺复兴时期的建筑艺术中找不到插手或者置喙的地方,因为文艺复兴时期建筑艺术理论的各个方面已经被彻底地研究过,得出的结论也是很周全地建立起来的。不仅如此,文艺复兴时期艺术家的做法与热衷古董、考古的人的观点刚好是背道而驰的,文艺复兴时期的艺术家总是把古典设计手法按照自己的理解,无视常规地应用到自己的实践中去。因此这些爱好古董的人士便把注意力转移到中世纪留下来的丰富的细节,而这些建筑细节恰好还没有被深入地研究开发过,因此他们在浪漫主义艺术思潮的驱使下,很自然地把自己的热情都转向了那里。本来是很偶然出现的浪漫主义倾向,现在一下子成了有意识的艺术追求。浪漫主义潮流中那些崇尚诗意的理论家也被迫绕过文艺复兴这个时期,因为这个时期的传统距离当时还不够久远,还不是用神秘而光荣这样的词汇来探讨这个时期的时候;因为它还不能像希腊建筑或者哥特建筑那样被赋予某种诗意的联想,因此它还不能被拿来象征某种思想或者理念,因此也就很自然地被崇尚诗意的浪漫主义理论家所抛弃。由于在浪漫主义艺术理论中,所有的风格都会被赋予某种象征意义,因而文艺复兴时期的建筑风格就被用来象征各种不受人喜爱的思想。而文艺复兴时期的建筑艺术产生过程给人的感觉是像平铺直叙的白话文一样,没有什么隽永的诗意。因此这个时期的建筑艺术形式也自然是没有什么诗意的了。[1]

这就是具体物质的建筑形式被转变成"文学艺术"思想的过程。我们暂时还不能说这种文学艺术的思想在建筑艺术的体验中毫无道理可言。任何一种艺术体验都包括两个方面:一个是直接的,另一个是间接的。直接的元素包括我们的感官体验与对形式的简单认知;这是通过视觉或听觉等感官直接领悟一件艺术作品,完全是根据我们直接与艺术品的接触,根据我们的本能对艺术品的价值作出判断。在这个直接体验之外,还有一种体验是这个艺术作品在我们的头脑中唤起某些联想,我们的主观意识开始对艺术作品

---

[1] 参见利塔比(Lethaby)先生在最近的一篇文章写的:"我认为,对于那些理解古希腊建筑风格和哥特风格的人们来说,他们一定会认同下面的说法:文艺复兴时期的建筑风格真的可以被称为无聊枯燥的风格……哥特风格见证了一个国家的猎人、工匠和运动员的训练过程;而文艺复兴时期的建筑艺术则见证了学究和廷臣的生活……"这段话出自一本建筑历史书,这本建筑历史著作在谈到整个文艺复兴时期的成就的时候,前后不足八页(还没有讨论巴比伦时期的建筑历史所用到的篇幅那么长)。这个事实,加上作者所说的内容,足以让我们得出以下的结论:文艺复兴时期的历史,在充满了浪漫主义思想的历史学家看来,是不值得深入探讨的,它被轻易地忽略过去,至于被讨论的部分也多半是被误解。当利塔比先生继续自己关于文艺复兴时期的论述时,他抱怨说这个时期的建筑都是些"建筑师"的建筑,大意是说,这类建筑不那么容易被转换成诗意的幻想或者历史的浪漫联想等文学语言,而是必须要先掌握一定的建筑艺术原则和知识,才能理解和欣赏这种建筑艺术。事实上在这里,文艺复兴时期的建筑被当作这个时期的社会状况来看待,而把这些建筑当作是艺术作品来看待的那些人,则被误解成为"建筑师"。在今天,像英国的这些有学问又有知名度的建筑艺术理论家们,随便一个人写文章都会用同样的口气来教训、开导我们。他们甚至在通俗读物上发表文章也这样讲,好像是这样的观点已经成为人人皆知的一般常识似的。看来要想清除浪漫主义的谬误还有相当长的路要走。——见 W. R. 利塔比,《建筑艺术》,1912 年出版,第 232 至 233 页。

and beyond this, there are the associations which the work awakens in the mind—our conscious reflections upon it, the significance we attach to it, the fancies it calls up, and which, in consequence, it is sometimes said to express. This is the indirect, or associative, element.

These two elements are present in nearly every aesthetic experience; but they may be very differently combined. Literature is an art which deals preponderatingly with 'expression.' Its appeal is made through the indirect element. Its emphasis and its value lie chiefly in the significance, the meaning and the associations of the sounds which constitute its direct material. Architecture, conversely, is an art which affects us chiefly by direct appeal. Its emphasis and its value lie chiefly in material and that abstract disposition of material which we call form. Neither in the one case nor in the other is the method wholly simple. Mere sound in poetry is an immediate element in its effect. And some visual impressions in architecture are bound up almost inextricably with elements of 'significance': as, for example, the sight of darkness with the notion of gloom, or of unbroken surfaces with the notion of repose. Nevertheless, the direct elements of poetry—its sound and form—are valuable chiefly *as means to the significance*. They are employed to convey refinements of meaning, or to awaken trains of association, of which mere unassisted syntax is incapable. They enrich or sharpen our *idea*. The sounds delight us because, in them, the sense is heightened; and formal rhyme, by linking one phrase with another, adds a further intricacy of suggestion. But the merely formal, merely sensuous values of poetry are fully experienced when we read a poem in an unknown language; and the experiment should assure us that in literature the direct elements are valuable, almost solely, as a means to the indirect, and that the method of the art is strictly associative. In architecture, on the other hand, so small is here the necessary importance of mere significance, that a building whose utilitarian intention is crudely ignoble, and which is thus symbolic of ignoble things, may easily affect us, through its direct elements, as sublime. Literature may possess abstract architectural properties—scale, proportion, distribution—independent of its significance; architecture may evoke a poetic dream, independent of its forms; but, fundamentally, the language of the two arts is distinct and even opposite. In the one we await the meaning; in the other we look to an immediate emotion resulting from the substance and the form.

The reason of this difference is obvious. The material of literature is *already* significant. Every particle of it has been organised in order to convey significance, and in order to convey the same significance to all. But for the material of architecture, no system of accepted meanings has been organised. If, therefore, we derive associative values from its forms, those values will be determined wholly

进行思考，在思想意识中，我们给作品附加了某些特别的意义，我们认识到它带给了我们幻想，自然也认识到它有时又代表了某种理念。这第二种体验就是间接的体验，是通过联想引申出来的。

任何一个审美体验都包括这两方面的因素，但是这两个因素在每一次审美体验中所发挥的作用总是不同的。文学艺术的主要作用在于它"所表现的内容"方面，因此它就更加喜欢间接的体验；它的侧重点与价值主要是通过文学艺术作品唯一的直接因素，声音，来重点传达作品中的意义、象征、联想。与此形成鲜明对照的建筑艺术则是完全通过直接感受来体验的艺术作品。它的侧重点与价值在于建筑的材料与建筑造型的组合方式。在两种情况中，我们不能说哪一种要比另一种简单。其实都不简单。诗歌中的声音有非常直接的作用；而建筑中的某种视觉表现形式则不可避免地引发某种"象征意义"，例如，眼睛看到阴影就自然联想到阴森森的黑暗，看到连续的表面就会立刻想到安详。不仅如此，诗歌中的能够直接感受到的两个因素，声音与形式，作为*象征手段*的表现方式非常重要。它们能够帮助加强主题思想的精炼，唤起一连串的联想，这些仅靠诗歌句法是远远不够的。它们丰富了诗歌的中心思想，并且使得中心思想更加清晰。诗歌的声音让我们心生愉悦，因为诗歌朗诵强化了我们听觉，形式的节奏韵律增强了连续不断的精美篇章，让我们对后续的诗句有所期待。当我们用不熟悉的外语来朗诵一篇诗歌的时候，我们仍然能够完全感受到它单纯的形式美、单纯的悦耳音节。这个实验让我们确信，在文学艺术中直接的感受非常重要，几乎是不用借助其他力量便可直接唤起那些间接的感受；它也使我们确信，文学艺术的方法就是一门唤起联想的艺术。在另一方面，建筑艺术中象征的意义则根本没有那么重要，建筑里面的实用性非常残酷地告诉人们建筑本来的卑微出身，建筑也就成了卑微杂事的象征，但是建筑通过自己的直接作用会影响到我们，让我们看到了建筑艺术的崇高。文学艺术或许具有某些抽象的建筑品质，如尺度、比例关系、构图分布等因素，这些因素与文学艺术的意义没有关系；建筑艺术也可能让人产生诗情画意般的联想，这与建筑的形式也没有什么关系；但是从根本上讲，这两种艺术语言是截然不同的，甚至是对立的。一个是传达给我们它们所包含的象征意义，另一个则是让人们从实物的体量与形式中直接获得某种亲身感受。

产生这种差别的原因很明显。构成文学艺术的材料本身就已经带有各自的一定意义的，作品中的所有组成部分组织到一起就是为了传达某种特定的含义，所有的组成部分都是为了表达那个共同的主题含义。但是，在建筑艺术中，组成建筑的那些材料没有什么大家公认的固定含义。如果我们从建筑形式中联想到某种意义，这些被联想到的意义完全取决于当时的境况与每个人的个性。我们对于建筑的解读彼此非常地不同。某一个人，或者某一

by the accidents of our time and personality. Our readings will disagree. Thus, while each individual, or generation, may add to the direct pleasures of architecture a further element of associative delight, this associative element is not fixed or organisable; it does not contain the true intention or typical value of the art, and cannot be fitted to contain them.

Now since language, meaning, and association play so large a part in our practical life, and form the very texture of our thought, there has been little danger at any time that the *significance* of literary art should be overlooked. There has never been—save perhaps to a slight degree in the eighteenth century—an 'architectural fallacy' in literature, though it has often been the case that the minor element of value—the sensuous element of literature—is totally forgotten. But this same habitual preoccupation with 'significance' which has kept literature vital has, in architecture, led us to lay undue weight on what is there the secondary element, and to neglect its direct value, its immediate and typical appeal. This, then, is the 'literary fallacy' in architecture. It neglects the fact that in literature meaning, or fixed association, is the universal term; while in architecture the universal term is the sensuous experience of substance and of form.

The Romantic Movement is a phase, precisely, of this literary preoccupation. It is the most extreme example of the triumph of association over direct experiences which the history of culture contains. Its influence upon taste can never be quite undone; nor need we wish it. Romanticism, as a conscious force, has brought with it much that is valuable, and holds the imagination of the age with an emphatic and pervasive control. But the danger is great lest a spirit which has rendered intelligible so many ancient and forgotten beauties, and created so many that are new, may, in its impetus, render ineffective for us some less insistent types of art, towards the perfection of which the tradition of centuries has austerely worked. Such an art is the architecture of the Renaissance. Here, then, if we indulge at all in literary ideas, let us at least be sure that they do not obscure from us the value of the style.

One fact should be stated in defence. These 'literary' ideas ought not to be the *primary* value of a material art; they are, nevertheless, its *ultimate* value. For, since man is a self-conscious being, capable of memory and association, all experiences, of whatever kind, will be merged, after they have been experienced, in the world of recollection—will become part of the shifting web of ideas which is the material of literary emotion. And this will be true of architectural experience. It may begin as a sensuous perception, but as such it is necessarily more transient and occasional than its remembered significance, and more isolated and particular than when fused

代人,在感受具体建筑形式美感的时候,会产生一定的联想,因而产生某种心灵上的愉悦,但是这种联想不是固定不变的,也不是组织有序的;它不能构成艺术的目的或者达到某种预知结果的手段。到目前还没有什么有效的方法,来帮助我们事先知道,什么样的做法定然会带来什么样的预期效果。

如此说来,语言、语言所传达的意义以及从字面上延伸出来的联想,这三者在我们的实际生活中扮演着重要的角色,构成了我们思想的载体。因此我们可以很肯定地说,假如把载体之外所具有的所谓象征意义忽略掉,其实也没有什么了不起的。在文学艺术中从来没有听说过什么"建筑艺术带给文学艺术的谬误",或许十八世纪可以除外,因为有那么一点点痕迹表明,那时的文学艺术似乎与建筑有着一点瓜葛。事实上,文学中那些次要的因素,如比例关系、尺度等等,常常为人忽略。但是在建筑艺术领域里情况大不相同。决定文学艺术生命的'象征意义'这个让人先入为主的概念,给了建筑艺术中那些次要的东西以根本不恰当的过于重要的地位,却忽略了建筑艺术中最直接的因素,即忽略了真正引起我们感官兴趣的实体。这就是建筑艺术中的"从文学艺术带来的谬误"。这种谬误忽略了一个基本事实,就是说,文学中所表达出来的意义是文学艺术中普遍存在的基本属性,它所产生的任何意义都是固定不变的,任何人阅读一篇文字,它所表达的意义是相同的;但是建筑艺术则相反,它的普遍存在着的基本属性是让人们亲身体验的具体实物和具体形式。

浪漫主义艺术运动正是这种文学艺术所主导的观念在发挥影响的一个时期。它是一个间接的联想胜过直接的体验的极端时期。它对于人们审美意识与判断产生了巨大的影响,而这种影响是无法清除的。实际上也没有必要加以清除。浪漫主义作为一种主观意识的力量,其实也带给艺术界很多有价值的东西,它不容否认地左右了一个时代普遍的艺术想象力。但是,我们也面临着一种巨大的危险。这种艺术精神让很多被遗忘的古代作品重见天日为人所知,同时也创造出许多全新的艺术作品。它同时以同样的力量驱使我们不能执著地完善某些艺术作品,不能让其有机会达到完美的程度。古代流传下来的那些完美的艺术作品正是这样严肃执著、朴实地经过几个世纪的演变才完成的。文艺复兴时期的建筑艺术就是这样一种艺术。因此,我们可以陶醉于某些文学作品的理念,但是至少我们不应该忘记古典建筑艺术风格的价值。

有一个事实可以说出来作为我们为自己辩护的理由。文学艺术中的那些理念绝不应该成为真实造型艺术的*主要*价值标准,但是它们的确应当成为造型艺术的*终极*价值标准。每一个人都自己的思想意识,都有自己的记忆与联想,个人的全部经验都会在联想中融合到一起,形成了一个记忆中的世界。这个由各种记忆、理念组成的思想网络不是线性的,而是叠加、错落、纠缠在一起的组织。这些记忆和理念刚好又是文学艺术所要借以表达那些情感的基本材料。对于建筑艺术的体验也将会是如同这里我们刚刚说的那样。开始的时候可能只是感官的直觉,但是那种感官获得的直觉很短暂,转眼就会忘记,而且也是很随机、偶然的,还不能仅仅通过这种直觉来获得并且记住建筑艺术的意义所在;那些感官的直觉还都是孤

by reflection with the rest of our remembered life. Its significance outlives it in the mind. There is, therefore, so to say, a literary background to the purely sensuous impression made upon us by plastic form, and this will be the more permanent element in our experience. When we renew the sensuous perception of the work of art, in addition to the immediate value this perception may have for us, there will be, surrounding it, a penumbra of 'literary' and other values. And as our attention to the sensuous properties relaxes, it is to these that it will naturally turn. In so far, then, as the literary values of the work of art enrich our complete experience of it, they are clear gain. And in so far as the Romantic Movement has stimulated our sensibility to such literary values, that also is a clear gain. It would be absurd to demand (as in some of the arts enthusiasts are constantly demanding) that we should *limit* our enjoyment of an art to that delight which it is the peculiar and special function of the art to provide. To sever our experience into such completely isolated departments is to impoverish it at every point. In the last resort, as in the first, we appreciate a work of art not by the single instrument of a specialised taste, but with our whole personality. Our experience is inevitably inclusive and synthetic. It extends far beyond the mere reaction to material form. But its nucleus, at least, should be a right perception of that form, and of its aesthetic function. It is reasonable, then, to claim that the aesthetic enjoyment which is proper and special to a given art should be the first and the necessary consideration, and that in relation to this the quality of a style should primarily be appraised. Whether or not that peculiar enjoyment can be enriched and surrounded with others of a different and more general nature must be a secondary question, and one with which the criticism of a given art, as such, need have no concern. When, therefore, our architectural critics condemn the Renaissance style on this secondary ground before they have fairly considered its claims on the primary ground, this, we may fairly say, is unsound and misleading criticism, criticism tending to obscure real values and diminish possible enjoyments, criticism vitiated by the Romantic Fallacy.

立、分散的,都是很具体的一个一个的片断,还没能通过思考而与整个生活经历联系起来。建筑艺术留给我们的意义远不止这些具体的感官直觉,它留在我们头脑里的记忆要比直接的感受久远得多。所以我们说,单纯地从建筑造型那里获得的感官体验一定存在着相应的文学形式出现的人文背景知识,二者的结合构成了我们对艺术体验的永恒认识。当我们再次对艺术品有感官体验时,那么我们知道,在这种直接感受之外,人文背景知识以及与之相关的其他经验都会如影随形。当我们的直接感受不再那么强烈的时候,那么我们对艺术作品的体验自然就向围绕在具体作品周围的那些人文背景知识倾斜。所以,只要文学艺术能够帮助我们完成对艺术作品的欣赏,能够丰富我们的体验,那这种文学作品对于建筑艺术来说就是有价值的。只要浪漫主义艺术思潮的内容是能够激发我们对于这类文学作品的兴趣,认识这些文学作品的价值,那么从这层意义上讲,浪漫主义也是有价值的。有人主张把我们对艺术作品的欣赏仅仅局限在具体作品依靠本身带给我们直接、有限的体验与享受,这种强加给自己的限制简直是不可思议的(很多热爱艺术的狂热人士总是提出这样的要求)。把我们对艺术作品的欣赏生硬地分割成不同的方面,并且把这些不同的方面对立起来,这种做法会使我们对艺术的欣赏随时随地陷入困境。对于一件艺术作品的欣赏,到了最后的结果与最一开始的感受,其实在性质上是一样的,我们不是通过某一个特别的方面,而是通过我们自己的真实个性来感受。我们的体验是包括多方面的,是综合的,它远远超出建筑形式带给我们的直觉。但是建筑艺术的体验,其核心在于正确地感知它的形式,正确地理解它的美学功能。这样,我们有理由得出下面的结论:对于任何艺术作品恰当的审美享受是第一位的,是不可或缺的,因此,作品的艺术风格就是最主要的欣赏对象。至于这种享受是否可以借助于其他艺术或者背景知识来获得强化、丰富等等,这些都是次要问题,而针对这些具体艺术作品而出现的评论、理论,都不应该纠缠这些次要的东西。因此,当我们看到有些建筑艺术理论不是从建筑艺术的本质出发对建筑艺术作品加以研究,而是根据这些次要的东西来谴责、批评文艺复兴时期的建筑风格的时候,我们知道这种理论是经不起检验的,也是在误导人们。这种理论只会掩盖住建筑艺术作品的真正价值,让我们错失在欣赏作品时所应当获得的享受。这样的建筑艺术理论是被浪漫主义谬论所毁掉的理论。

THREE

The Romantic Fallacy (*continued*) : *Naturalism and the Picturesque*

I. Romanticism has another aspect. We have seen that it allows the poetic interest of distant civilisation to supplant the aesthetic interest of form. But the romantic impulse is not attracted to history alone. It is inspired by the distant and the past; but it is inspired, also, by Nature. For, obviously, those qualities which romanticism seeks, these Nature possesses in the highest degree. Nature is strange, fantastic, unexpected, terrible. Like the past, Nature is remote. Indifferent to human preoccupations and disowning human agency, Nature possesses all the more forcibly an imaginative appeal. Thus, in the last century and earlier, together with the ballad-revival and the historical fiction, came, far more powerful than either, a new poetry of Nature. Under the influence of this poetry, Nature's unconsidered variety became the very type and criterion of beauty, and men were led by an inevitable consequence to value what is various, irregular, or wild, and to value it wherever it might be found. As in the cult of the past, so, too, in this cult of the 'natural,' it was literature, the true instrument of the Romantic Movement, that led the way.

It is evident that architecture and the criticism of architecture have reflected this poetic change. The formal garden, necessarily, was the first object of attack. In the Renaissance taste the garden was an extension of the main design. It was a middle term between architecture and Nature. The transition from house to landscape was logically effected by combining at this point formality of design with naturalness of material. The garden was thus an integral, an architectural, element in the art. But when Nature, through poetry, acquired its prestige, the formal garden stood condemned. Unpleasing in itself, because 'unnatural,' it was in addition a barbarous violence, a ruthless vandalism upon pools and trees. It was an offence against Nature all the more discordant because it was expressed in Nature's terms. Thus, before the impact of Naturalism shook traditional design in actual architecture, the formal garden was already gone. Eighteenth century philosophers, seated under porticoes still impeccably Greek, were enabled comfortably to venerate Nature—or, if not Nature, at least her symbol—as they watched their ancestral but unromantic gardens give place to a 'prospect' of little holes and hills. At their bidding a change was wrought throughout Europe, as sudden as it was

第三章
浪漫主义艺术理论的谬论（续）：崇尚自然与追求绘画构图式的造型

我们在前面已经看到浪漫主义对诗学的兴趣，看到它把遥不可及的远方文明嫁接到具体的艺术作品形式上的做法，但是，它还有另外的一面。它不是仅仅对于历史表现出兴趣，它不仅从遥远的东西和过去的东西里寻找灵感，而且它也从大自然中寻找灵感。显然是因为属于自然范畴里的东西都具有浪漫主义艺术家们所苦苦追求的各种品质的最高境界。自然界是一个陌生奇怪的世界、是一个充满幻想精彩的世界，也是一个不可预知、恐怖的世界。它和过去的历史一样，给人的感觉很遥远。它根本不理会人们是怎么想的，也与人类的各种活动毫无关系，因此大自然对于人们的好奇心又具有巨大而神秘的吸引力。在过去的那个世纪以及在那以前，伴随着传奇叙事故事的流行，伴随着历史题材的小说的流行，人们又发现了赞美大自然的诗篇，而且对大自然的赞美大大地超过了对传奇故事和历史故事的传诵。由于受到诗歌艺术的影响，大自然中的各种天然去雕饰的表现形式成为艺术创作的依据和评判标准，人们不可避免地要从多样化、不规则、带有野趣等美学价值来鉴赏一件艺术作品，甚至把艺术作品所表现的内容是否能够很容易地在现实中找出对应的东西，作为评判的标准。如同对过去历史的崇拜一样，对大自然的崇拜也是因为文学艺术而产生。文学艺术是浪漫主义运动中的真正推手。

建筑艺术与建筑艺术理论也反映出这种诗歌艺术所产生的影响。规则的园林艺术是第一个遭受到攻击的对象。在文艺复兴时期，园林设计被看作是建筑主体设计的延伸，它是建筑艺术作品与大自然之间的一个过渡环节。从建筑物到大自然的过渡，当然以采用天然材料最为得体，这也最符合人们正常的思维逻辑。建筑周围的园林景观是艺术创作的一部分，是隶属于建筑的。但是，当大自然借助于诗歌的力量，成为艺术创作的主宰，规则几何形式的园林景观忍受着诅咒、谴责。因为"不自然"，当然是不美观的了。不但不美观，而且对于山水树木粗野地施加暴力，无情地加以破坏。最无法让人忍受的是，这些对于自然的暴力居然借用的是自然的名义。所以，在崇尚自然的审美观念动摇传统的建筑理论之前，规则几何形式的园林景观设计早已被清除掉了。十八世纪里的哲学家们坐在地地道道的古希腊风格的门廊下，惬意地崇拜着自己心目中的大自然，眼看着从祖先手里继承下来的古典园林中规则、缺乏浪漫色彩的部分被拆掉，取而代之的是小丘小坑。在不能直接欣赏大自然的情况下，也要间接地陶醉一下大自然的象征。这种改变突然之间席卷整个欧洲，不但迅速，而且彻底。每一条街道似乎都没有被放过，直达通畅的道路被扭曲了，原来平坦的路面也变得颠簸不平。

complete. In a moment every valley had been dejected, the straight made crooked, and the plain places rough.

The change in architecture was not slow to follow. Here, as the last chapter showed, a romantic sense of history, treating styles as symbols, could look with equal favour on the Gothic and the Greek, and had provoked a romantic revival of both. But the romantic sense of Nature weighted the balance in favour of the mediaeval. The Gothic builders belonged to the 'nobly savage' north, and had built against a background of forest and tempest. The Greeks stood for reason, civilisation, and calm. More than this, a certain 'natural' quality belonged to the Gothic style itself. Like Nature, it was intricate and strange; in detail realistic, in composition it was bold, accidental and irregular, like the composition of the physical world. Among the causes of the Gothic revival, the poetry of Nature, that cast on all such qualities its transforming light, may certainly be given an important place.

The influence of the sense of Nature upon building did not exhaust itself in the taste for Gothic. In England there grew up a domestic architecture which attaches itself to no historic style and attempts no definite design. It is applied, like the Georgian manner before it, indifferently to the cottage and the great house. But while the Georgian taste sought to impart to the cottage the seemly distinction of the manor, the modern preference is to make the manor share in the romantic charm of the cottage. In Latin countries this architecture is not found; its place is wholly taken by a resurrection of the 'Styles.' But in England, where the hold of style is slighter and the sentiment of landscape more profound, the rustic influence in taste has been extreme. It favours an architecture which satisfies practical convenience, and, for the rest, relies on a miscellany of sloping roofs and jutting chimneys to give a 'natural' beauty to the group. Save for a certain choice in the materials and some broad massing of the composition, the parts bear no relation to one another or to the whole. No such relation is attempted, for none is desired. The building grows, without direction, from the casual exigencies of its plan. The effect intended, if not secured, is wholly 'natural.' The house is to take the colour of the countryside, to lie hidden in the shadows and group itself among the slopes. Such in fairness is its ideal, realised too seldom. So far as this architecture takes any inspiration from the past, it looks to the old farm-buildings long lived in, patched, adapted, overgrown: buildings, so unconscious in their intent, so accidental in their history, as almost to form part of the Nature that surrounds them, and for whose service they exist.

What measure of beauty may belong to such an architecture will later be considered. It is irrelevant here to insist on the unfortunate effect it is calculated to produce when reiterated, with how monotonous a variety, on either side

很快，建筑艺术的改变也随后跟进。我们在上一章看到，浪漫主义看待历史是把各种风格当作是代表不同意义的象征。对于持这种观点的人来说，哥特风格和古希腊风格都是受欢迎的，都激发起复兴这两种风格的热情。但是，浪漫主义对待自然的态度，则让他们更偏爱中世纪的哥特风格。哥特风格的建造者是"高贵、野蛮"的北方人，他们建筑的背景是森林与暴风雨。而古希腊人则代表着理性、文明与祥和。事情还不仅仅如此。哥特建筑风格本身包含着某种"自然"的品德，它如同自然一样精巧诡异。细微处非常简单，但是组合到一起后，整体上就变得非常有气魄、随心所欲，充满了偶然性和不规则性。这和自然界的现象有着异曲同工之妙。在引起哥特复兴的各种因素中，对自然的赞美诗篇无疑起到了关键的作用。自然的力量无所不在。

对于自然的崇尚理所当然地影响到建筑艺术，而且这种影响力在哥特风格上的作用丝毫不减。英国本土发展出一种属于自己文化的居住建筑，不属于任何历史风格，也没有固定的设计语言。它一视同仁地对待小小农舍和豪门大宅，在这一点上，它与之前的乔治时期建筑有类似之处。但是，在乔治时期，人们的喜好是把明显属于豪门大宅的某些特征加到小农舍建筑的上面，而现在人们的喜好则是把豪门庄园装扮得富有小农舍的浪漫色彩。在拉丁地区那些国家里没有出现这种建筑，结果在那些地区就只好是把"过去的某种建筑风格"翻出来炒冷饭。但英国对于建筑风格似乎很冷淡，而对自然景色却是十分地钟情和陶醉，因此在英国人的审美爱好上，粗犷不规则的倾向则是被推向了极致。他们喜欢的建筑造型在满足方便实用的基础上，讲究大大小小的斜屋顶的堆积，屋顶上伸出许多烟囱，让建筑群形成一种"自然美"的感觉。建筑群除了在建筑材料的特定选择与整体构图上的平衡之外，其中的各个部分彼此之间没有任何关系，各个部分与整体也没有任何关系。设计人也从来没有努力去追求这些关系，因为在他们看来根本没有必要。建筑的造型是根据它的用途，从那些必然的平面设计中衍生出来的，不需要别人给它指点方向。设计人所追求的造型完全是"自然"的流露，至于能否达到这个效果则无法保证。即使达不到，也在这个方向上努力过了。房子在色彩上要与周边的乡村一致，要退到影子里面去，要按照山坡的走势组织群落。这类主张很好，也很理想，但是很少能够实现。如果说这类的建筑需要从过去寻找设计的灵感，那么它很自然地就会去乡村找那些有人居住了很久的老房子，修修补补、增建扩建改造过多次，而且略显拥挤。这类房子是在无意识的情况下生长起来的，没有计划与发展方向，在整个存在的过程中一切都是很偶然地发生着，和周边的大自然的演变过程一致，好像是在为大自然服务似的。

这类建筑的美学价值到底是怎样的，它们好看在哪里？关于这些我们将在后面用相当的篇幅来探讨，在这里暂且不谈。纠缠于马路两边单调的建筑形式是因为简单的重复而产生的必然结果之类的问题，在现在也没有什么意义。这些说法就算是都正确，我们仍然可以

of a continuous street. But certainly, whatever be its merits, the habit of taste which it implies is hardly favourable to an understanding of the Renaissance. Order and subtleties of proportion require an habitual training in the eye. The Greeks, as some of the 'optical' corrections of the Parthenon have revealed, responded here to distinctions of which today even a practised taste will be almost insensible. The Renaissance inherited their ideal, if not their delicacy of sense. But a 'natural' architecture, so far from affording such practice to the eye, raises a prejudice against order itself; because whatever qualities a 'natural' architecture may possess are dependent on the negation of order. A taste formed upon this violent and elementary variousness of form, conceives a Renaissance front as a blank monotony because that, by contrast, is all it can discern. What wonder, then, if it accepts the verdict of the poetry of Nature, and declares the Renaissance style to be a weary and contemptible pomp, while it endows its own incompetence with the natural 'dignity' of the fields and woods.

Two duties, then, were required of architecture when the poetry of Nature had done its work. First, it must disguise, or in some way render palatable, the original sin of its existence: the fact that it was an artificial thing, a work of man, made with hands. To this end Nature herself might seem to have intended a variety of creeping, and ultimately overwhelming, plants, by means of which much of the architectural art of England has been successfully rendered vain.[1] To eradicate the intellectual element of design, to get rid of the consistent thought which means formality, is thus the first or negative condition of a 'natural' architecture. Its second aim is more positive. When once the evil spirit of conscious art has been exorcised, the door can be opened to a pandemonium of romance. The poetry of Nature can infect architecture with all her moods: idyllic in the rustic style we have described; fantastic and wild in every kind of mediaeval reminiscence or modern German eccentricity.

It is of the essence of romantic criticism that it permits literary fashion to control architectural taste. This is the cardinal point to which once more we are brought back, and on which once more we may insist. That the architectural judgment is made in unconsciousness of the literary bias is immaterial. A literary fashion is easily conceived of as an absolute truth, and the unconsciousness of a prejudice only adds to its force. For the power of literature extends far beyond its

---

[1] The habit of smothering fine architecture in vegetation is peculiarly English. The chapel of Trinity College at Oxford—to take an example out of a thousand—is habitually indicated to visitors as an object of special admiration *on account of* a crude red creeper which completely conceals it, together with the fact that it is, or would be, one of the most graceful works of architecture in that city. *Naturam furca expellas.* ... But our romantic professors have evidently abandoned the struggle and exchanged Horace for Wordsworth.

肯定地说,这种审美观念所透露出来的思维方式,对于理解文艺复兴时期的建筑艺术基本上是没有什么帮助的。建筑中的秩序与其中比例上的微妙关系是需要长时间对眼睛的训练才能掌握的技艺与审美习惯。古希腊人在帕台农神庙上所展现的各种"视觉上"的矫正手段,对于今天那些即使是最能把握住当前审美取向的人们来说,也是无法感受到的。文艺复兴时期的人们继承了古希腊的美学中秩序的理念,至于是否承袭了各种相应的手段与技艺则还无法断定。追求建筑艺术"自然效果"的人们,在自己的眼中当然是反对任何秩序的东西,因为"自然"的建筑造型要在建筑艺术中排除任何看起来有规律的东西。这种以排斥规律性的东西为诉求,以形式的多样化为标准的审美趣味,在看到文艺复兴时期的建筑正立面造型的时候,就会产生出单调的感觉,因为这是他们能够做出的唯一认识。如果这一种审美观念接受了"大自然"的诗意作为审美评判标准,并以此作为根据来宣判文艺复兴时期的建筑艺术为乏味、卑下的浮夸风气,而我们却看到他们自己在那里水平低下、笨手笨脚地追求大自然中平川田野、山林树木的"高贵气息"的时候,也就不会感到有什么奇怪的了。

当大自然的诗意成为审美原则的时候,它要求建筑艺术必须完成两项工作:第一项就是要把自己的原罪掩盖起来,即使掩盖不了,也要做出一副讨好的嘴脸出来。建筑与生俱来有一个原罪,那就是,建筑作品无论怎样讲,都不外乎是一个完全由人工建造出来的东西,根本不是自然形成的。它是人们用自己的双手,根据自己的需求建造出来。为了达到掩盖建筑这个原罪的目的,大自然已经亲自创造了随处可见、多种多样的素材,以供人们选用,而且到最后甚至有点喧宾夺主的架势。大自然创造的素材就是树木植物。在利用植物渲染建筑方面,英国的建筑艺术绝对称得上是登峰造极。[1]因此,在建筑设计中把人类智慧的痕迹抹去,把理性连根拔起,把追求建筑固有形式的思想清除干净,这就成为"自然风格"建筑艺术的第一要务。到了它的第二个任务的时候,它就不再是从负面提出要求来排除什么了,而是给出一个正面的建议。因为在第一项要务中已经把恶魔清除掉了,所以第二项就是敞开了建筑艺术的大门,迎接浪漫主义艺术的各路神仙粉墨登场。大自然的诗意凭借自己的喜怒哀乐在感染着建筑艺术:有我们在前面说过的粗犷的田园风格,有对中世纪各式怀旧事物充满的不着边际的幻想,或者在北方德国正在流行的当代的怪癖爱好。

浪漫主义艺术理论的本质就是把文学艺术中流行的风气转移到建筑艺术上来,并左右建筑艺术的审美观念。现在我们又被带回到这一关键的问题上。关于这一点,我们也想借此机会再次强调一下。浪漫主义理论就是在无意识的情况下把文学艺术上的偏好应用到建筑艺术的评判上面,这种做法是完全错误的。文学艺术中的某些流行观念很容易地被人们接受,并奉为绝对的真理,而这个观念的形成过程又是在无意识中形成的,它的这种无意识

---

[1] 把一座本来很精美的建筑用植物青藤包裹起来的确是非常能够代表英国人的一种做法。比如我们随便举一个例子,在牛津大学三一学院的小礼拜堂,在向来访的客人介绍这座建筑的时候,一个常常被拿来当作是它最特殊的特征便是它被那些蔓延生长的红色爬山虎全部覆盖住了,同时再加上一句,说它是整个牛津最为优雅的建筑物。这只是成千上万的例子之一。垛干草的叉子可以*宣告大自然已被赶走*……具有我们的浪漫主义观念的教授们显然不在乎这些绿叶凋零以后该怎么办,他们就是认定要用浪漫主义时期的华兹华斯(Wordsworth)诗句来替代古罗马诗人贺拉斯(Horace)的诗句。

conscious students; by a swift contagion it determines, even in illiterate minds, the channels of their thought, the scope of their attention, and the values to which they will respond. It leads men to say, at a given epoch, summarily: 'The artificial or the formal is less worthy than the natural,' without any necessary analysis of what these abstract terms involve. Their aesthetic attention to the concrete case is obstructed by the phrase; and architecture serves as a mere symbol of the idea.

But this, the central point of the Romantic Fallacy, must be guarded from misunderstanding. The influence of literature upon the arts of form exists at all times, and is often beneficial. Romanticism is a permanent force in the mind, to be neither segregated nor expelled. It is only in the manner of its operation that the fallacy occurs. The arts of form have their native standards, their appropriate conventions; standards and conventions founded in experience, and necessary to render them effective in any undertaking, howsoever inspired. When for any reason tradition, which is the vehicle of those standards and conventions, wavers or decays, then the literary influence will, in all likelihood, impose inappropriate standards of its own. The necessary balance between formal and significant elements, which in every art is differently poised, is then overweighted. Overcharged with literary significance and atrophied in its design, the art of form loses the power to impress; it ceases, in any aesthetic sense, to be significant at all.

Thus, in transporting romance from poetry to architecture, it was not considered how different is the position which, in these two arts, the romantic element must occupy. For, in poetry, it is attached *not to the form but to the content*. Coleridge wrote about strange, fantastic, unexpected, or terrible things, but he wrote about them in balanced and conventional metres. He presented his romantic material through a medium that was simple, familiar, and fixed. But in architecture this distinction could not be maintained. When the romantic material entered, the conventional form of necessity disappeared. 'Quaint' design and crooked planning took its place. For here form and content were practically one. And, further, the romantic quality of the material was, in architecture, extremely insecure. The 'magic casements' of Keats have their place in a perfectly formal and conventional metric scheme that displays their beauty, and are powerful over us because they are imagined. But the casements of the romantic architecture, realised in stone, must lack this reticence and this support. They were inconvenient rather than magical, and they opened, not on the 'foam of perilous seas,' but, most often, upon a landscape-garden less faery than forlorn.

Certain images of architecture in their proper context, formal and poetic, are romantic. Remove them from that context, and render them actual, and it becomes

性又反过来强化了这一观念。因为文学艺术的影响力远远超出文学本身的专业范围,在它横扫一切的传染范围内,即使是那些目不识丁的人,都会跟着受到很大的影响。它左右了人们的思想方法、思考范围以及判断价值标准。在任何一个具体的时期里,人们都会概括地说出这样的话:"人造的东西,或者规则的东西,总不如自然的东西好看。"他们在说这话的时候,根本是没有经过大脑思索的,也没有去想过这个抽象的说法到底是什么含义。这些人的审美判断在遇到具体问题的时候立刻被这句话给搞模糊了;建筑艺术成了某一种理念的纯粹象征。

我们在这里对于自己给浪漫主义艺术谬论的核心思想所下的定义必须有一个清楚的认识,绝不可以让它遭到误解和滥用。文学艺术带给造型艺术的影响从来就没有间断过,而且总是有益处的。浪漫主义是人类头脑里固有的一种力量,是不能被分离出来的,也不能被清除出去的。浪漫主义本身没有什么毛病,问题出在它发挥自己的影响力的过程中。就在这个过程里,浪漫主义的谬误出现了。造型艺术有自己的规律和标准,有自己的习惯做法。这些从实践中积累的标准与习惯,在艺术创造过程中仍然会不间断地发挥着作用,无论艺术家的灵感、动机、目的是什么。当承载这些标准与习惯做法的传统变得过时,或者不再发挥作用的时候,文学艺术的那些评判标准便开始施加自己的影响。这时在造型艺术里强调象征意义的必要性便被夸大。文学艺术的特性被过分地强调,而造型艺术本身的规律却被忽视,因此造型艺术就失去了力量。艺术作品失去力量,也就失去了其中的任何意义,这在任何审美体验中都是如此。

在把诗歌艺术中的浪漫主义转移到建筑艺术里来的时候,这些理论家们根本不考虑这些浪漫主义的因素在这两种不同的艺术领域里所扮演的角色到底有什么不同。要知道,在诗歌艺术里,浪漫主义因素所依附的*不是形式方面的语言,而是诗歌的内容*。柯勒律治(Coleridge)在自己的诗歌中描写了不少奇异、怪诞、意想不到的题材,有时甚至是令人恐怖的题材,但是,他在具体的写作中采用的手法还是传统的描写手法,也注意到平衡关系。他是用简洁、熟悉的常见手法来表现他的浪漫主义主题。但是,这种区别在建筑艺术中是无法做到的。当浪漫主义侵袭到建筑艺术领域里的时候,传统的形式就必然被排挤得消失了。"奇巧"的设计以及弯弯曲曲的规划成为主流。在建筑艺术领域,它的形式和内容是同一个东西,建筑形式就是建筑内容,建筑内容就是建筑形式。不仅如此,建筑艺术中的浪漫主义的成分极其不稳定,没有自己的安全感。济慈(Keats)诗歌中的那"神奇的窗扉",无论在诗歌的形式上,还是在诗歌的传统手法上,都完美地成为优美诗歌的一部分,对于我们读者来说,它有无法抵抗的美感与感染力,这完全是因为它是出自于我们的想象力。但是,浪漫主义的建筑中的窗扉则不同,它完全是用石头制作而成的,缺乏任何想象空间与情感上的支撑。建筑中的这些窗户失去了所有的神奇魔力,又缺乏传统窗户的方便实用。而这些窗扉开启的时候,也不是看到"可怕的海浪与海浪的泡沫",最有可能看到的是凄凄惨惨的一座小小花园,远不是一个神话传奇故事。

某些建筑的形象在恰当的文字描写中都是很浪漫的,不管这个建筑形象是规则的,还是如同图画般构图不规则的。但是,假如我们把它们从文字中转移到现实中,真实地建造

evident that there is nothing inherent in the architecture itself that can evoke an imaginative response. Again, there are actual works of architecture that by the lapse of time are almost fused with Nature, and by the course of history almost humanised with life. These, too, are romantic. But if they are repeated anew, it becomes evident that the romantic element was adventitious to the architectural value. The form itself, which must inevitably be the object both of architectural art and criticism, is found to be valueless altogether, or valued only by a vague analogy of thought. And this, in effect, is the case with the conscious architecture of romance. Sharply concrete, divested of the charm of age, it lacks alike the material beauty and the imaginative spell. The formal basis is lacking which alone can give it power.

II. But the prejudice against the 'unnatural' style of the Renaissance was something more than an association of architecture with *poetical* ideas. As that, indeed, it began. But we shall underrate its force, and falsely analyse its ground, if we do not recognise in it, also, an association of architecture with *ethical* ideas. The poetry of Nature furnished the imagery of the gospel of freedom. The Romantic Movement, with its theory of Natural Rights, gave to Nature a democratic tinge. The cult of Nature had its say on conduct: it was a political creed. It was more; for, in proportion as orthodoxy waned and romanticism gathered force, a worship of Nature—for such, in fact, it was—supplanted the more definite and metaphysical belief. A kind of humility, which once had flowed in fixed, Hebraic channels, found outlet in self-abasement before the majesty, the wildness and the infinite complexity of the physical creation. Of all the changes in feeling which marked the nineteenth century, none perhaps was profounder or more remarkable than this, and none more dramatic in its consequences for art. The instinct of reverence, if science dislodged it from the supernatural world, attached itself to the natural. This sentiment, which for the agnostic mind was a substitute for religion, became for the orthodox also the favourite attitude of its piety. A vague pantheism was common ground between the Anglican Wordsworth, the rationalist Mill, and the revolutionary Shelley. Nature, unadorned, was divine herself—or, at the least, was God's garment and His book; and this, not in the elegant and complimentary sense in which Addison might have so regarded her, but with a profound power to satisfy the mystic's adoration. The argument assumed a different plane. To be 'natural' was no longer a point merely of poetic charm—it was a point of sanctity. With Ruskin, for example, the argument from Nature is always final. 'Canst thou draw out Leviathan with a hook?' To improve on Nature's architecture were alike impertinence. It is even suggested that forms are beautiful precisely in relation to the frequency with which Nature has

出来，那么我们立刻就会认识到一点，那就是，建筑物本身其实并不具有任何能够激发起我们想象力的成分。真实的建筑物随着时间的推移，逐渐变成大自然的一部分。随着历史的演变，建筑成为生活的一部分，获得了它自己的人格。这本身就是很浪漫的事情。但是，把这些都重新再来一次，让这些建筑都变成崭新的建筑物，那么，从前的那些浪漫的东西就不见得一定能跟着再次发生。这说明建筑中的浪漫成分不是必然的，这样的建筑形式本身从建筑艺术实践和建筑艺术理论两方面来说都是没有什么价值的，或者仅仅在把它与某种思想含含糊糊地联系到一起的时候才会具有那么一点点价值。但是，建筑形式本身却又是实践与理论无法避免的研究对象。上面说的这个现象便是从主观意识出发来追寻所谓的浪漫主义建筑所必然出现的局面。浪漫主义的建筑作品实际上非常具体实在，没有任何虚幻的成分，又缺少岁月磨练的魅力，因此它既没有具体的实物美，也缺少想象的魅力。这种建筑的形式基础缺少一种重要元素，而它缺少的元素恰恰是给建筑物带来力量的东西。

二、反对文艺复兴时期建筑艺术，指控这些建筑缺乏"自然美"的做法实际上完全是一种不顾事实的偏见。这种偏见并不单纯地只是把建筑艺术与*诗歌*的审美观念结合到一处。事实上，它只是这样开始的。如果我们不能同时认清躲藏在建筑艺术与诗歌艺术关系背后的建筑艺术与*道德观念*的关系的话，那么，我们将低估前一种关系的影响力，也会错误地对其根源进行分析。大自然的诗意为我们提供了什么是随心所欲的最好示范，浪漫主义运动又给大自然涂上一层民主的色彩，提出了不可剥夺的自然权力的概念。关于人们的行为准则，崇拜自然的这一派也有话要说：这是一种政治理念。并且随着过去的正统观念逐渐消失，浪漫主义的影响逐渐扩大，对自然的崇拜便有了更多的信徒，获得了更多的一般抽象意义上的影响力。在大自然的神圣庄严面前，在变化多端、形式多样化面前，某种谦卑心理在自叹不如的情况下不断滋生起来。从前这种谦卑的心理只有一个固定的渠道，就是用希伯来语写成的经典。标志着十九世纪各种变化的东西中，恐怕没有任何一种变化要比这种心理上的变化带来更深远的影响，也没有任何其他的改变比这种心理上的改变对于后来的艺术产生如此戏剧性的影响。因为科学的发展把人们对超自然崇拜的敬畏心理清除出去，那么现在出现的这种自卑心态则顺理成章地向大自然表现自己的谦卑。这种情感对于不可知论者来说已经取代了他们原先的宗教，成为这时的正统，并且人人都乐于认同这种情感。信仰英国圣公会教派的华兹华斯、理性主义者米尔（Mill）以及富有革命精神的雪莱（Shelley），他们三人的共同基础就是一种模糊不清的泛神论思想。天然去雕饰的大自然本身就是很神圣的东西，至少它是上帝的外套，也是上帝启迪人类的书籍。艾迪生（Addison）曾经赞美过的那个优雅与友善的大自然或许与这里的大自然有所不同，但是，我们面前的这个大自然仍然有着无比的力量，仍然具有神秘的号召力，让人对它顶礼膜拜。只是膜拜的理由有所不同。"自然的效果"不仅仅是诗意上的给人的吸引力，也是一种陶冶净化情操的神圣力量。比如，拉斯金（Ruskin）就认为大自然给出的答案总是最终的答案，没有争辩的余地。"你能够用鱼钩把海底的怪兽钓上来吗？"大自然营造的一切都是完美的，试图改变大自然的作品就是亵渎神圣的粗鲁行为。他甚至认为，人造的形式只有当它与自然形式的节奏韵律合拍的时候才是美的。他不仅把大自然是上帝启发人们的书籍这一纯属随意编造的神学概念引申一步，推论出研究自然形态的神圣价值，而且认为研究人类的本能是大逆不道的

employed them. And not only does he place a sacramental value on the study of Nature deduced from an arbitrary theological doctrine that it is God's 'book,' but he makes it a sin to study the human instinct, as though Nature's 'book' had expurgated man, and the merit of creation ceased at the fifth day. Doubtful logic this—and scarce orthodox theology! Yet there is little doubt that Ruskin's reiterated appeal to the example of Nature to witness against the formal instincts of man, did far more to enforce the prejudice against the 'foul torrent of the Renaissance' than he effected either by detailed reasoning or general abuse. In the face of all this poetry and rhetoric, in the face of all the sermons that were eloquent in stones, it is not surprising that Naturalism became the aesthetic method, and the love of Nature the most genuine emotion of our age. The emotion was as universal as it was genuine. A rich harvest of invention rewarded this attentive humility in the empirical sciences; the generation was encouraged by Emerson to 'hitch its waggon to a star'; the discipline of Nature, poetically inspired and religiously sanctioned, was pragmatically confirmed. Once more in the changes of civilisation, to 'live according to Nature' became a creed.

But to 'live according to Nature' means also, incidentally, to build and to garden according to Nature. And since the sublimity of Nature—its claim to worship—lay in its aloof indifference to man and in its incalculable variety, to build and garden according to Nature meant, as the progress of art soon demonstrated, to have a house and garden which betrayed, so far as possible, no human agency at all; or, at least, such human agency as might be manifested must be free from one specifically human quality—the 'self-contemplating reason.' This, with its insistence on order, symmetry, logic, and proportion, stood, in the ethics of Nature, for the supreme idolatry.[1]

On the one side was Nature: the curves of the waves, the line of the unfolding leaf, the pattern of the crystal. All these might be studied, and in some way architecturally employed—no matter how—so long as the knowledge and the love of them were evident. On the other stood the principles of Palladio, and all the pedantry of rule and measure, made barren by the conscious intellect. The choice between them was a moral choice between reverence and vanity. This was the refrain of *The Stones of Venice* and all the criticism 'according to Nature.'

The cult of Nature has a venerable history; but it is interesting to notice the

---

[1] This may perhaps furnish a philosophic basis for the advice once offered by a French nobleman, when consulted as to the most propitious method of laying out a garden in the then novel Romantic Manner: 'Enivrez le jardinier et suivez dans ses pas.' The 'self-contemplating reason,' temporarily dethroned by this expedient, is, for Ruskin, a constant source of political tyranny, architectural pedantry and spiritual pride.

罪恶,好像大自然早在创世纪的第五天里就已经把人类的创造力从人类的各种本能中清理掉了似的。逻辑似乎不通,神学理论也非正统。但是,拉斯金不断地引用自然界的例子来作为自己的武器,攻击人们所创造出来的形式。他的这种攻击言论不是给人以清晰的理性推导,也没有产生普遍的影响,只是极大地加强了人们心中已经形成的"文艺复兴时期肮脏的恶流"这样的偏见,这一点是毫无疑问的。面对这些诗歌、文学修辞方面的巨大影响,面对那些用石头书写出来的华丽庄严的神谕,当我们看到崇尚自然成为我们这个时代的美学方法、对于自然的热爱成为我们这个时代最真挚的情感的时候,也就不足为怪了。这种真挚的情感同时也遍及每一个角落。在经验科学的积累方面,这种对自然的谦卑自然得到了大自然的奖励,获得了大量的收获;这一代人在爱默生(Emerson)的鼓励下,满怀信心地"要把马车拴到星星上";而在诗歌的启发下,在宗教般地遵守中,自然的原理实际上得到不断的证实。在社会文明的不断变化中,"遵照自然的启示来生活"成为一种坚定的信条。

但是,"遵照自然的启示来生活"很偶然地也恰好包含了根据自然规律来建造房屋与设计花园的意思。由于大自然具有无限的崇高性,需要人们的顶礼膜拜,同时对人们的所作所为却又无动于衷,它展示给人们的形式又是变化多端,无法预料的,因此,所谓的根据自然规律来建造房屋和花园就意味着要排除任何人工的痕迹;即使不可能完全排除,至少也要把带有人类推理成分的那些内容清除掉,让这些建筑和花园与人类具有的自我反省与推断能力脱离关系。而艺术的发展过程表明这根本就是自欺欺人,我们在后面的讨论中会进一步揭示这一点。人类的自我反省与推断能力所强调的内容就是秩序、对称、逻辑以及各部分之间的比例关系,这种推理能力在自然规律中是受到最高敬意的。[1]

我们现在所面对的问题有两个方面,一方面是大自然:波浪式的曲线、舒展开的树叶、水晶的图案。所有这些都是可以研究的,也可以在建筑中加以运用,但是必须要明显地看得出设计者对自然的热爱与对自然知识的掌握。另一方面则是帕拉第奥的设计原理,这些原理讲的都是各种规则与尺度,都是很枯燥乏味的理性推导与计算。在二者之间作出选择是一个道德问题的抉择,是选择崇高还是选择浮华空虚的问题。把这个问题提高到道德层面上来,是《威尼斯的石头》那本书以及所有鼓吹"根据自然来设计"的建筑理论一直喋喋不休鼓噪的口头禅。

崇拜自然的思想本身有着悠久的历史,但是,当我们注意一下它的演变过程会发现有

---

[1] 这句话可以为我们提供一个哲学注解,帮助我们来理解下面这个忠告。当有人问到在设计花园的时候,什么是最恰当的布局方法,这时一位法国贵族给出最符合当时浪漫主义手法的忠告是:"把园丁用酒灌醉,然后你就跟着他后面走,这样的花园布局最完美。"这种简单操作的方法可以暂时排除"自己一个人凭空想出来的理由",但是,在拉斯金看来,"自己一个人凭空想出来的理由"正是政治暴君、建筑艺术学究和在精神上自以为是等现象的永久根源。

change it has here undergone. For Nature, as the romantic critics conceive it, is some thing very different from the Nature which their Stoic predecessors set up as an ideal, and very different also from Nature as it actually is. For the element in Nature which most impressed the Stoics was law, and its throne was the human reason. To 'follow reason' and 'to live according to Nature' for Marcus Aurelius were convertible terms. The human intellect, with its inherent, its 'natural' leanings towards order, balance, and proportion, was a part of Nature, and it was the most admirable and important part. But Nature, in the ethical language of her modern aesthetic devotees, stands most often in definite contradistinction to the human reason. They were willing to recognise authority 'in the round ocean and the living air,' but few remembered with Wordsworth to add: 'and in the mind of man.' The architect's work must be a hymn to creation, must faithfully reflect the typical laws and imitate the specific character of all that Nature presents. But the typical law and specific character of humanity, to impose order and rhythm on our loose, instinctive movements and proportion on our works—this is the unworthy exercise of 'self-contemplating Greeks,' the mark of 'simpletons and sophists.' While all things in nature fulfil their own law, each after its kind, man alone was to distrust his law and follow that of all the others; and this was called the example of Nature. Yet, since even so some choice is in practice forced upon him, the sole result of 'following Nature' is to sanctify his own caprice. Nature becomes the majestic reminder of human littleness and the insignificance of other people's thought. It is difficult to treat with total seriousness a phase of opinion so fatally paradoxical. Yet it sank deep into the public taste; and even now a discernible taint or moral reproof colours the adverse criticism of formal architecture; and a trace of conscious virtue still attends on crooked planning, quaint design and a preference for Arctic vegetation unsymmetrically disposed.

The creed of Nature entailed two consequences: first, a prejudice against Order and Proportion, and, therefore, against the Renaissance—for however deeply Order and Proportion may characterise the laws of Nature, they are far to seek in its arrangement; secondly, an emphasis on representation, on fidelity to the natural fact. This was soon made apparent in painting—first, in the microscopic realism of the Pre-Raphaelites; later, with more regard to the facts of vision, in impressionism. Architecture—an abstract or, at the least, a utilitarian art—might have been expected to escape. But it contained one element which exposed it to attack: it contained architectural sculpture. It followed, therefore, that this element, which admitted of representation and could be pressed directly into the cult and service of Nature, should become supreme. 'The only admiration worth having,' it is said in *The Seven Lamps*, 'attaches itself wholly to the meaning of the sculpture

一个很有趣的变化。浪漫主义理论家所鼓吹的自然与古希腊时期提倡禁欲主义的斯多噶（Stoic）学派所提倡的自然有着很大的不同，与真实的自然也不尽相同。让斯多噶派最为印象深刻的地方是大自然的规律性，而掌控这些规律的是人类的理性。对于古罗马帝国的皇帝马尔库斯·奥列里乌斯（Marcus Aurelius）来说，"遵循理性"与"按照自然规律生活"是同一件事，二者是可以互换的。人类的智慧，它的传承以及它对秩序、平衡、比例关系的掌握，这些都是大自然过程的一部分，而且是其中最重要、最受人尊敬的部分。但是在现在那些鼓吹崇尚自然主义的理论家看来，自然好像总是在与人类的理性相抵触，好像故意在作对似的。他们只希望承认"环绕着的大海与有生命的空气"所体现的权威，却忘记了华兹华斯在那后面还有一句："也要认识到人类理智"的权威。建筑师的设计作品必须是对上帝创造世界的赞美，必须诚惶诚恐地反映出大自然提示给人们的法规，模仿大自然提供的式样。能够代表人文精神的典型法则与特征就是给我们的作品中那种混沌散漫、凭直觉行事的方式施加秩序和节奏，施加比例关系，但是，这些都是"希腊人自己在思考"，都是"一些傻瓜和诡辩"的东西，根本不值得一提。自然界的一切事物都可以遵循自己的发展规律，唯独人类是不可以的，他们必须遵循其他事物的规律，这就是自然法则。然而，在某些不得已的具体情况下，当没有其他选择的时候，所谓的"遵循自然法则"不过就是把自己的办法加以合理化而已。自然就是用来不断地提醒人们是多么的渺小，其他人的想法是多么的微不足道的。这类的思维方式和做事原则很难让人严肃看待他们的主张。但是，这些观念早已深深地渗入到民众的普遍审美意识当中，甚至直到今天，它依然带着明显的误解与偏见，用道德的谴责口吻来抨击任何规则一点的建筑设计语言；人们的道德意识仍然让他们更加偏爱一些弯曲的规划设计布局、奇异的造型和把来自北极的植物不规则地布置在建筑周围。

对大自然的崇拜直接产生两个后果：第一个，产生敌视秩序与比例关系的偏见，因此便敌视文艺复兴文化。因为尽管大自然的规律从深层意义上来讲是以秩序与比例关系为特征的，但是在表面上是不大容易察觉的。第二个后果，它导致了对于自然细节的模仿与表现，而且是力求极其真实的描绘。这一点在美术界很快便显露现出来，最初是前拉菲尔学派（Pre-Raphaelites）那种细微局部的刻画，后来是追求视觉效果的印象派。建筑艺术因为是抽象的或至少是实用的大概会躲过这一类的影响吧，人们或许会这样想。但是，建筑里面因为包含了一种元素使得它也无法逃避这种影响。这种元素就是建筑中的雕塑与雕刻。由于雕塑的存在，雕塑又必须为大自然服务，必须表现出对自然的崇拜，结果使雕塑成了建筑中的最重要的成分。《建筑艺术中的七盏灯》（The Seven Lamps）一书中就说："建筑中最值得敬佩的地方就是建筑中的雕塑和色彩所表现出的意义。""体量的比例关系完全是抽劣不入流的东西。"这时建筑中的雕塑不但是脱离了与建筑的恰当关系，而且是成为建筑的目的与评判标准。根据同一个逻辑，这时的雕塑也必须是写实的。如果说，建筑艺术有任何意义

and the colour of the building.' 'Proportion of masses is mere doggerel.' And not only was sculpture thus thrust out of its true relation and made the chief end and criterion of architecture, but it was required, by the same argument, to be realistic. But architecture, if it means anything, means a supreme control over all the elements of a design, with the right to arrange, to modify, to eliminate and to conventionalise. Here, instead, arrangement becomes 'doggerel' and convention a blasphemy. In this, it will be noticed, the romanticism of Nature reached a conclusion exactly parallel to the romanticism of History. The latter, as we saw, becoming antiquarian, emphasised detail at the expense of the whole, and allowed architectural detail to deteriorate into a stylistic symbol. So, in this case, sculpture takes the place of architecture and deteriorates into realism. All this was necessarily fatal to the Renaissance style. Here there was little sculpture, and that little for the most part was conventional. Artificial in detail, artificial in design, here was an 'unnatural' architecture. Further condemnation could not be required.

Ⅲ. No fashion could have so securely established itself that was rooted in preferences altogether irrational or even new. Naturalism in architecture is partly a poetical taste; partly it is an ethical prejudice, and in each case it has been shown to be fallacious. But naturalism is also frankly aesthetic: a preference not merely of the fancy or the conscience, but of the eye. It may have entered modern architecture by a kind of false analogy, and may still derive from poetry a half-unreal support; but it has a solid footing of its own. For the place of what is unexpected, wild, fantastic, accidental, does not belong to poetry alone. These are the qualities which constitute the *picturesque*—qualities which have always been recognised as possessing a value in the visual arts. And one cause of offence in Renaissance architecture is precisely its lack of this picturesqueness of which Nature is so full. For the sake of this merit to the eye, how much decay has been endured and awkwardness forgiven! In a theory of architecture, what place then, if any, can be found for this true merit of the picturesque? What was, in fact, its place in the architecture of the Renaissance? To these questions an answer should be given before the romantic criticism of architecture can be fairly and finally dismissed.

If the wild and the accidental are absent from Renaissance architecture, it is certainly not because the men of that period were blind to their attraction. The term *pittoresco* was, after all, their own invention. It stood, on its own showing, for the qualities which suggest a picture, and are of use in the making of it. Picturesque elements—elements that are curious, fantastic, accidental, had been sought after in the painting of Italian backgrounds almost from the first. Their presence gave a special popularity to such subjects as the *Adoration of the Kings*, depicted, as by convention they habitually were, with strange exotic retinues and every circumtance

的话，它的一个意义就在于全面控制它上面的各个组成部分，可以按照建筑规律加以组合排布、调整、取舍，让建筑中的所有部分符合常规。然而在浪漫主义建筑艺术理论中，建筑布局成为拙劣不入流的东西，常规成了亵渎神圣的做法。这样一来，我们看到，崇尚自然的浪漫主义与崇尚历史的浪漫主义殊途同归，二者非常地一致。历史浪漫主义导致的结果是爱好古董，是不顾整体关系地追求古代的细部，让细部成为代表建筑风格的标签、象征符号。而在这里，自然浪漫主义的建筑，雕塑取代了建筑，成为写实主义的东西。这些都是与文艺复兴艺术不相容的东西，它们必置文艺复兴艺术风格于死地。在文艺复兴时期的建筑艺术中，雕塑的成分不是很多，几乎看不到，而仅有的那些雕塑又是最常规的，没意思。建筑细部都是人类设计出来的，建筑设计也是人工的，整个"建筑"的艺术是那样的"不自然"，还需要进一步加以谴责吗？

三、完全非理性的时髦理论是不可能具有如此强大的影响力的。甚至全新的理论也不可能。建筑艺术中的崇尚自然主义有一部分是属于诗歌艺术的，有一部分是属于道德偏见的。我们已经看到这两方面的谬误所在。但是，自然主义同时也是一种美学主张：它不完全是诗意的幻想与良心的说教，它也是视觉审美的判断准则。它可能是因为一种错误的类比关系而进入当代建筑艺术领域，它甚至仍然从诗歌艺术中获得半真半假的支持；这些都不影响它具有自己坚实的基础。对于一种带有某些意外、狂野、怪异、偶然等特征的艺术现象来讲，它不见得一定要属于诗歌艺术范畴。这些特征就是我们所说的图画般的造型，它们从来都是具有视觉艺术的特点的。文艺复兴时期的建筑艺术就是因为缺乏这种图画般构图才招致崇尚自然一派艺术理论家的攻击的，我们知道，大自然绝对不会缺乏这种图画般的构图。可怜我们的眼睛，它们忍受了多少腐化堕落的形式啊，又要原谅多少令人尴尬难堪的构图与造型啊！在建筑艺术理论中，这种图画般构图的学说在其中到底处于一种什么位置？它在文艺复兴时期的建筑艺术当中又处于一种什么样的位置？在我们彻底否定浪漫主义艺术中的建筑理论之前，必须就这两个问题做出回答。

如果我们说文艺复兴时期的建筑艺术中缺少狂野与偶然的元素，这绝对不是因为那个时代的人对这些不敏感，没有注意到它们的视觉魅力。事实上，概括这个艺术特征的意大利文术语 pittoresco 就是文艺复兴时期的发明。这个字的意思就是表示一些元素具有某些性质，让我们看到它们的时候一下子便能联想到一幅画，它们在一幅图画的形成中起到重要的作用。这些犹如图画构图般性质的元素在意大利绘画艺术中，几乎打一开始便受到画家们的追捧。这些元素据有令人好奇、富有幻想、出其不意等特点。比如《东方三博士在耶稣基督诞生的时刻前来朝拜》（*Adoration of the Kings*）之类题材的画作中，这种构图与艺术手法非常流行。习惯上，在这种题材的绘画作品中，画家细致地描绘出不少令人惊奇的来自远方的异族人物和随扈，场景也多半是幻想中的东方城镇或者乡村。所以到了十七世纪中期，当图画般构图这个艺术术语开始流行的时候，它并不是说绘画领域里刚出现的某种新手法或者新技巧，更多的是用在艺术分析评论里面，说明人们这时刚刚清楚地认识到艺术领域里的

of the fancied East. Thus the word itself, when, soon after the middle of the seventeenth century, it came into use, marked not so much a new virtue in painting as a new analytic interest, taking note for the first time of a permanent character in the art. Nor were these romantic elements limited to landscape and costume. They took the form, often enough, of inventions of fantastic architecture. And this is the more significant since in the Renaissance painters and architects are almost one fraternity, and the two arts were frequently conjoined.

But their sense of the freedom appropriate to the painted architecture is in strong contrast to the strictness they imposed upon themselves in the concrete art. The nearer art approached to the monumental, the more this self-denying ordinance became severe. Whatever surrounds us and contains our life; whatever is insistent and dominating; whatever permits us no escape—that, they felt, must be formal, coherent, and, in some sense,serene. Real architecture, by its very scale and function, is such an art. It is insistent, dominating, and not to be escaped. The wild, the fantastic, the unexpected in such an art could not therefore be allowed to capture the design. That, if we may judge from their work, was the principle in which Renaissance architects put faith.

This principle, like all the principles of Renaissance architecture, rested on a psychological fact. The different effects which art is able to produce, however various and incommensurable they may radically be, are commensurable at least in this: that each in some degree makes a demand on our *attention*. Some works of art affect us, as it were, by infiltration, and are calculated to produce an impression that is slow, pervasive, and profound. These seek neither to capture the attention nor to retain it; yet they satisfy it when it is given. Other works arrest us, and by a sharp attack upon the senses or the curiosity, insist on our surrender. Their function is to stimulate and excite. But since, as is well known, we cannot long react to a stimulus of this type, it is essential that the attention should, in these cases, be soon enough released. Otherwise, held captive and provoked, we are confronted with an insistent appeal which, since we can no longer respond to it, must become in time fatiguing or contemptible.

Of these two types of aesthetic appeal, each commands its own dominion; neither is essentially superior to the other, although, since men tend to set a higher value on that which satisfies them longest, it is art of the former kind which has most often been called great. But they do both possess an essential fitness to different occasions. Wherever an occasion either refuses or compels a sustained attention, a right choice between the two types will be a first condition of success. Fantastic architecture, architecture that startles and delights the curiosity and is not dominated by a broad repose, may sometimes be appropriate. On a subdued

早已存在的一种固有的特征。作品中展现的具有浪漫色彩的元素并不局限于人物的服饰或者田野场景，还包括很多完全是幻想出来的建筑，而且建筑在画面上的分量相当地大。这一点非常重要，因为文艺复兴时期的画家和建筑师是属于同一个行会的，大家以兄弟相称，而且这两种艺术常常是彼此不分，同时出现的。

但是，在绘画作品中表现出来的那种无拘无束富于幻想的建筑，与当时的艺术家和建筑师在真实生活里建造出来的严谨建筑作品相比较，二者形成了十分强烈的对比。建筑艺术越是接近纪念性，它就越是否定属于自身固有的性质。凡是那些围绕在我们身边与生活密不可分的东西，凡是那些恒久不变与强势主导的东西，凡是那些让我们想甩也甩不掉的东西，在文艺复兴时期的人们看来，它们都必须是堂堂正正、规规矩矩、有条有理、和谐一致的，从某种意义上讲，追求的也是一种平静。真实世界里的建筑从尺度和作用两方面来讲，正是这样的一种艺术，它恒久不变，它左右着人们的行为，它也让人无法摆脱。在建筑艺术中，那些带有一点狂野、不切实际的幻想、让人出乎意料的东西是不被建筑设计所接受的。如果我们从文艺复兴时期的建筑实例中能够做出某一种判断的话，那么这个判断就应当被视作是文艺复兴时期建筑师们所信奉的一条基本原则。

这个原则如同这个时期建筑艺术中的其他原则一样，它是基于心理学上的一个现象。艺术作品所产生的各种效果，无论它们彼此之间是多么的不同，多么的互不相干，但是有一个结论可以应用到所有的艺术效果上的：它们都在不同程度上希望获得我们对它们的注意。有的艺术作品是通过把自己浸润渗透到我们的心灵中，然后对我们产生作用。这类作品绞尽脑汁地创造出一种缓慢侵入我们的心灵的方式，不动声色地慢慢扩散自己的影响，而且是潜移默化的深远影响。这类作品根本没有打算惊动我们以便获取我们的注意力，它们也没有打算抓住我们的注意力以后便紧紧地不松手。但是，这类作品具有一种力量，只要我们注意到它们，那么我们就会立刻从中获得满足与享受。另外一类作品则是千方百计地试图捕获我们的注意力，不遗余力地向我们的感官发起猛烈攻击，或者刺激我们的好奇心理，大有不达目的决不罢休的架势。这类艺术作品的作用就是刺激我们，让我们兴奋起来。但是我们大家都清楚地知道，我们对于这类的刺激所产生的反应不会持久，我们对它们的注意力很快便会消失。因为持续地对这类艺术作品做出回应，过了不多久，我们就会变得麻木，再持续下去的话，一定是疲倦，甚至会产生抵触情绪。

对于这两类艺术作品的审美体验，每一类都有自己的适用范围，不能说哪一类一定就比另外一类高明。但是通常人们会认为，能够给自己带来长久一点的美感享受的艺术作品会具有更高一点的价值，因此，前一种的艺术作品会被更多的人认为比后一种的作品略胜一筹。但是，这两类艺术作品的确是各有各的长处，各有各的适用场合。在一个给定的情况下，当时的具体情形需要的是人们持续的注意力呢，还是对于这种持续注意力的排斥呢？这时候必须要在这两种艺术类型之间作出一个选择，这是成功的第一步。富于幻想特征的建筑艺术，那些能够勾起我们好奇心、让我们感到兴奋的建筑艺术，那些不需要保持雍容镇定姿态的建筑，有时候正是非常合适的建筑风格。在小尺度的建筑物上、在花园的某一个角落

scale, and hidden in a garden, it may be pleasant enough; but then, to be visited and not lived in. At a theatrical moment it will be right. It may be gay; it may be curious. But it is unfitted, aesthetically, for the normal uses of the art, for it fatigues the attention; and architecture once again is insistent, dominating and not to be escaped.

The practice of the Renaissance was controlled, if not by this reasoned principle, at least by an instinctive sense for its application. Even in the picture—since this, too, must have its measure of attention—the 'picturesque' element is made subordinate; it is subdued to that wider composition of line and tone and colour which contains it. And the complete picture itself is, or should be, subordinate once more to the formal scheme of the architecture, where it fills an appointed place. Consequently, the 'accidental' element, in the final result, is adequately submerged within the formal; it gives, without insistence, the charm of strangeness and variety to a general idea which it is not suffered to confuse.

This the Renaissance allowed; but the Renaissance went further. It was not only in painting that the picturesque could be favourably included; it was not only in its farms and hill-town buildings, pictorial as their beauty is. The Renaissance ended by reconciling the picturesque with classic architecture itself. The two were blended in the Baroque. It is not the least among the paradoxes of that profoundly great style that it possesses, in complete accord, these contrary elements. To give the picturesque its grandest scope, and yet to subdue it to architectural law—this was the baroque experiment and it is achieved. The baroque is not afraid to startle and arrest. Like Nature, it is fantastic, unexpected, varied and grotesque. It is all this in the highest degree. But, unlike Nature, it remains subject rigidly to the laws of scale and composition. It enlarged their scope, but would not modify their stringency. It is not, therefore, in any true sense accidental, irregular, or wild. It makes—for the parallel is exact—a more various use of discords and suspensions, and it stands in a closely similar relation to the simpler and more static style which preceded it, as the later music to the earlier. It enlarged the classic formula by developing within it the principle of movement. But the movement is logical. For baroque architecture is always[1] logical: it is logical as an aesthetic construction, even where it most neglects the logic of material construction. It insisted on coherent purpose, and its greatest extravagances of design were neither unconsidered nor inconsistent. *It intellectualised the picturesque.*

---

[1] I am speaking throughout of baroque architecture at its best. Naturally, in some cases there is charlatanism, or an ignorant attempt, to imitate the forms without perceiving the theory of the art. But the essence of the modern 'picturesque' taste in architecture is its absence of theory, its insistence on the *casual.*

里，出现一个这类建筑可以让环境变得非常愉快；但是，这类建筑只是适合人们来这里走走，而不是居住在其中。在类似娱乐的场合里，这类建筑非常恰当。它们可以带给人们欢乐，也有引人入胜的细节。但是对于平时的日常使用，从审美角度来说，这类建筑就不合适了，因为它们很容易引起注意力的疲倦。这时，建筑又回到了它的恒久不变、强势主导、让人无法摆脱的性格上来。

文艺复兴时期的艺术创作都是经过理性考虑之后，有节制地进行，如果不是严格的理性分析，至少也是凭直觉在运用着这样的理性分析原则。即使是在绘画创作中，具有"图画般构图"性质的元素也是在充当配角的角色，它必须要服从于更高一层的构图需要，与整体构图中的线条、调子、色彩等元素配合。因此完成的整体构图应当是服从于建筑艺术的形式规律，配合建筑的形式主题，恰当地出现在它应该出现的位置上。这样一来，那些"偶然、意外"的元素在最终的构图中被限定在恰当的位置，服从整体的需要；它们在大的原则控制之下，会因为自己的奇异而带给人们一点乐趣，带来一点变化，同时又不会过分地坚持渲染这些奇特的东西。人们不会因为它们的奇特而感到困惑、茫然。

这就是文艺复兴时期的艺术所允许的一些变化，但是，绝不止这些。不仅仅是这个时期的绘画艺术包含了追求图画效果的构图元素；不仅仅乡村、山坡上的建筑允许包括这种图画效果的构图元素，这时的建筑师把古典建筑语言与追求图画般构图效果结合到一处，成就了我们称之为巴洛克风格的建筑艺术。这种伟大的艺术风格把最不可思议的两种矛盾着的东西结合到一起，而且是结合得如此完美。巴洛克风格最了不起的试验就是把追求图画效果的构图上升到无以复加的地步，同时又让这种艺术满足建筑艺术自身的规则，它做到了。巴洛克风格一点也不会害羞，它很乐于吸引人们的注意，刺激人们的感官。它与大自然的相同之处在于，它有很多富于幻想的成分，常常出乎人们的预料，变化莫测，也时常出现一些怪诞的东西。在程度上也不是轻描淡写，而是无所不用其极。但是，巴洛克艺术与大自然还有很大的不同，它服从于严格的比例关系与整体的构图效果，它在自己能够发挥的空间内发挥得淋漓尽致，但是同时也严格遵守自己的界限。因此，可以这样说，巴洛克艺术从来都不是意外偶然的东西，不是真正意义上的不合规矩与放肆狂野。它在应该嘈杂的时候，就很放得开地大声喧哗，也充分利用人们的悬念等手段来强化这种效果，但是它与在它之前出现过的那些简单、安静的建筑风格保持着非常密切的关系。仔细观察，我们就会发现它们还是非常相似的，可以说保持着平行的关系。就好像一首音乐作品的后面部分与前面部分之间所保持的那种关系。它在自己追求动感的原则下，丰富了古典主义建筑的形式语言。巴洛克建筑中的动感是具有很强逻辑性的动感。巴洛克建筑艺术从来都是具有逻辑性的[1]：即便是在建造的时候会忽略建筑材料使用方面的逻辑性，但是在美学方面、在审美判断方面，它从来都是坚持自己的逻辑。它十分强调和谐一致的总的目标，在采用最为夸张的设计手段的地方，从来没有放松对于整体效果的把控，一切都没有放松局部与整体之间的一致性。巴洛克艺

---

[1] 我在说到巴洛克建筑艺术的时候指的是它的最优秀作品。在一些情况下，巴洛克艺术中当然有一些属于骗人的把戏，也有一些属于无聊的胡搞，在不知其所以然的时候胡乱模仿一些形式。但是，当代流行的"追求绘画构图效果"的建筑艺术从根本上讲就是这一派别，其实根本没有什么理论可言，所以它只能是不断地强调*随意而为*。

That the baroque style should be supreme in the garden and in the theatre—the two provinces which permit design its greatest liberty—was to be expected. The fountains and caryatides at Caprarola, the stage conceptions of Bibbiena and Andrea Pozzo, are unsurpassed. But the baroque could satisfy no less the conditions of a monumental and a permanent art. The colonnade of St. Peter's, Bernini's St. Andrea, the Salute at Venice, the front of the Lateran, are 'exciting' architecture: they startle the attention; they have the vivid, pictorial use of light and shade; the stimulus of their effect is sharp. In all this they achieve the immediate merit of the picturesque. Yet their last and permanent impression is of a broad serenity; for they have that baroque assurance which even baroque convulsion cannot rob of its repose. They are fit for permanence; for they have that massive finality of thought which, when we live beside them, we do not question, but accept.

Here, then, in the painting and architecture of the Renaissance, is an example of the fit employment of the picturesque. But these restrictions were not destined to be respected. The cult of Nature, by its necessary hostility to convention, modified that treatment of the picturesque and destroyed in it those saving qualities which can reconcile it with a 'dominating and insistent art': the qualities of reserve, finality, and repose.

While the Renaissance was in its vigour, the romantic view of Nature was no enemy of classic architecture. Of this the painters give us evidence enough. The painting of Claude Lorrain poeticised Nature in a luminous Virgilian mood, to which his vision of classic architecture, so far from being foreign, was the almost necessary complement. Without the austere quiet of his temples, Nature, in its tranquillity, might seem less human than he dreamt; without their Corinthian state, less sumptuous. Poussin, more sylvan in his interpretation, is not less classic in his forms. The more dramatic nature-painters—Salvator and the rest —did not press the wildness of their inspiration beyond its natural confines. It is perhaps only with Piranesi that a new spirit begins to show its force. In Piranesi, the greatest master of the picturesque in art, Nature holds architecture in its clasp, and, like the 'marble-rooted fig tree,' shatters and tortures it in its embrace. The consequences which were in due course to follow from the union are foreshadowed in the earliest phase of this master's art. He conceived a vision of infernal dungeons, without meaning, exit or hope; architecture, surrendered to the picturesque, was doomed in two generations to fall to the chaos without achieving the grandeur of Piranesi's 'Carceri.' Piranesi's etchings were multiplied rapidly and widely circulated; and the effect of their picturesque power on the imagination of the eighteenth century

术风格追求的是理智控制下的图画般构图。

巴洛克建筑艺术风格在花园和剧场中应该是具有大显身手的机会的,因为这两个地方允许艺术家能够最大自由地发挥自己的想象力。意大利中部城市卡普拉洛拉(Caprarola)里的喷泉与女像柱(caryatides),红衣主教毕比耶纳(Bibbiena)和安德烈·珀佐(Andrea Pozzo)的舞台设计,这些都无人能比。但是,巴洛克艺术在纪念性建筑方面、在永恒性艺术方面的潜能也毫不逊色。圣彼得大教堂前面的柱廊,伯尔尼尼设计的圣安德烈教堂(St. Andrea),威尼斯的安康圣母圣殿(the Salute),拉特兰宫殿(Lateran)的正立面,这些都是令人兴奋的建筑艺术:它们无一例外地都在力图引起人们对它们的注意;它们富有活力,对于光影的运用非常具有戏剧效果;它们带给人们的感官刺激非常强烈。总而言之,这些建筑都显然取得了图画般构图的效果。但是它们留给人们的最终印象却是晴朗宁静,这是因为巴洛克建筑艺术具有一种自信,即使是包括了一些具有动感、起伏的元素,也不影响整个建筑的沉稳,这一点非常适合表达永恒性。正是因为巴洛克建筑艺术具有这种宏大气魄,具有坚定的主张与信念,因此当我们面对这些建筑的时候,我们不会产生任何质疑,我们只有接受它们的权威。

我们在上面的论述中看到了文艺复兴时期的绘画艺术与建筑艺术是如何恰到好处地把图画般构图的想法与具体实践相结合的做法。但是其中的某些自我约束的做法并不能得到别人的认同与尊敬。崇拜自然的浪漫主义一派,出于它们对于传统常识性做法的敌视,把巴洛克艺术风格中的图画般构图等元素加以改变,把其中可以让作品避免轻浮的一些做法毁掉。被浪漫主义清除的东西正是那些"稳重自信、恒久不变的艺术成分",这些成分具有一定的矜持、坚毅、沉着性质。

文艺复兴时期的艺术即使是在自己最富有活力的时刻,也不会把崇尚自然的艺术爱好看作是古典主义艺术的敌人。这一点在画家留给我们的作品中足以作证。克劳德·洛兰(Claude Lorrain)的风景画就具有诗歌般的意境,他刻画的自然景色具有维吉尔诗歌中的那种光明灿烂,画中的古典风格建筑与这种自然景色搭配得恰到好处,互相衬托。这里的建筑语言都是最为熟悉的样式,一点也没有格格不入的风气。如果没有这些朴素安静的神庙建筑物,那么画面中的大自然也不会获得这种宁静与安逸,可能会更像是在梦幻中,而不是显现在那样活生生的气氛中;如果画中没有了科林斯建筑(Corinthian)的华丽,他的风景画也就失去了现在我们看到的奢侈与气派。普辛(Poussin)在自己的画中更加注重于树木的刻画,但是在整体构图上仍然遵守着古典构图原则。即使是其他追求画面戏剧效果的自然风景画家们,例如萨尔瓦多(Salvator)等人,他们在自己的创作中也不会为了表现自然的奔放而不顾艺术规律,忘记了艺术表现中应当遵守的分寸。大概只有在皮拉内西(Piranesi)的艺术创作中,这种浪漫主义艺术思潮才比较明显地显露出它的影响。在追求图画般构图这一思潮影响下进行创作的艺术家当中,皮拉内西应该是最伟大的艺术家。在他的画中,大自然把建筑物玩弄于股掌之间,就像生长在石头上的一棵大榕树的树根,在拥抱缠绕石头的时候,也同时把它粉碎,让它经历磨难。皮拉内西画中的建筑与自然的紧密结合自然对他后期的作品产生影响,而实际上这种风格在他早期的作品中就已经是有迹可循的了。他设想出一种地狱般的景象,阴暗的监狱,没有什么具体的实质意义,也看不到从中逃出的出口在哪里,更看不到任何希望;后来有两代艺术家,根据浪漫主义的构图原则所创造出来的建筑作

was decisive. Thus the way was made ready for the work of literature, and the new poetry of Nature when it came was reinforced by an existing fashion. Painting and literature were now as one. The taste for the picturesque, defensible, enough in those two arts, could not be long constrained within their limits. A picturesque architecture was required—an architecture untrammelled by those restraints which even the baroque style had hitherto observed. The philosophy of the Revolution favoured this impulse of the arts. True, it wrapped itself at first in a Greek mantle and David contrived a Doric background for its sages and tyrannicides. But 'natural' rights and a creed of anarchy could not for ever ally themselves with the most austere, the most conventional of styles. The philosophy of freedom invoked for architecture, as for life, the magic charm of Nature. But the material of architecture, no less than that of politics, was unfitted to receive its impress. For, in these obdurate forms, variety must prove tedious and licence lose its fascination.

But such an argument is incomplete. Picturesque building, it may be replied, in so far as it is insistent, curious and wilfully capricious, like the modern style of Germany or the fantastic style of the Gothic revival, may be thus unfit. But architecture which aims at the picturesque *need not* be insistent. There is a romanticism of conceits: the romanticism of Chambord, or the poetry of Donne. But there is also a romanticism of natural simplicity: the romanticism of Wordsworth and of a 'rustic' architecture. Architecture, in fact, can be picturesque without affectation, and various without disquiet. Why should not this be favoured? Where is the fault in that domestic type of architecture, in which we see a variety of form conditioned solely by convenience? Here will be repose, because the picturesqueness is unstudied, fitting the house to unselfconscious nature. No insistent appeal is thrust on the attention, for no deliberate appeal exists. This, in our time, is the true rival to the Renaissance style. It is this architecture which has so firm a hold in England, which seems to us so good to paint and good to live in. Poetry and sentiment are in its favour; it indolently provides pleasure to the eye. Leave it to be overgrown and it will be soon 'transformed by the enchantment of Nature to the likeness of her own creations.' Its beauty is secure from fashion, for it is elementary and genuine.

This is true; but how much shall we be willing to forego for the sake of this inoffensive, this sometimes charming, architecture? With what is it contrasted? It is usually implied that the alternative is *mere* formality. Formality, too, has its inherent, its, perhaps equal, charm. But it has more. It is the basis of *design*.

品，没有一件能与皮拉内西的铜版画《监狱》(Carceri)中所刻画的建筑相比。这组铜版画中的建筑物拥有前所未见的宏大气魄。这套作品不断地被复制出售，对于十八世纪的人们关于图画般构图的认识具有决定性的影响力，对于当时人们的形象也同样具有决定性作用。因此，当文学作品开始接受浪漫主义影响的时候，它们追寻模仿的对象就是从美术作品中获得的印象；在人们崇尚大自然的时候，这些存在于美术作品中的浪漫主义艺术思潮立刻强化了这种倾向。绘画艺术与文学艺术在这时走到了一起，成为一体。这两种艺术中追求图画般构图的艺术倾向是可以理解的，但是很快就不再局限于它们自己本来的地界，延伸到其他的艺术领域里了。建筑也必须是遵守图画般的构图才行，原先的很多约束被抛弃，甚至连巴洛克时期都遵守的一些原则也被抛弃。法国大革命的理想欢迎这样一种艺术思潮与艺术思想。在最开始的时候，大革命的理想披了一件古希腊的外套，法国艺术家大卫(David)在自己的画作中，勉强地为那些圣贤或者暴君安排了希腊多利克建筑与柱式。但是，天赋的人权思想与对于无政府主义的追求，让这些人不可能长久地保持那些最严肃、最常规的建筑风格的。对于自由的信仰导致了他们的建筑艺术崇尚自然，这与他们对待生活的态度是一致的。但是，建筑艺术的物质性却不允许接受这种不符合常规的做法，建筑的这种自身规律一点儿也不会让自己向政治运动屈服。在顽固的建筑形式中，出现过多的变化会让建筑作品变得异常琐碎，也就从而失去它的魅力。

但是这样的论据还不够充分。或许有人会说，对于一些要求具有永恒性的建筑来说，能够激起人们好奇心和故意为变化而变化的图画般构图有可能是不大合适，如当今在德国出现的那些现代风格的建筑，或者哥特复兴中出现的花里胡哨的东西。可是追求图画般构图的建筑师不需要具有永恒性。浪漫主义艺术有自己华丽的追求，例如香波尔城堡(Chambord)建筑和多恩(Donne)的诗歌创作。浪漫主义也有自然清新、简洁单纯的创作，如华兹华斯的诗歌和乡村纯朴的建筑。建筑可以是图画般构图的创作，同时又避免所有装腔作势的虚假成分；可以富于变化，同时又不失宁静。这有什么不好吗？在居住建筑形式上，我们见过很多根据使用上的方便而出现一些形式上的变化与多样性，这又有什么错呢？这些都会重归于安然稳重的，因为图画般的构图不是经过认真研究才搞出来的东西，它是在无意识的情况下让住宅变得好用而已。至于是否能持久地引人入胜、抓住人们的注意力实际上根本不在考虑之内，因为这种风格中根本不存在经过仔细算计、认真考虑之后而产生出来的吸引人的东西。在当今，这种建筑的确是文艺复兴时期建筑风格的真正对手。这类建筑在英国很有地位和影响，为很多英国人所接受。这类建筑物看上去很好看，很入画，住在里面也很方便实用。它们具有很强烈的诗意，很能唤起人们对它们的情感，对于人们的视线也不刻意地去争夺。它让自己的外貌自由生长，"很快就跟随着大自然一道生长，接受大自然的改造，最后成为有如造化本身创造出来的东西一样"。这类建筑的形式美非常自信，没有一点时髦的东西在里面，因为它的产生过程是最最基本的真实过程。

这些都没有错。但是，有多少次在我们设计建筑物的时候是单纯地为了它们的这种与世无争的性格，有时甚至是有点妩媚的特征的呢？这种性格与特征又是与什么相比较之下才显露出来的呢？没有说。但是通常的潜台词是，它是在与规则形式的建筑风格相比较，而且是*只有*规则形式的建筑风格与之对立。规则的建筑形式一样也有自己迷人的地方，而且

Everything in architecture which can hold and interest the intellect; every delight that is complex and sustained; every subtlety of rhythm and grandeur of conception, is built upon formality. Without formality architecture lacks the syntax of its speech. By means of it, architecture attains, as music attains, to a like rank with thought. Formality furnishes its own theme and makes lucid its own argument. 'Formal' architecture is to the 'picturesque' as the whole body of musical art to the lazy hum and vaguely occupying murmur of the summer fields.

All this is sacrificed; and perhaps even that little merit is not gained. Time and decay, colour and the accidents of use, the new perspective from the unforeseen angle of chance vision, may be trusted to give picturesqueness to the austerest architecture. Confusion will not lose its charm because there once was thought. Design is no implacable enemy of the picturesque; but the picturesque *ideal* is at variance with tradition and repugnant to design.

Our concern is here with one point only. It is not, certainly, that the picturesque is without merit; the merit of it is indeed too obvious. It is that, as an ideal, the picturesque renders taste obtuse, or suffers it to remain so. Like a coarse weed, not unbeautiful in itself, it tends to stifle every opportunity of growth. The modern taste for picturesqueness—as the old painters suffice to prove—brought with it nothing that was new. Nature, and man's work, is full of a picturesque beauty that has never passed unnoticed. But the aesthetic content of the picturesque is not constructive and cannot be extended. Nevertheless, it is upon this quality, so low in the scale, so unhopeful for future creation, and so unhelpful for an understanding of the formal past, that modern taste has been concentrated. This is the novelty and the prejudice.

There is a beauty of art and a beauty of Nature. Constuction, when it relaxes the principles of design, does not become Nature; it becomes, more probably, slovenly art. Nature, for a living art, is full of suggestion; but it is none the less a resisting force—something to be conquered, modified, adorned. It is only when the force of art is spent, when its attempt is rounded and complete, that Nature, freed from the conflict, stands apart, a separate ideal. It is thus the last sign of an artificial civilisation when Nature takes the place of art. Not without reason, it was the eighteenth century at its close—that great, finished issue and realised pattern— which began the natural cult. For a single moment, while the past still imposed its habit upon thought, disaster was arrested. The cult of Nature was a convention like the rest, and sought a place within the scheme. But the next step was the suicide of taste. Taken in isolation, made hostile to the formal instincts of the mind, Nature led, and can only lead, to chaos; whence issued a monstrous architecture: *informe*

也毫不逊色。但规则建筑所具有的性质不仅仅是这个,还有很多其他的方面。规则的形式是*造型设计*的基础。建筑艺术中任何能够吸引人们理性与智慧的成分,任何能够带给人们复杂的持久的愉悦、任何能够体现微妙的韵律节奏与构思的宏大的成分,都是建立在规则形式基础之上的。离开了这些规则的形式,建筑就在自己的语言中失去了应有的句法。正是通过这些句法,建筑艺术与人的思想结合到了一起,这和音乐里所包含的音乐语言与人的思想结合到一起是一样的。规则的形式有自己的主题和属于自己的逻辑。把"规规矩矩正式的"建筑艺术与图画般构图的建筑艺术作一个比较,那就好比是全部的音乐艺术与在夏天的旷野里远处传来的一些断断续续、漫不经心哼出的小曲儿的比较。

所有的这一切都被忽略了,甚至连其中最小的优点也都被否定。岁月的痕迹和衰败的外貌,色彩斑驳与当初没有设想过的用途,加上看问题的新角度,都让追求图画般构图的新建筑成为新宠,让人们忘记了过去那些严肃的建筑艺术。认识上的混乱不会掩盖艺术作品固有的魅力,因为艺术作品中所曾经包含有的思想仍然存在。堂堂正正的设计从来就不是图画般构图的敌人,但是追求图画般构图效果的这一理念不认同传统价值,誓与理性设计不共戴天。

我们在这里只关心一点。我们当然不是要否定图画般构图的优点,它的优点实际上是非常显而易见的,这不是问题所在。我们要强调的是,作为一种理念,在追求图画般构图的时候,造成了审美品位的偏差,变得不敏感了。如果继续保持这样的态度,就会让审美判断受到负面影响。这就好比草坪中的杂草,它本身并不是特别难看,只是在阻碍草坪的正常生长。当今流行的追求图画般构图的艺术潮流并没有什么新内涵,古代的艺术家可以证明这一点。大自然的也好,人工设计的也好,到处都有这种图画般的情景,而我们对此也看得一清二楚。追求图画般构图的做法在美学内容上没有什么建设性意义,它不能拓展美学上的新领域。不仅如此,这种美学品位不够高远,层次很低;它过于沉迷于过去而不是向前展望,而对于理解过去曾经有过的规则形式却又没有任何的帮助,现代的审美观念就是建立在这些特征上面的。这就是一种时髦和追求与众不同的新奇而已,是一种偏见。

艺术的美和自然的美是两种不同的东西。即便是在设计原则放松的情况下,当建筑可以自由发挥的时候,它也不会变成自然的一部分。它最可能的结果是成为一种邋遢的艺术作品。大自然对于有生命的艺术来说具有无限的启发与暗示,但是,它归根到底是一种抵抗的力量,必须要经过对它的征服、改造、修饰等过程。只有当艺术的力量作用在它上面,只有当作用在它上面的艺术达到自己当初的目的的时候,大自然才与原来的不同,不再有任何内在的冲突,而成为一种独立的理想。这是人们创造出来的文明中所剩下的最后一个特征,之后,大自然便取代了艺术。当十八世纪结束的时候,这个问题已经经过了辩论,已经是有了答案的问题,并且开始出现在实践中了。这个时候出现对于自然的崇拜也就没有什么值得奇怪的了。当过去的传统仍然作用于我们思想的时候,大的灾难可以得以暂时的避免。对于大自然的崇拜和其他的思潮是一样的,都是一种习惯做法,都是在一个大的目标下寻求自己的位置。但是由此发展出来的审美观念则是一种品味上的自杀行为。这种观念是受到脱离整个环境的某个片断启发而生,天生的带有断章取义的特征,同时又与人们倾向于规则缜密的思维逻辑相敌对,结果崇拜自然只能带来思想上和实践上的混乱,也因此从中产

ingens, cui lumen ademptum. Thus it was that by the romantic taste the artificial was scorned, though art, whatever else it is, is necessarily that; and it was scorned simply because it was not natural, which no art can hope, by whatever casuistry, to become.

生出怪兽般的建筑:没有了形式,让我们的目光放在什么地方呢(informe ingens, cui lumen ademptum)。因此我们认为,完全是因为浪漫主义的审美观念才使得人工建造的东西遭到人们的嘲笑与鄙视。而艺术作品无论从任何角度进行解读,无论从这些解读中得到怎样不同的结论,最终仍然无法改变它们是人工创造的东西这一最简单的事实。艺术创造仅仅是因为不够自然而遭到嘲笑,但是无论怎样论辩,艺术创作是绝无可能成为自然的一部分的。

FOUR
The Mechanical Fallacy

Such, in broad outline, were the tendencies, and such, for architecture, the results, of the criticism which drew its inspiration from the Romantic Movement. Very different in its origins, more plausible in its reasoning, but in its issue no less misleading, is the school of theory by which this criticism was succeeded. Not poetry but science, not sentiment but calculation, is now the misguiding influence. It was impossible that the epoch of mechanical invention which followed, with singular exactness, the close of the Renaissance tradition, should be without its effect in fixing the point of view from which that tradition was regarded. The fundamental conceptions of the time were themselves dictated by the scientific investigations for which it became distinguished.

Every activity in life, and even the philosophy of life itself, was interpreted by the method which, in one particular field, had proved so fruitful. Every aspect of things which eluded mechanical explanation became disregarded, or was even forced by violence into mechanical terms. For it was an axiom of scientific method that, only in so far as phenomena could so be rendered, might any profitable results be expected from their study. To this rule the arts proved no exception. But they were affected by the prevailing theories in two contrary directions. In many minds, aesthetics, like all philosophy, became subordinated to the categories of materialistic and mechanical science. On the other hand, those who valued art tended more and more to claim for each art its separate consideration. For, since the essence of the scientific procedure had been the isolation of fields of inquiry—the subjection of each to its own hypothetical treatment—it was natural that the fine arts, also, should withdraw into a sphere of autonomy, and demand exemption from any values but their own. 'Art for art's sake,' for all its ring of aestheticism, was thus, in a sense, a motto typical of the scientific age; and Flaubert, who gave it currency, was an essentially scientific artist. But the fine arts employed their autonomy only to demonstrate their complete subservience to the prevailing scientific preoccupation. Each bowed the knee in a different way. Thus Painting, becoming confessedly impressionistic, concerned itself solely with optical facts, with statements about vision instead of efforts after significance. Literature became realistic, statistical, and documentary. Architecture, founded, as it is, on construction, could be rendered,

第四章
建筑技术决定论的谬论

我们在前面从宏观的角度概括了一个大的趋势,即一些理论家从浪漫主义艺术思潮中获得灵感而产生出来的建筑艺术理论。最近还有一个新的建筑理论学派,它与浪漫主义艺术理论的起源截然不同,在论理的过程中更具有说服力,在误导人们的思想方面则是如出一辙。这一派的理论非常成功地影响了一大批人。这个理论的巨大影响力在于它并不强调诗学艺术,而是着重强调科学;它不是宣扬什么个人的情感而是使用科学的运算与演绎。一个以机械发明著称的时代,讲究精确严密的时代,它伴随着文艺复兴时期传统的结束而诞生。如果说文艺复兴时期的传统在这个时代的世界观形成过程中没有发挥任何影响,那简直是不可思议的事情。这个时代的那些最基本的概念本身,正是科学研究方法的直接产物。

用这个理论来解释生命中的每一个活动,甚至生活哲学本身,被证明很有成效,尤其是在某一个特定的领域里更是这样。有些原本不可能用技术的理由加以解释的方方面面被忽略了,或者粗暴地把它们归纳到技术的范畴之内,强行地用技术理由加以解释。因为科学方法的公理告诉人们,只要它能够准确地描述一个现象,那么认真地研究这个科学方法很有可能在其他方面也有帮助。基于这样的认识,艺术领域也不可能置身事外。但是艺术所受到的影响来自两个彼此矛盾的流行理论:很多人在当时认为,美学如同各种哲学学派一样,是从属于物质与技术占主导的科学范畴的;另一方面的观点则是认为艺术领域内的各个分支彼此越来越不同,因此必须针对每一个分支,分别进行单独的考虑。科学研究的方法从根本上讲,就是把自己的研究范围与周围分离开来,针对每一个内容提出针对性的假说,然后再加以证明。所以很自然地在艺术领域也有人认为,艺术应该划分出明确的属于自己的领域,要求自成体系,把其他与自己无关的价值排除在外。所以,在一个讲究科学的时代,无论自己的艺术流派是什么,大家不约而同地喊出了"为了艺术而艺术"的口号。至于鼓吹这个观念的福楼拜(Flaubert)本人,他根本就是一位非常讲究科学的艺术家。但是纯粹的美术在使用自成体系这个概念的时候,它仅仅是在试图证明自己完全附属于当时最为流行的科学观念。每一种艺术分支都在采用自己独特的方式,卑躬屈膝于所谓的科学。绘画因此成了视觉印象,仅仅注重视觉上的效果并力求把这个效果表现出来,却忽略了艺术作品本身所具有的意义。文学作品成了写实的纪录与统计,成了资料性质的东西。建筑艺术更是因为建造技术的缘故,比其他艺术走得更远,成了纯粹科学技术的一个方面,建筑艺术的创作也就变成了工程师的计算。凡是在工程技术具有决定性影响的地方,这里的技术就从手段变为了目的。在一个时代的各种思想观念中,当它们的价值观与达到目标的手段相混淆的时候,情形就更是如此。

even more readily than the rest, in the terms of a purely scientific description; its aims, moreover, could easily be converted into the ideals of the engineer. Where mechanical elements indisputably formed the basis, it was natural to pretend that mechanical results were the goal; especially at a time when, in every field of thought, the nature of value was being more or less confused with the means by which it is produced.

Now, although the movement of thought we have just described was in no way allied to the Romantic, and may even, in a measure, be regarded as a reaction against it, yet one characteristic at least the two had in common, and that was an inevitable prejudice against the architecture of the Renaissance. The species of building which the mechanical movement most naturally favoured was the utilitarian—the ingenious bridges, the workshops, the great constructions of triumphant industry, proudly indifferent to form. But, in the 'Battle of the Styles,' as the antithesis between Gothic and Palladian preferences was at that time popularly called, the influences of science reinforced the influences of poetry in giving to the mediaeval art a superior prestige. For the Gothic builders were not merely favourites of romance; they had been greatly occupied with the sheer problems of construction. Gothic architecture, strictly speaking, came into existence when the invention of intermittent buttressing had solved the constructive problem which had puzzled the architects of the north ever since they had set out to vault the Roman basilica. The evolution of the Gothic style had been, one might almost say, the predestined progress of that constructive invention. The climax of its effort, and its literal collapse, at Beauvais, was simply the climax and the collapse of a constructive experiment continuously prolonged. In no architecture in the world had so many features shown a more evidently constructive origin, or retained a more constructive purpose, than in the Gothic. The shafts which clustered so richly in the naves were each a necessary and separate articulation in the structural scheme; dividing themselves into the delicate traceries of the roof, construction is still their controlling aim. The Greek style alone could show a constructive basis as defined; and, for a generation interested in mechanical ingenuity, the Gothic had this advantage over the Greek, that its construction was dynamic rather than static, and by consequence, at once more daring and more intricate. Thus Gothic, remote, fanciful, and mysterious, was, at the same time, exact, calculated, and mechanical: the triumph of science no less than the incarnation of romance. In direct contrast with this stood the architecture of the Renaissance. Here was a style which, as we have seen, had subordinated, deliberately and without hesitation, constructional fact to aesthetic effect. It had not achieved, it seemed not even to have desired, that these two elements should be made to correspond. Where the Renaissance

FOUR

我们以上所说的这种思想认识与浪漫主义艺术思潮应该没有什么关系,在某些方面甚至是反对浪漫主义倾向的,但是它们二者之间却有一个共同之处,而这个共同点正是让它们形成了一种偏见,来共同敌视文艺复兴时期的建筑艺术。技术派的理论家最喜欢的具有代表性的建筑类型当然是实用型的构筑物——巧妙的桥梁建筑、工业厂房、巨大的工业用的设备装置等,这些建筑物在自己的建造设计当中,对它们的最终形式都是采取一种高傲漠视、无所谓的姿态。当哥特风格与帕拉第奥风格的争论甚嚣尘上的时候,当时人称"风格大战"交战正酣的时候,科学的影响力介入进来了。科学的力量强化了诗意一方的战斗力,从科学角度加强了中世纪艺术的影响力,使之身价大涨。哥特时期的建造者们不仅仅具有浪漫的情怀,而且他们还特别地注重建造技术。来自北方的建筑师们长期以来一直被如何在罗马人留下的巴西里卡(basilica)布局的形制上加建拱顶的问题所困扰。当他们采用了一排排飞券来解决拱顶稳定的问题时,严格讲起来,这就是所谓的哥特建筑风格诞生的时刻。或许可以说,哥特建筑风格的演变过程就是这个技术问题的解决过程。这个拱顶的技术性实验的高潮在博韦主教教堂(Beauvais)的拱顶成功地建造起来而后又坍塌垮掉。这个实验过程极其漫长。在世界建筑历史上,还没有一种建筑艺术像哥特建筑艺术那样,它上面的很多艺术特征完全是出自建筑技术,同时又保持着原有的技术作用。许多柱子汇集到一起形成了中庭两旁的柱墩,柱墩上的每一个柱子都分别与整体结构中的某些元素有着直接对应的关系。柱子上升到屋顶以后,从不同方向与那里的分隔线条结合起来,形成我们看到的图案,哥特建筑中的建造技术仍然是它形式的主导因素。古希腊建筑风格从理论上讲也反映了它的建造技术,但是对于崇尚技术发明的那一代人来讲,哥特建筑有着比古希腊风格明显的优势,因为它的建造技术是具有强烈的动感的,不像希腊那样是静止的。动感的结果就是更加具有艺高胆大的吸引人的地方,更加精巧。因此,来自遥远的地方、华丽精美又具有相当的神秘感的哥特建筑同时又是精美绝伦、严格的力学计算、精湛的建造技术:这不但是科学的胜利,也是浪漫主义艺术的化身。而眼前的文艺复兴时期的建筑立刻与之形成鲜明的对比。我们看到的文艺复兴的建筑风格则是让建造技术服从于美学的追求,而且是毫不犹豫地这样做。这种风格的建筑作品中看不出有什么把美学追求与建造技术相结合的痕迹,更不要说在这方面的成功实例了。文艺复兴时期的建造者想到什么形式,便无所顾忌地加以采用,即使与建造技术相背离也无所谓。同样的思维逻辑,文艺复兴时期的人们对于建造技术丝毫没有兴趣,即使是那些必不可少的技术手段,只要是与艺术家的构思不相符,他们就千方百计地把这些痕迹覆盖掉,就算是没有明显的理由,也要把它们隐藏起来。建造技术被视作是建筑艺术的仆人,建筑艺术也就毫不客气地把它当作仆人、奴隶来对待。这些构件被用来充当自己的技术角色的同时,并不给它们对外界露面的机会。这还不够,这些技术构件所产生的形式被用来作为单纯的装饰用途,比如各种线脚、壁柱已经没有任何结构作用,就如同

builders wanted the effect of a constructional form, they did not scruple to employ it, even where it no longer fulfilled a constructive purpose. On the other hand, with equal disregard for this kind of truth, those elements of construction which really and effectively supported the fabric, they were constantly at pains to conceal, and even, in concealing, to contradict. Constructive science, which so long had been the mistress of architecture, they treated as her slave; and not content with making mechanical expedients do their work while giving them no outward recognition, they appropriated the forms of a scientific construction to purely decorative uses, and displayed the cornice and pilaster divorced from all practical significance, like a trophy of victory upon their walls. And, in proportion as the Renaissance matured its forms and came to fuller self-consciousness in its methods, this attitude towards construcation, which had already been implicit in the architecture of ancient Rome, with its 'irrational' combination of the arch and lintel, became ever more frank, and one might almost say, ever more insolent. Chains and buttresses in concealment did the work which some imposing, but unsound, dome affected to contribute; facades towered into the sky far above the churches, the magnitude of whose interiors they pretended to express, and buildings which, in reality, were composed of several stories, were comprehended within a single classic order.

  It is useless to minimise the extent to which such practices were typical of the Renaissance. Although it is only in Italy, and in the seventeenth century, that the most glaring examples are to be found, yet the principles which then reached their climax were latent, and even in many cases visible, from its earliest period. They are inherent in the point of view from which the Renaissance approached the question of aesthetics. And, on the continuous plane of increasing 'insincerity' which the style, as a whole, presents, it would be unreasonable and arbitrary to select this point or that as the limit of justifiable licence, and to decry all that came after, while applauding what went before. This, none the less, is the compromise which is fashionable among those critics who feel that concessions must be made, both to the strictures of the 'Scientific' criticism on the one hand, and to the acknowledged fame of the 'Golden Age' of architecture on the other. But such a procedure is misleading, and evades the real issue. It is, on the contrary, imperative to recognise that the Renaissance claimed and exercised this licence from the first, and to make the closest examination of the doctrines which that claim involves. The relation of construction to design is the fundamental problem of architectural aesthetics, and we should welcome the necessity which the Renaissance style, by raising the question in so acute a form, imposes for its discussion. But the issue is not such a simple one as the 'scientific' criticism invariably assumes.

  We must ask, then, what is the true relation of construction to architectural

挂在墙上的优胜奖杯一样。随着文艺复兴时期的建筑在形式方面的不断成熟,最自我的形式也逐渐充满自信,这种对于建筑技术的忽略就更加明显。过去在古罗马时期还很含蓄的'非理性'做法到了这时便是堂而皇之地尽情流露,比如过梁与拱券的组合便是一例。我们可以这样说,这时的建筑艺术包含有一种傲慢在其中。铁链和隐藏着的扶壁必须用来加固穹顶产生的侧向推力。正立面上的塔楼高耸入云,比教堂主体要高出很多。教堂内部宏伟的大厅以及周围的建筑实际上有好多层,但是建筑师硬是把它表现为一层的高度,便采用了放大尺寸的建筑柱式。

在文艺复兴时期,这类做法多得不计其数,我们没有必要对此加以掩饰,或者轻描淡写。尽管代表这类做法的主要著名作品大都是集中在意大利,集中在十七世纪,尽管这种做法在此之前已经是十分常见,但是这项原则在它取得最辉煌成就的时候却并不为人们所察觉。它是文艺复兴时期的人们对待审美观念的一个基本态度,是审美观念中固有的成分,以至于人们对它反而是视而不见了。这种艺术风格作为整体,它是沿着'违反结构逻辑'的发展方向在不断地深化,如果我们在这个演变过程中挑选一点作为这种艺术实践的分界点,对于发生在这一时间点之后的作品痛加指责,而对于在此之前的则是鼓掌喝彩,这显然是毫无道理可言的。但是,当今很多理论家却都是这样做的,而且是很乐此不疲。这些理论家认为有必要划分出两个特殊区域,一方面满足"科学的"艺术理论所要求的严密逻辑性,另一方面又能照顾到建筑艺术"黄金时代"的辉煌成就。但是这样的划分只能误导民众,只能是迷失问题的关键所在。我们真正应当注重的关键问题,恰恰与上面的做法相反。我们应该承认文艺复兴时期的艺术实践,从一开始就认定,并且一直在实践着这种"违反结构逻辑"的做法,然后认真地、仔细地研究那个时代之所以这样做的理由。建造技术与造型设计之间的关系是建筑美学问题的核心所在,我们应该感到有一种紧迫感,文艺复兴时期的人们通过自己的建筑艺术形式,十分尖锐地把这个问题摆在我们面前,迫使我们必须作出回答。我们应该欢迎这样一个挑战和机会。这个问题并不像"科学的"艺术理论所想象得那样简单。

我们一定会提出这样一些问题:建造技术与建筑艺术之美到底是一种怎样的关系?

beauty; how did the Renaissance conceive that relation; and how far was it justified in its conception?

Let us begin by attempting, as fairly as we may, to formulate the 'scientific' answer to the first of these questions; let us see where it leads us, and if it leads us into difficulties, let us modify it as best we can, in accordance with the scientific point of view.

'Architecture,' such critics are apt to say, 'architecture *is* construction. Its essential characteristic as an art is that it deals, not with mere patterns of light and shade, but with structural laws. In judging architecture, therefore, this peculiarity, which constitutes its uniqueness as an art, must not be overlooked: on the contrary, since every art is primarily to be judged by its own special qualities, it is precisely by reference to these structural laws that architectural standards must be fixed. That architecture, in short, will be beautiful in which the construction is best, and in which it is most truthfully displayed.' And in support of this contention, the scientific critic will show how, in the Gothic style, every detail confesses a constructive purpose, and delights us by our sense of its fitness for the work which is, just there, precisely required of it. And he will turn to the Doric style and assert the same of that. Both the great styles of the past, he willl say, were in fact truthful presentations of a special and perfect constructive principle, the one of the lintel, the other of the vault.

Now, in so far as this argument is based on the Greek and mediaeval practice of architecture, it is an argument *a posteriori*. But it is clearly useless to reason dogmatically *a posteriori*, except from the evidence of all the facts. If *all* the architecture which has ever given pleasure confirmed the principle stated in the definition, then the argument would be strong, even if it were not logically conclusive. Admitting, then (for the moment), that the description given of Greek and mediaeval architecture is a fair one; admitting, also, the Greek pre-eminence in taste, and the acknowledged beauty of the Gothic, the argument from these is clearly not, in itself, an adequate condemnation of a different practice employed by the Romans and the Renaissance, which has enjoyed its own popularity, and whose case has not yet been tried.

But we may suppose our scientific critic to reply that he does not base his case on authority, but on the merits of his definition: that his argument is, on the contrary, *a priori*, and that he cites Greek and mediaeval architecture merely as an illustration. Can we say that the illustration is a fair one? Is it a sufficient description of the Greek and Gothic styles of architecture to say that they are 'good construction, truthfully expressed' ? Is it even an accurate description?

文艺复兴时期的人们又是怎样认识这个关系的？在他们的认识中又有多少是具有充分的说服力呢？

首先，让我们先来针对第一个问题，设想一下"科学的"答案会是怎样的吧。我们会尽可能地做到不偏不倚、公平合理；再让我们看看这个答案会把我们带到什么地方；如果他把我们带到了一个不利的境地，那么我们看看能否再根据科学的观点，对于这个不利的结论进行一番修正，让它变得好一些。

坚持科学观念的理论家们大概会这样说，"建筑就是建造房子。它之所以成为艺术的根本特征就在于它必须解决结构的技术问题，而不是靠光和影子的视觉图案。因此，在评价一个建筑的时候，它的这个独特性质决不能被忽略，因为这是它之所以成为与其他艺术明显不同的艺术的决定性因素。每一个艺术分支都是根据它自身的特殊性质来进行评价与判断的，建筑的评价判断标准正是应该建立在它的这种与建筑结构规律的关系上面。概括地说，如果一个建筑物的建筑结构与构造能够最真实地反映在它的建造过程与最终结果上面的话，那么，这个建筑的建造手段就是最好的，也因此它的建筑艺术也就是美的。"为了支持他们的这个说法，这些理论家们会借用哥特建筑风格来说明，强调其中的每一个细节都在传达着建造过程中的作用，说明这些细节恰恰是因为它们恰到好处地满足了自己特殊的作用而出现在那个特定的地方，而我们的感官感受到了这一点之后便产生了一种愉悦。这些理论家们还会借用古希腊多立克（Doric）风格的建筑来论证同样的观点。这两种建筑风格都是过去历史上出现过的伟大的艺术风格。理论家们会说，这两种风格真实地表现了各自的特殊性与建造技术原理，一个表现的是梁柱结构体系，另一个是拱券结构体系。

上面这个根据古希腊和中世纪的实例来论证自己观点的做法有一个问题，它只是叙述了一个后天的经验（a posteriori），是马后炮，不是问题之所以如此演变的真正原因。根据个别的经验来当作教条进行推理是没有说服力的，除非这个结论能够概括所有的情形。如果所有能够让我们感受到愉悦的建筑艺术都无一例外地证实了这个结论，就算它在逻辑上还不是完全判断的最后定论，那么这个论点显然将会是无法批驳的。退一步讲，即便是这个关于古希腊与中世纪建筑艺术的论述是公允的，即便是古希腊的建筑艺术很杰出，哥特建筑艺术的美得到民众的公认，这几点也并不能成为谴责其他艺术创作的理由，还不足以用来指责古罗马和文艺复兴时期建筑艺术。要知道，后面这两个时期里的建筑也是受到民众的极大喜爱的。至于这两个风格的是与非，到目前还没有经过审判和裁决呢。

秉持科学理念的理论家们肯定不服气，他们会辩驳说自己的论点并不是在强词夺理，完全是根据自己理论概念的真实性与说服力在做出的判断；他们的结论或者论点不仅仅是后天的个别经验，不是马后炮，而是一个具有普遍意义的判断，是判断建筑艺术优劣的前提条件（a priori）；他们引用古希腊和中世纪的风格来说明自己的观点，不过是列举两个例子来说明问题而已。他们引用的这两个实例能够证明他们自己的论点吗？他们有充分的理由说明古希腊与中世纪的建筑艺术风格的确是"优秀的建筑技术，真实地反映了建筑技术"的建筑吗？抛开理由的充分与否不说，这两种建筑真的反映了建造技术吗？

Are they, in the first place, 'good construction'? Now, from the purely constructive point of view—the point of view, that is to say, of an engineer—good construction consists in obtaining the necessary results, with complete security and the utmost economy of means. But what are the 'necessary' results? In the case of the Greek and Gothic styles, they are to roof a church or a temple of a certain grandeur and proportion; but the grandeur and proportion were determined not on practical but aesthetic considerations. And what is the greatest economy of means? Certainly not the Doric order, which provides a support immeasurably in excess of what is required. Certainly not the Romanesque, or earliest Gothic, which does the same, and *which delights us for the very reason that it does so*. Greek and mediaeval construction, therefore, is not pure construction, but construction for an aesthetic purpose, and it is not, strictly speaking, 'good' construction, for, constructively, it is often extremely clumsy and wasteful.

Can we now describe it as 'construction truthfully expressed'? Not even this. For the Greek detail, though of constructional origin, is expressive of the devices of building in wood; reproduced in stone, it untruthfully represents the structural facts of the case.

And if by 'truthfully expressed construction' it is meant that the aesthetic *impression* should bring home to us the primary constructive facts (a very favourite *cliché* of our scientific critics), how are we to justify the much applauded 'aspiring' quality of Gothic, its 'soaring' spires and pinnacles? In point of structural fact, every dynamic movement in the edifice is a downward one, seeking the earth; the architect has been at pains to impress us with the idea that every movement is, on the contrary, directed upwards towards the sky. *And we are delighted with the impression.*

And not only does this definition, that the beauty of architecture consists in 'good construction truthfully expressed,' *not* apply to the Greek and mediaeval architecture, not only does it contradict qualities of these styles which are so universally enjoyed, but it *does* apply to many an iron railway-station, to a printing press, or to any machine that rightly fulfils its function. Now, although many machines may be beautiful, it would be a *reductio ad absurdum* to be forced to admit that they all are; still more that they are essentially more beautiful than the Greek and Gothic styles of architecture. Yet to this conclusion our definition, as it stands, must lead us.

Clearly, then, when Greek and Gothic buildings are cited in support of the view that the essential virtue of architecture lies in its being 'good construction truthfully expressed,' we must take objection, and say, either these styles, and, *a fortiori*, all others, are essentially bad, or our definition must be amended. The scientific

FOUR

首先一点，这两种建筑是否是反映了"优秀的建造技术"的建筑呢？单纯地从建造技术的角度来看这个问题，也就是从工程师的角度来看这个问题，一个建筑物如果能用最经济的手段来满足最基本的使用要求，同时又有足够的安全可靠性，那么，这个建筑物从单纯的技术角度来讲就是优秀的建筑物。满足最基本的使用要求指的是什么呢？古希腊和中世纪的建筑风格主要希望解决的问题无非是给神庙和教堂加建一个屋顶，同时表现出建筑物的宏伟气魄并且达到某种特定的比例关系。表现出这种宏伟气魄与某种特定的比例关系已经不是技术或者实用上的要求了，它们完全是出于美学的考虑。而经济的手段又是怎样体现的呢？绝对不会是多立克柱式，因为古希腊的多立克柱式所使用的石头不知超出技术要求多少倍。罗马帝国时期的风格和早期哥特风格的建筑也是如此。*我们喜爱这些建筑风格完完全全因为它们保持着目前的这个不合理的样子。*古希腊和中世纪的教堂建筑并不完全是单纯地追求建筑技术，而是一种美学实践，是为了满足特定的美学追求。它们也不是"优秀"的建造技术的代表，严格说起来，它们在结构上都常常是笨拙臃肿的，而且十分浪费建筑材料。

我们能说这些建筑物真实地表现了它们的建造技术吗？恐怕这个问题的答案也是否定的。古希腊建筑上面的细节的确是与建造技术有一定的关联，但是那些都是原型木建筑上的细节，现在借用石头把它们复制出来，它们并不是石头结构所必需的细节。

如果说"真实地反映出建造技术"指的是建筑物上出现的关键建造技术特征所带给人们的美学感受（崇尚科学的艺术理论家们特别喜欢这种论调），那么我们怎样来解释哥特建筑上面为人们津津乐道的"高耸、升腾"的感觉呢？怎样解释那些具有"一飞冲天"气势的尖塔和大大小小的尖顶呢？因为从结构角度来讲，建筑中能够表现出建筑物受力规律的构件应该是指向地面才符合逻辑的呀，怎么现在都指向上方呢？显然，哥特建筑的建造者们是在绞尽脑汁地把所有的建筑构件统统指向天空。而*我们喜欢它最后的这个样子。*

我们现在看到，"真实地表现出建造技术"的建筑才会产生出的建筑美这个定义不仅仅不能说明古希腊、中世纪建筑艺术的本质，不仅仅与那两种建筑艺术风格中为广大民众所喜爱的艺术特征相矛盾，而且，它刚好适用于诸如铸铁搭建的铁路车站、印刷机器或者任何普通的机器设备。我们不否认有些机器的确是很美观，但是强迫我们接受"所有的机器都美观"则是强人所难，显然是荒谬的结论（a reductio ad absurdum）。不但如此，如果这个定义成立的话，它不可避免地会得出这样一个结论：一切机器都要比古希腊、中世纪的建筑艺术还要美观。

显然，用古希腊和中世纪的建筑来证明建筑艺术的根本性质在于"真实地表现建造技术"这一定义是站不住脚的，对于这个观点和举例证明的做法，我们必须表示反对。我们必须明确地大声指出，要么是这两种建筑风格，或者加重一点分量说，所有的建筑风格，它们都是错误的风格，要么就是这个定义自己是错的，必须对它进行修正。讲究科学这一派的艺术

criticism would presumably prefer the latter alternative. Those of its supporters who *identify* architectural beauty with good and truthful construction (and there are many) it must disown; and we may suppose it to modify the definition somewhat as follows:

Beauty, it will say, is necessary to good architecture, and beauty cannot be the same as good construction. But good construction is necessary as well as beauty. We must admit, it will say, that in achieving this necessary combination, some concessions in point of perfect construction must constantly be made. Architecture cannot always be ideally economical in its selection of means to ends, nor perfectly truthful in its statement. And on the other hand, it may happen that the interests of sincere construction may impose some restraint upon the grace or majesty of the design. *But good architecture, nevertheless, must be, on the whole, at once beautiful and constructively sincere.*

But this is to admit that there are two distinct elements—good construction and beauty; that both have value, but are irreducible to terms of one another. How then are we to commensurate these two different elements? If a building have much of the second and little of the first—and this, many will say, is the case of Renaissance architecture—where shall we place it, what value may we put upon it, and how shall we compare it with a building, let us say, where the conditions are reversed and constructive rationality co-exists with only a little modicum of beauty? How is the architect to be guided in the dilemma which will constantly arise, of having to choose between the two? And, imagining an extreme case on either side, how shall we compare a building which charms the eye by its proportions and its elegance, and by the well-disposed light and shade of its projections, but where the intelligence gradually discovers constructive 'irrationality' on every hand, and a building like our supposed railway station, where every physical sense is offended, but which is structurally perfect and sincere? Now, the last question will surely suggest to us that here, at any rate, we are comparing something that is art (though, it may be, faulty art) with something that is not art at all. In other words, that from the point of view of art, the element of beauty is indispensable, while the element of constructive rationality is not. The construction of a building, it might conceivably be suggested, is simply a utilitarian necessity, and exists for art only as a basis or means for creating beauty, somewhat as pigments and canvas exist for the painter. Insecure structures, like fading pigments, are technical faults of art; all other structural considerations are, for the purposes of art, irrelevant. And architectural criticism, in so far as it approaches the subject as an art, ought perhaps to take this view.

But there the scientific criticism should certainly have its reply. Granting,

理论家们大概会选择后一个建议,把定义作一番修改。这个修改过的理论必须要把那些认为建筑美也就等同于"真实地表现出建造技术"的好建筑的那些人排除在外(持那样观点的人为数不少)。这个修改过的理论或许会这样讲:

美是一个好建筑必不可少的条件,美并不等同于好的建筑技术所建造出来的房子。但是好的建筑建造技术是获得建筑美的不可或缺的要素。这个理论会接着说,我们必须承认,为了取得这些不同要素的完美组合,在建造技术方面的要求就不得不做出一些让步,不能过于追求建造技术方面的十全十美。建筑艺术在追求自己的目的的时候,也并不总是确保自己采用的手段是最经济的,在处理方式上也不是毫无瑕疵,也不是总能做到里外如一。话又说回来了,如果真的按照建造技术的要求做出来的话,建筑的外观和设计的构思说不定还会受到负面影响呢。*但是,无论怎么说,好的建筑从整体上讲,它既是美的,同时在技术上又是真实的。*

但是,这个修改过的理论承认了两个完全不同的建筑要素:好的建造技术和建筑美。它们有各自不同的价值取向,相互之间无法增减转换。那么在这两种截然不同的元素之间怎样找出适用于彼此的衡量标准呢?如果一个建筑中包含了很多第二种元素,却有很少的第一种元素(我们可以说,这正是文艺复兴时期建筑所处的状态),我们把它划分在哪一类呢?我们怎样来衡量这类建筑的价值呢?我们怎样把这类建筑同与它刚好相反类型的建筑进行衡量比较呢?所谓的相反类型就是第一种元素所占的比例甚高,而第二种元素比例甚低。在这种进退维谷的境地中,我们该给建筑师怎样的忠告,让他们在下次遇到同样问题的时候能够作出恰当的选择?现在再让我们设想一个极端的情况,这两种元素都各自走向极限。假设一个建筑凭借自己完美的比例关系和优雅的造型,凭借着自己的光和影在凹凸之间所形成的图案,让我们的眼睛大大地享受了一番,但是,我们的智慧却让我们逐渐地发现这个建筑内部的建造技术非常地"违反理性",几乎所有的地方都不符合建造技术方面的理性原则;我们再假设另外一个建筑,例如火车站,那里的每一处让我们在视觉上都不顺眼,感到不舒服,但是理智告诉我们,它的每一处在建造技术上都是无懈可击的。我们怎样把这两个建筑进行一番比较呢?这个问题实际上可以换一个说法,就是一件艺术作品(也许不是优秀的艺术)与一件与艺术无关的东西怎样进行比较呢?从艺术角度来看问题,美是不可或缺的,但是建筑技术中的理性元素则不是。从这里我们很自然地会引申出一个推论,那就是一座建筑的建造技术只是必要的工具和手段,它只是为艺术提供了一个可以创造出"美"的平台和基础,就好比是画家手里的颜料与摆在他面前的画布。建筑结构如果不坚固,就和褪色的颜料一样,是技术上的缺陷,在这种情况下再谈别的与技术相关的话题就已经没有任何意义了。如果我们仍然把建筑当作一门艺术的话,那么我们的建筑理论就应该采取类似于这样的一种观点。

坚持科学观念的艺术理论家们对于这一点当然有话要说。他们大概会这样讲,即便是

it will say, that beauty is a more essential quality in good architecture than constructive rationality, and that the two elements cannot be identified, and admitting that the criticism of architectural art should accept this point of view, there is still a further consideration. It will claim that architectural beauty, though different from the simple ideal of engineering, is still beauty *of structure*, and, as such, different from pictorial or musical beauty: that it does not reside in patterns of light and shade, or even in the agreeable disposition of masses, but in the structure, in the visible relations of forces. The analogy between construction and the mere material basis of the painter's art, it will say, is false: we take no delight in the way a painter stretches his canvas or compounds his pigments, but we do take delight in the adjustment of support to load, and thrust to thrust. It is no doubt legitimate to add decorative detail to these functional elements; they may be enriched by colour or carving; but our pleasure in the colour and the carving will be pleasure in painting or sculpture; our specifically architectural pleasure will be in the functions of the structural elements themselves. It is in this vivid constructive significance of columns and arches that their architectural beauty lies, and not simply in their colour and shape, as such, and so far as the structural values are absent, and the eye is merely charmed by other qualities, it is no longer architectural beauty that we enjoy. Only, these functional elements must be vividly expressed, and, if necessary, expressed with emphasis and exaggeration. The supporting members must assure us of their support. Thus, the Doric or the Romanesque massiveness, while it was in a sense bad science, was good art; yet its beauty was none the less essentially structural. Thus, the printing press or the railway station will now appropriately fall outside our definition because, although truthfully and perfectly constructed, and fit for their functions, they do not *vividly* enough express what those functions are, nor their fitness for performing them. Structurally perfect, they are still structurally unbeautiful. On the other hand, the arches and pilasters of many Renaissance buildings may be agreeable enough as patterns of form, but are no longer employed for the particular structural purpose for which apparently they are intended, and so, in diminishing the intelligibility and vividness of the whole structure, diminish at the same time its beauty. Thus, the one group fails because, though functional, it is not vivid; the other because, though vivid, it is not functional.

　　Such, or somewhat such, would be the statement of a 'scientific' view of the relation of construction to architectural design, as we should have it when divested of its more obviously untenable assertions and stated *in extenso*. In the modern criticism of architecture, we are habitually asked to take this view for granted, and the untenable assertions as well; and this is accepted without discussion, purely owing to the mechanical preconceptions of the time, which make all criticisms

FOUR

在一个好建筑中,它的美要比追求建造技术中的纯粹理性要来得重要,即便是建造技术与美不能彼此画上等号,即便是我们也可以假设建筑艺术理论应该采取这样的观点,但是建筑艺术中还有一个事实我们不得不加以考虑。这些理论家会说,建筑艺术美尽管不同于最简洁的工程技术设计,但是它归根到底毕竟还是在欣赏一座具有庞大**构架**建筑物的美,也正是因为这一点,使得它不同于图画或者音乐中所具有的那种美:建筑美不在于光和影的图案,甚至不在于均衡的体量组合,建筑美是在于它的大的结构框架,在于结构框架上所反映出来的各种受力关系。把建筑技术成分与画家手里的材料相比较是不恰当的,是错误的。我对于画家如何把自己的画布固定在框子上毫无兴趣,我们对于画家如何调混自己的颜料也没有什么兴趣,但是我们对于建筑中出现的那些支撑是如何抵抗重力,对于那些横向加固的构件是如何抵抗水平方向上的侧推力等等,却是饶有兴趣,并从中获得极大的享受。我们不是说绝对不能在这些发挥力学作用的构件上施加某些装饰,当然可以在上面施加一些颜色或者雕刻一个造型,但是,那些装饰部分到头来,已经不是建筑美,我们从那些装饰上感受到的美其实又回到了绘画或者雕刻艺术的美,与建筑美有着本质的不同。我们对于建筑艺术的享受还是在于那些正在正常地发挥着作用的结构构件上面。建筑的美正是在于这些具有生命力的柱子与拱券上面,它们正生动地发挥着自己的作用。建筑美并不在于它的色彩和形状。只要是离开了建筑结构部分,我们的眼睛实际上是在欣赏别的东西,并不是在欣赏建筑艺术之美。只有当这些与结构有关的元素在生动地发挥着自己的作用,只有当建筑把这些结构元素充分地表达出来,甚至在表达手段上多少带有一点点夸张,这时我们才在欣赏建筑艺术之美。建筑的支撑元素一定要让我们感觉到它的支撑能力才行。因此,古希腊的多立克柱式,或者古罗马时期的柱子才变得很粗大,从科学角度讲或许不合理,或者没必要,但是从艺术角度讲,这些则是好的艺术作品,无论怎样讲,它们最终毕竟是建筑结构中的元素。这样说来,印刷车间和铁路车站大概就会被我们的新定义排除在外了,因为这些东西虽然在建造技术上很理性,也很真实地反映了内部的使用与技术构件的作用,但是它们并没有对这些内容加以*生动地*表现,从而让我们感受到这些构件的作用到底是什么,也没有表现出这些构件的确是在正常地发挥它们自己的作用。它们的结构很理性,但是不美。在另一方面,文艺复兴时期建筑中的一些拱券、壁柱等元素,看起来与空间的形态很协调,但是,它们并不具有真实的结构作用,实际情况与它们在表面上传达给我们的信息是不一致的,结构上是虚假的。在利用虚假的手段让建筑结构失去理性与生命力的同时,也就失去了它的美。因此,头一类建筑物不美,是因为它们虽然结构是真实的,但是表现手段不够生动;另一类建筑物不美,是因为它们虽然手段生动,但是结构并不发挥自己真正的结构作用,是虚假的。

这应该就是崇尚科学技术的艺术理论家所坚持的关于建筑艺术与建造技术的"科学"观点吧。就算是不很准确,也应该相差不远。它又多花了些笔墨把一些明显站不住脚的内容从原先的定义中删掉。在当今流行的建筑艺术理论中,理论家把这个观点当作是理所当然的东西,他们假设我们都会自然而然地接受这个观点,最好是连同原先那些站不住脚的观点一起接受下来。他们让我们接受这些观点,却又连接受的理由都不愿意给出,这完全是受到了当时对于技术的迷信与盲从这样一种社会风气的左右,认为只要是借用"技术"作为理

on the score of 'structure' seem peculiarly convincing. Such a view, even in the modified form in which we have stated it, sets up an ideal of architecture to which indeed the Greek and mediaeval builders, on the whole, conformed, but to which the Romans conformed very imperfectly, and to which the Renaissance, in most of its phases, did not conform at all. It cuts us off, as it seems, inevitably, from any sympathy with the latter style. Before accepting this unfortunate conclusion, let us see whether the ideal is as rational and consistent as it sounds.

In the first place, it is clear that the vivid constructive properties of a building, in so far as they are effectively constructive, must exist as *facts*. The security of the building, and hence also of any artistic value it may possess, depends on this; and a support which seemed to be adequate to its load, but actually was not, would, as construction, be wrong. But in so far as they are vivid, they must exist *as appearances*. It is the effect which the constructive properties make on the eye, and not the scientific facts that may be intellectually discoverable about them, which alone can determine their vividness. Construction, it may be granted, is always, or nearly always, in some sense, our concern, but not always in the same sense. The two requirements which architecture so far evidently has are constructive integrity in fact, and constructive vividness in appearance. Now, what our scientific critics have taken for granted, is that because these two requirements have sometimes been satisfied at the same moment, and by the same means, no other way of satisfying them is permissible. But there has been no necessity shown thus far, nor is it easy to imagine one, for insisting that these two qualifications should always be interdependent, and that both must invariably be satisfied at a single stroke. Their value in the building is of wholly disparate kind: why, then, must they always be achieved by an identical expedient? No doubt when this can be done, it is the simplest and most straightforward way of securing good architectural design. No doubt when we realise that this has been done, there may be a certain intellectual pleasure in the coincidence. But even the Greeks, to whom we are always referred, were far from achieving this coincidence. When they took the primitive Doric construction, and raised it to a perfect aesthetic form, the countless adjustments which they made were all calculated for optical effect. They may not have entailed consequences *contrary* to structural requirements, but at least the optical effect and the structural requirements were distinct. The Renaissance grasped this distinction between the several elements of architectural design with extreme clearness. *It realised that, for certain purposes in architecture, fact counted for everything, and that in certain others, appearance counted for everything. And it took advantage of this distinction to the full.* It did not insist that the necessary fact should itself produce the necessary appearance. It considered the questions separately, and was

由就会让理论成为不容置疑的权威，大家都会自然而然地信服。我们在前面以理论家的口气所叙述的这个观点，就是勾画出一幅理想中建筑艺术的图画，从整体上讲，古希腊和中世纪的建筑刚好与理想画面完全一致；古罗马时期的不够完美，但也算是基本上一致；而文艺复兴时期的建筑艺术绝大多数是不符合这幅图画所勾勒出的标准的。这个新定义好像让我们无言以对，让我们找不到一点同情文艺复兴时期建筑艺术的理由。在我们以沮丧的心情来被迫接受这个结论之前，先看看这个被科学艺术理论家们勾勒出来的理想画面，到底是不是真的那样推理缜密、无懈可击呢。

首先要说的是，根据这个定义，建筑艺术中那些活生生的与建造技术相关的元素，它们必须是已经有效地建造出来的东西，必须是*客观存在*的了。建筑的坚固与安全性，以及其他附加的艺术价值都取决于这一点。一个建筑的承重体系如果看上去没有问题，但是实际上却不能满足力学要求，那么这个建筑物从技术上讲就是坏的。但是，这些建筑物又必须能够生动地表现出这一点，那它就必须是取决于*视觉*的东西。它是建造技术作用于眼睛所产生的效果，并不是依赖于我们的理性与智慧去研究才能够发现的东西。表现的生动与否完全取决于人们的眼睛。建造技术方面的事情当然是我们所关心的，从某种意义上讲，我们总是对于建造方面的事情有所关心，但是侧重点则是根据具体情况而随时有所不同。到这里，我们明显地看到了建筑艺术中所必不可少的两件事：一件是建造技术的坚固，这是客观事实方面的性质；另一件是建造手段在外观上给人的感觉是否生动地表现了这种技术上的坚固。我们的崇信科学的理论家们因为看到在某些案例中，这两种情况刚好因为同一种技术手段而同时发生，就理所当然地认为不应该有其他的方式来满足这两项要求。但是到目前为止，我们还没有看到有什么必要一定要做到只采用一种技术手段来同时满足这两项不同的要求，还没有看到有什么必要一定要让这两件不同的事情互相联系在一起，也无法想象出有任何理由一定要这样做。这两件事在建筑艺术中具有不同的价值与作用，为什么一定要做到"一箭双雕"，为什么一定要强求把两件事情通过同一种技术手段来同时解决呢？当然，如果能做到这样，无疑是最理想的，这是完成一件优秀设计作品最简单、最直接的途径。这一点是毫无疑问的。而且真的能够通过一种技术手段来达到两种目的，我们的理性与智慧也一定会从中感受到那种灵性与愉悦，这一点也是毫无疑问的。但是，即使是不断被拿来当作模范经典的古希腊建筑也远远没有做到这一点啊。古希腊人把过去原始多立克柱式的建筑形制拿来进一步完善的时候，他们经过无数次的实验，把这个柱式的形式调整到了极其完美的程度，他们所作的从来就是在追求完美的视觉效果而已。虽然他们的做法并没有导致任何对于结构不利的情形，但是他们对于视觉上的造型与结构上的要求还是保持着清醒的区分的。文艺复兴时期的艺术家们对于这种客观存在着的区别有着更清醒的认识，他们在自己的建筑设计中，对于不同的元素分别采用不同的手段加以处理。*在这个时期的建筑艺术中，艺术家们认识到，在某些情况下，建造技术的理性胜过一切，同时在另外某些情况下，建筑艺术造型必须是压倒一切的。*他们把这两种不同的情况与对应手段发挥到了极致。他们并不会强迫自己必须做到一点，就是要让那些不可或缺的必要技术措施同时也承担艺术上十全十美的角色。这时的艺术家会把这两方面的问题分别加以考虑，也满足于分别处理的做法。他们不需要戴着脚镣来跳舞。这些艺术家创造出来的建筑看起来生机勃勃，给人坚固牢靠的感觉，同时也采取一切必要的技术手段来保证建筑物的确能够做到坚固牢靠。再让我们

content to secure them by separate means. It no longer had to dance in fetters. It produced architecture which *looked* vigorous and stable, and it took adequate measures to see that it actually *was* so. Let us see what was the alternative. Greek architecture was simply temple architecture. Here, architectural art was dealing with a utilitarian problem so simple that no great inconvenience was encountered in adjusting its necessary forms to its desired aesthetic character. Nor was there any incongruity between the aesthetic and practical requirements of a Gothic cathedral. But the moment mediaeval building, of which the scientific criticism thinks so highly, attempted to enlarge its scope, it was compelled to sacrifice general design to practical convenience, and was thereby usually precluded from securing any aesthetic quality but the picturesque. And even so it achieved only a very moderate amount of practical convenience. Now the Renaissance architecture had to supply the utilitarian needs of a still more varied and more fastidious life. Had it remained tied to the ideal of so-called constructive sincerity, which means no more than an arbitrary insistence that the structural and artistic necessities of architecture should be satisfied by one and the same expedient, its search for structural beauty would have been hampered at every turn. And, since this dilemma was obvious to every one, no one was offended by the means taken to overcome it.

And not only was the practical range of architecture thus extended without loss to its aesthetic scope, but that scope itself was vastly enlarged. In the dome of St. Peter's we see a construction, the grandeur of which lies precisely in the self-contained sense of its mass, and the vigorous, powerful contour which seems to control and support its body. Yet actually the very attempt to give it this character, to add this majestically structural effect to the resources of architectural art, meant that Michaelangelo ran counter to the scientific requirements of a dome. The mass which gives so supreme a sense of power is, in fact, weak. Michaelangelo was forced to rely upon a great chain to hold it in its place, and to this his successors added five great chains more. Had he adhered, as his modern critics would desire, to the Byzantine type of dome, which alone would of itself have been structurally sufficient, he must have crowned St. Peter's with a mass that would have seemed relatively lifeless, meaningless, and inert. Structural 'truth' might have been gained. Structural vividness would have been sacrificed. It was not, therefore, from any disregard of the essential constructive or functional significance of architectural beauty that he so designed the great dome, but, on the contrary, from a determination to secure that beauty and to convey it. It was only from his grasp of the relative place for architecture of constructional fact and constructional appearance, that he was enabled, in so supreme a measure, to succeed. And it was by their sense of the same distinction that the architects of the Renaissance, as a

来看看不这样做的话，还有什么其他的办法没有。古希腊的建筑只是一些很简单的神庙建筑。在这里，由于建筑艺术所需要解决的技术问题过于简单，因此在调整造型、追求美学理想的时候，不会遇到什么技术上的困难和制约。哥特建筑风格的那些大教堂在建造技术上与使用功能上也不存在什么冲突，都很容易同时满足两方面的要求。但是，让我们再好好看看中世纪的建筑艺术，这些被崇尚科学的理论家们推崇备至的建筑艺术，当它们在使用功能上必须要包含其他一些内容的时候，我们就会发现它们在艺术追求方面常常捉襟见肘，顾此失彼，往往不得以采用一些应付手段与权宜之计来牺牲艺术的完美，最后把很多美学上的考虑都放弃，仅守住图画般构图这一条。即便如此，中世纪的建筑仅仅满足一些不是很复杂的使用要求。到了文艺复兴时期，这个时期的建筑使用功能要求复杂，人们的生活也多样化，对建筑提出更多的要求。假如这个时期的人们果真如同理论家们所鼓吹的那样，盲目坚持建造技术必须真实地反映在外观上，也就是盲目地坚持力争做到通过一种技术手段既解决建造问题，同时又创造出美观的造型，那么，他们的建筑技术之美就一定是寸步难行。因为当时人人都知道这个问题，所以当文艺复兴时期的艺术家提出分别解决两个问题的手段时，没有任何人觉得这个办法有什么不妥。

这样一来，在建筑艺术没有受到任何损失的同时，它又极大地满足了使用功能对建筑提出的诸多需求。不仅如此，艺术范围不但没有受损，而且也得到相当大的拓展。比如在圣彼得大教堂的穹顶建造中，我们所感受到的那个宏伟、高大的气势正在于这个穹顶的巨大体量以及它完全依靠自己的力量雄踞建筑群之上，它带有弹性的曲线给人以旺盛的生命力和强大的力量的感觉，让人觉得这种力量足以支撑自己。努力追求这样的效果就是把这个建筑物中结构技术上的元素化作了视觉艺术的基础，这个追求在事实上却是让米开朗基罗必须要违反穹顶建造技术中的某些技术制约。这个看上去很有力量的巨大穹顶在技术上实际上是非常薄弱的。米开朗基罗迫不得已采用一条巨大的铁链子来捆住穹顶的根部来使得它不至于塌落下来。而他的后继者又另外再增加五条铁链子来捆绑这个穹顶。如果米开朗基罗当初真的像那些理论家要求的那样，墨守成规地遵循穹顶的技术要求，那么他只能建造出一个类似拜占庭风格的穹顶。穹顶在技术上也许是变得真实了，也够坚固了，但是圣彼得大教堂也因此会变得毫无生气，失去所有我们现在看到的东西，这个穹顶也就没有什么意义了，只是一个单纯的穹顶技术再一次实现而已，这个建筑会非常的沉闷。结构技术上的"真实"目的达到了，但是结构体系给人那种活生生的生命力却失去了。因此，米开朗基罗在设计他的这个大穹顶的时候，并不是完全地抛弃建筑艺术中那些结构元素的作用和表现力，而是相反，他就是要把那种活生生的力量表现出来。因为他有选择性地在建造技术与建筑外观之间取得一种相对平衡，再通过极端的手段来表现这种结构关系，所以他成功了。文艺复兴时期的建筑师们正是基于对技术与艺术这两方面的深刻认识，不仅仅给建筑带来了美观的造型，而且让这些古代的艺术形式通过灵活的手段重新为普通的生活带来一种向上的风气，让人们获得了尊严。这不是一个人、两个人的实践，而是那一代人所形成一个艺术流派。这些

school, not only enriched architecture with new beauty, but were able to dignify the current of ordinary life by bending to its uses the once rigid forms of the antique. And this they did by basing their art frankly on the facts of perception. They appealed, in fact, from abstract logic to psychology.

A similar defence may be entered for the Renaissance practice of combining the arch with the lintel in such a way that the actual structural value of the latter becomes nugatory, and merely valuable as surface decoration, or for its elaborate systems of projections which carry nothing but themselves. If we grant that architectural pleasure is based essentially upon our sympathy with constructive (or, as we have agreed, *apparently* constructive form), then no kind of decoration could be more suitable to architecture than one which, so to say, re-echoes the main theme with which all building is concerned. In Renaissance architecture, one might say, the wall becomes articulate, and expresses its ideal properties through its decoration. A wall is based on one thing, supports another, and forms a transition between the two, and the classic orders, when applied decoratively, represented for the Renaissance builders an ideal expression of these qualities, stated as generalities. The fallacy lies with the scientific prejudice which insists on treating them as particular statements of constructive fact wherever they occur. And, if the Renaissance architects, on their side, sometimes introduced a decorative order where on purely aesthetic considerations the wall would have been better as an undivided surface, or if they introduced a decorative order which was ill-proportioned in itself, or detracted from the spatial qualities of the building—which was, in fact, unsuccessful *as decoration*—this we must view as a fault rather of practice than of theory. And their tendency to abuse their opportunities of pilaster treatment must be held to spring from an excessive zeal for the aesthetics of construction, the nature of which they understood far more exactly and logically than their modern critics, who, while rightly insisting on the fundamental importance of structure not only in architectural science, but in architectural art, overlook the essentially different part which it necessarily plays in these two fields, and who imagine that a knowledge of structural fact must modify, or can modify, our aesthetic reaction to structural appearance.

To this position the scientific criticism would have a last reply. It will answer—(for the complaint has often been made)—that this *apparent* power and vigour of the dome of Michaelangelo depends on the spectator's ignorance of constructive science. In proportion as we realise the hidden forces which such a dome exerts, we must *see* that the dome is raised too high for security, and that the colonnade falls too low to receive the thrust, and that, in any case, the volume of the colonnade is inadequate to the purpose, even were the thrust received.

艺术家这样做的依据就是他们自己直觉的坦白流露。他们所信奉的不再是抽象的逻辑,而是民众的心理感受。

我们可以用同样的理由来解释,在文艺复兴时期的建筑艺术中,为什么拱券与横梁会出现我们所看到的那种关系。在文艺复兴时期的建筑中,很多横梁是出现在拱券之上的,从结构技术上讲,这个横向的过梁是累赘,没有结构作用的,完全是视觉上的装饰,就是说,那些细节丰富的横梁并不支撑任何东西,它们只负担自己的荷载而已。如果我们认定对于建筑艺术的享受根本在于对它建造技术的欣赏(或者说这种对于技术元素的外在表现,这一点我们同意),那么,建筑中采用的任何装饰元素都应该围绕着建筑物的主要结构形式而展开,各种装饰物体都是在呼应结构体系才行。在文艺复兴时期的建筑艺术中,墙体不再是承重元素了,而是一种表现工具,来表现艺术家心中的理想,因此艺术家可以采用与自己理想相一致的装饰手段来打造这片墙。但是墙身给人的感受只有一件事情,那就是它在支撑着上面的重量,同时形成了上下两部分之间的力量传递。当古典柱式在文艺复兴时期被拿来用作装饰手段的时候,正说明这个时期的艺术家在象征性地表达那种结构技术的原始本意,正在表达着一种普遍的概念。所谓的科学观念就是带有一种强烈的偏见,它的谬误所在就是认为这种梁柱体系,只要是出现,就只能是具体地用在建造技术中,在别的地方出现就是违反结构逻辑的。这显然是荒谬的。如果一面墙体在没有任何装饰的情况下,它的效果会是最好的,但是艺术家却偏偏为了自己的审美爱好而强行放上一个古典柱式把它破坏了;或者说这个艺术家在应该放置柱子的地方放一个比例关系十分难看的柱子,让整个空间因此被破坏掉了,我们会说这是实践中的失误,不是理论上的错误。艺术家们往往在操作的时候,会把壁柱的比例关系夸大一些,这些都是出于他们对于视觉审美的理解,是从视觉的角度来解释建筑结构与技术的特征。这些艺术家关于这一点的认识要比当今理论家们的认识深刻得多,无论是在逻辑上,还是在造型细节上,都是如此。这些理论家们在强调建筑技术不但是科学,而且也是艺术这一方面也许是对的,但是他们忽略了技术在这里所扮演的两个角色是有本质上的差别的;他们错误地坚持认为,结构技术作为科学必须矫正我们对于建筑造型的审美习惯,也能够矫正那些审美习惯。

就我们上面说的问题,这些崇尚科学的艺术理论家们还会反驳说(这种批评的声音从来就没有间断过),米开朗基罗的这个穹顶之所以会让人产生一种力量与生命力的感受,那是因为有这种感受的人对于建筑技术方面的科学无知。当我们认识到这个穹顶所产生的巨大作用力,虽然我们用眼睛看不见,但是我们会立刻明白这个穹顶被举起来太高了,以至于它的稳定性很差,而它下面的柱子位置过低,根本不能平衡抵抗掉穹顶所产生的侧向水平推力。这排柱廊的体量无论从哪一个方面来讲,都不够,甚至连真实的作用在它上面的推力都抵抗不了。

This is one of the commonest confusions of criticism. Just as, in the previous question, the scientific view fails adequately to distinguish between fact and appearance, so here it fails to mark the relevant distinction between feeling and knowing. Forms impose their own aesthetic character on a duly sensitive attention, quite independently of what we may know, or not know, about them. This is true in regard to scientific knowledge, just as in the last chapter we saw it to be true in reference to historical or literary knowledge. The concavity or convexity of curves, the broad relations of masses, the proportions of part to part, of base to superstructure, of light to shade, speak their own language, and convey their own suggestions of strength or weakness, life or repose. The suggestions of these forms, if they are genuinely felt, will not be modified by anything we may intellectually discover about the complex, mechanical conditions, which in a given situation may actually contradict the apparent message of the forms. The message remains the same. For our capacity to realise the forces at work in a building *intellectually* is, to all intents, unlimited; but our capacity to realise them *aesthetically* is limited. We feel the value of certain curves and certain relations of pressure to resistance by an unconscious (or usually unconscious) analogy with our own movements, our own gestures, our own experiences of weight. By virtue of our subconscious memory of these, we derive our instinctive reactions of pleasure, or the reverse, to such curves and such relations. But the more complex forms of construction can address themselves only to the intelligence, for to these our physical memory supplies no analogies, and is awakened by them to no response. So, too, if there be an exaggerated disparity between the visible bulk of a material and its capacity for resistance, as for instance in the case of steel, it is perfectly easy to make the intellectual calculus of its function in the building, but it is quite impossible to translate it into any terms of our own physical experience. We have no knowledge in ourselves of any such paradoxical relations. Our aesthetic reactions are limited by our power to recreate in ourselves, imaginatively, the physical conditions suggested by the form we see: to transcribe its strength or weakness into terms of our own life. The sweep of the lines of Michaelangelo's dome, the grand sufficiency of its mass, arouse in us, for this reason, a spontaneous delight. The further considerations, so distressing to the mechanical critic, remain, even when we have understood them, on a different plane, unfelt.

This theory of aesthetic must indeed be dealt with more adequately in a later chapter, but even if our scientific assailant refuses to admit the distinction between knowing and feeling to be important, and claims—for to this it seems he is reduced—that aesthetic feeling is consequent on *all* we know, and that architectural beauty lies, in fact, in the intelligibility of structure, his position—and it seems to

这是艺术批评中最常见的混淆说法。就如同在上一个问题的讨论中我们已经揭露的那样，这类理论家不能正确、清醒地区分建筑技术上的真实与视觉上的真实。与此类似，在我们正在讨论的这个问题中，他们仍然不能正确地区分人们的感受与理性知识之间的差别。建筑形式在把自己的美学特征作用于我们的感觉的时候，在抓住我们的注意力的时候，与我们对于它们的理性了解与掌握到什么程度没有任何关系。在上一章里，我们分析过历史与文学艺术的知识对我们的影响，那些分析与结论对于这里的科学知识也是适用的。曲线的凹或凸、体量上的大关系、局部与局部之间的比例、基础与上层建筑、光和影，这些形式都在述说自己的语言，在向我们传达这些形式所代表的强壮还是软弱的含义，呈现给我们的可能是充满活力的形象，也可能是冷静沉着的神态。这些形式传达给我们的信息一旦被我们接收到了，我们之后无论怎样进行科学研究也很难加以改变。我们的智慧或许可以从这一大堆形式语言中找出其中的某些技术，它们从严格的技术角度上看或许与外观形式语言所传达给我们的信息相冲突，但是，仍然无法改变我们已经接收到信息。因为我们的*理性认识能力*在认识建筑技术方面是无限的，但是我们对于建筑的*美学感受能力*却是有限度的。我们对于某种曲线的走势，对于某种压力和抵抗力都有着下意识的反应，我们在潜意识里把这些形式与我们自己的身体动作、姿势以及我们对于重力的直接经验等联系起来，将建筑形式与我们的肢体动作在潜意识里进行了类比。我们潜意识里对于这些亲身经验的记忆驱使我们对于建筑形式上的某些曲线或者比例关系作出直接的反应，从而产生出一种相应的快乐或者悲伤的情绪。但是更复杂的建筑形式只能在理性分析的时候才能够理解，我们已经不会有直接的反应，因为我们对于复杂的形式没有什么记忆，因此不会有什么类比效应，也就不会产生什么相应的感受。如果建筑材料的体积与它所能承受的荷载，在人们的认知上有落差，例如钢材，而这种落差又被夸大的时候，经过理论计算，人们很容易理解这种材料在建筑中的作用，但是这种计算几乎是不可能被转化成人们的亲身经验的。我们自己对于这种令人迷惑的关系了解不多。我们看到的建筑形式，经过想象力的作用，对我们产生一定的生理上的作用。我们的美学体验完全取决于这种生理上的体验：换句话说，我们是把建筑形式的强弱转换成我们人生经历中的某些体验。米开朗基罗设计的穹顶，它外轮廓的曲线和它巨大的体量，正是因为自己的形式立刻让我们感到一种自然而然的欣喜。进一步的理性分析让崇尚技术的理论家感到非常痛苦，但是对我们来说，即便是我们已经理解其中的道理，这些分析也完全是另一个层面的问题，是我们感受不到的。

无疑，我们将会在本书的后面必须就这种美学理论进行深入充分的讨论。但是就目前的讨论结果来看，就算是那些崇尚科学的理论家们拒绝承认在心理感受与理性分析之间加以区别有着极其重要的意义，坚持认为审美感受归根到底还是我们全部知识产生作用的结果，坚持认定建筑艺术的美在于对建筑结构的理性认识，他们的问题实际上也已经被回答了。因为经过深入理性的分析之后，那些在穹顶根部把穹顶捆绑在一起的铁链也必须是我

be the last—is simply met. For if it is to be a case of full understanding, the chains which tie the dome are part of what we understand. Why are we to conjure up the *hidden* forces of the dome, and refuse to think of the chains which counteract them? But, granted the chains, the structure is explained, and the knowledge of the fact should give the scientific critic the satisfaction he desires. And if our pleasure lies in intellectually tracing, not the means by which the structure is made possible, but the relation of the structure to its purpose, then this pleasure would be derivable from the work of the Renaissance architect no less than from that of the mediaeval one. For, given that the end proposed by the former is understood to be different—and we have shown that it *was* different—from that proposed by the latter, then the different methods chosen in the two cases are no less exactly adjusted to their ends in the one case than in the other. No doubt when the aesthetic sense is atrophied, when the attention is concentrated upon scientific curiosity, when the Renaissance architect is conceived to have attempted something different from what he did attempt, then the dome of St. Peter's may induce nothing but an intellectual irritation. But then, this attitude to architecture, carried to its logical results, ignores its character as an art altogether, and reduces it simply to engineering; and we have already demonstrated the *reductio ad absurdum* which that involves.

Thus vanishes the argument from structure. The prestige which still, in all our thought, attaches to mechanical considerations, have given to so weak a case a perverse vitality. One central point should, however, be clear from this analysis. It may be restated in conclusion, for it is important. Two senses of 'structure' have been entangled and confused. Structure, in one sense, is the scientific method of 'well-building.' Its aim is '*firmness*.' Its end is achieved when once the stability of architecture is assured. And any means to that end are, scientifically, justified in proportion to their effectiveness. Structure, but now in a different sense, is also the basis of architectural '*delight*.' For architecture, realised aesthetically, is not mere line or pattern. It is an art in three dimensions, with all the consequence of that. It is an art of spaces and of solids, a felt relation between ponderable things, an adjustment to one another of evident forces, a grouping of material bodies subject *like ourselves* to certain elementary laws. Weight and resistance, burden and effort, weakness and power, are elements in our own experience, and inseparable in that experience from feelings of ease, exultation, or distress. But weight and resistance, weakness and power, are manifest elements also in architecture, which enacts through their means a kind of human drama. Through them the mechanical solutions of mechanical problems achieve an aesthetic interest and an ideal value. Structure, then, is, on the one hand, the technique by which the art of architecture is made possible; and, on the other hand, it is part of its artistic content. But in the

们理性分析的一部分。为什么要求我们把穹顶里看不见的受力关系必须考虑在讨论之内，却不允许我们把抵抗这些受力的铁链算作是解决这个问题答案的一部分呢？但是，如果我们接受这些铁链的话，那么穹顶的结构分析也就没有问题了，技术的理性分析也给了那些理论家所要求的答案。如果我们从建筑艺术中获得的快乐是来自于理性与智慧对于各种问题的理解，我们关心的不是采用什么具体的手段把它做出来的，而是关心建筑结构与建筑目的之间的关系，那么，我们从文艺复兴时期建筑艺术中获得的快乐绝不会比从中世纪的建筑艺术中获得的要少。我们已经在前面分析过，中世纪时期的建筑与文艺复兴时期的建筑对使用有不同的要求，因此针对两个时期的不同需要，两种建筑分别提出了具有各自说服力的解决方式，每一个建筑都不比另一个缺少任何要素。如果我们不考虑艺术造型与美学因素，如果我们把我们的全部注意力集中在所谓的科学好奇心方面，如果不能正确理解文艺复兴时期的艺术家在追求什么，那么，我们必然从圣彼得大教堂的穹顶上什么也看不到，只会注意到那里违反理性的某些东西。但是这种对待建筑艺术的态度根本就是忽略了建筑本来还是一门艺术这个条件，把建筑艺术简化为其中依赖于计算的工程技术部分，所以他们得出那种结论也就不足为怪了。我们在前面已经说明了这种结论的荒谬（reductio ad absurdum）。

以建筑结构为依据的理论到现在都破灭了。以建筑技术为依据的理论在我们所有人的思想中有着令人敬畏的地位，但是我们所看到的关键论述却都是不堪一击的。它的一个中心论点在我们的分析中应该被说明得很清楚了。因为这一点十分重要，我们在结论总结里，再强调一次。他们没有分清纠结在一起的关于建筑结构的"两种不同的认识"，这把他们的思想搞混乱了。建筑结构，一方面是科学的方法，是解决建造技术的手段。它的目标是在一座好建筑中，担负着确保建筑"坚固"的责任。只要一座建筑解决了稳定、坚固的问题，它的目的也就达到了，自己的使命也就完成了。只要能够最有效地解决科学技术的问题，那么任何具体的技术手段都应该是被接受的。另一方面，建筑结构还是建筑艺术给人带来愉悦的基础。因为美观的建筑不仅仅是线条和图案的组合，它是一个三维的艺术作品，而三维的艺术作品有着不同于其他艺术形式的特点。它是空间与体量的艺术，从中可以让人感受到各个部分间的相互关系，是各种制约力量相互平衡、妥协的结果，它把建筑材料按照我们自己能够感受到的最基本定律组合到一起。重量与支撑、困难与努力、虚弱与强壮，这都是我们生活经验中能够感受到的基本内容，这些内容与我们自己的某些情绪是密不可分的，例如轻松愉悦、狂喜欢乐、压抑郁闷等感受都因为那些不同的经验而引起。但是，重量与支撑、虚弱与强壮恰好又是建筑艺术的基本特征，它们通过建筑上的一些手段表现出来，这些表现在建筑上的特征也带给人们相同的感受。解决建筑中技术问题的那些技术手段，通过这种方式变成了我们的审美体验，具有了美学价值。因此我们说，建筑结构一方面通过自己的技术让建筑物得以实现，另一方面它又是建筑艺术中的重要组成部分。但是前者完全服从于力学定律，后者则是服从于心理学规律。正是它的这种两重性、两种作用，带给我们思想上的混乱。建筑结构在人们思想中产生的美学效果并不是跟随建造技术的发展而一起（pari

first case it is subject to mechanical laws purely, in the second to psychological laws. This double function, or double significance, of structure is the cause of our confusion. For the aesthetic efficacy of structure does not develop or vary *pari passu* with structural technique. They stand in relation to one another, but not in a fixed relation. Some structural expedients, though valid technically, are not valid aesthetically, and *vice versa*. Many forces which operate in the mechanical construction of a building are prominently displayed and sharply realisable. They have a mastery over the imagination far in excess, perhaps, of their effective use. Other forces, of equal moment towards stability, remain hidden from the eye. They escape us altogether; or, calculated by the intellect, still find no echo in our physical imagination. They do not express themselves in our terms. They are not powerful over us for delight.

In proportion as these differences became distinguished, the *art* of architecture was bound to detach itself from mechanical science. The art of architecture studies not structure in itself, but the effect of structure on the human spirit. Empirically, by intuition and example, it learns where to discard, where to conceal, where to emphasise, and where to imitate, the facts of construction. It creates, by degrees, a humanised dynamics. For that task, constructive science is a useful slave, and perhaps a natural ally, but certainly a blind master. The builders of the Renaissance gave architecture for the first time a wholly conscious liberty of aim, and released it from mechanical subservience. To recall the art of architecture to that obedience is to reverse a natural process, and cast away its opportunity. The Mechanical Fallacy, in its zeal for structure, refuses, in the architecture of the Renaissance, an art where structure is raised to the ideal. It looks in poetry for the syntax of a naked prose.

passu）改变的。二者相互间有关联,但不是一对一的那种直接、固定的关系。有些建筑结构手段在技术上是合理的,但在美学视觉上则不合理;反之亦然。在建筑的建造过程中,很多技术手段的确都把自己的力量表现出来,也让人感觉到这些手段的坚实可靠。这是在充分地掌握了人们想象能力的基础上发挥出来的一种技能,这种表现出来的力量实际上远远超出力学方面的实际需要。在另外的情况下,还有一些其他的受力则不被表现出来,一直藏在人们眼睛察看不到的地方。这一部分永远不会被我们感受到,即使是经过理性计算证明它的存在,但是在我们的想象中,根本没有任何回应。这些被藏起来的受力是我们经验中感受不到的东西,它们对于我们从建筑艺术中能否感受到建筑美没有什么作用。

  根据我们对建筑结构中所包含的两种认识所作的区分,建筑的艺术部分与建造技术、建筑科学是没有什么必然关系的。建筑艺术关注的对象不是建筑物的技术本身,而是建筑作用在人们心灵上的那种效果。这种效果是根据直觉、先例等经验所获得的,直觉加上先例告诉我们,对于那些真实的建造技术手段,在什么地方应该舍弃、什么地方应该遮盖起来、什么地方应该重点加以强调、什么地方应该进行模仿。这样做,建筑艺术成为了一种人性化的互动结果。为了这个目的,建造科学与技术方面的内容只是一个有点实用价值的奴隶而已,或许把它的身份再提高一点,算是一个同盟军,但是绝对不是瞎了眼睛的主人。文艺复兴时期的建筑艺术家第一次清醒地让建筑目的获得了自由解放,让它摆脱了技术的束缚。将建筑艺术再次沦为技术的奴隶,屈从于技术,是违反自然发展规律的。鼓吹建筑技术决定论是出于对于技术的狂热,而对于文艺复兴时期建筑艺术中把建筑结构上升到了理想状态的做法视而不见,对它采取否定的态度。这种技术决定论的谬误在于,它是在诗歌艺术中寻找不加修饰的散文句法。

The Ethical Fallacy

I. 'I might insist at length on the absurdity of (Renaissance) construction... but it is not the form of this architecture against which I would plead. Its defects are shared by many of the noblest forms of earlier building and might have been entirely atoned for by excellence of spirit. *But it is the moral nature of it which is corrupt.*'[1]

'It is base, unnatural, unfruitful, unenjoyable and impious. Pagan in its origin, proud and unholy in its revival, paralysed in its old age... an architecture invented as it seems to make plagiarists of its architects, slaves of its workmen, and sybarites of its inhabitants: an architecture in which intellect is idle, invention impossible, but in which all luxury is gratified and all insolence fortified; the first thing we have to do is to cast it out and shake the dust of it from our feet for ever. Whatever has any connection with the five orders, or with any one of the orders; whatever is Doric or Ionic or Corinthian or Composite, or in any way Grecised or Romanised; whatever betrays the smallest respect for Vitruvian laws or conformity with Palladian work —that we are to endure no more.'[2]

A new temper, it is clear, distinguishes this rhetoric from the criticism we have hitherto considered. The *odium theologicum* has entered in to stimulate the technical controversies of art. The change of temper marks a change, also, in the ground of argument: 'It is the moral nature of it which is corrupt.' Fresh counts are entered in the indictment, while the old charges of dulness or lack of spontaneity, of irrational or unnatural form, are reiterated and upheld before a new tribunal. Barren to the imagination, absurd to the intellect, the poets and professors of construction had declared this architecture to be: it is now repugnant to the conscience and a peril to the soul.

From the confused web of prejudice which invests the appreciation of architecture, we have therefore to disentangle a new group of influences, not

---

[1] *The Stones of Venice*, vol. III. chap. ii. § 4.
[2] *The Stones of Venice*, vol. III. chap. iv. § 35.

第五章
道德决定论的谬论

"我可以用上很长的篇幅来说明(文艺复兴时期)建筑在建造技术方面的荒唐性……但是,我要强调说明的是,我所反对的并不是这种建筑艺术的形式。这里出现的这种形式上的缺陷,实际上在过去很多尊贵的建筑形式中也都曾经出现过,这种视觉上的缺陷也可能已经完全被建筑中高尚的精神所弥补过来了。但是,这种建筑形式中的道德本质却是堕落的。"[1]

"这些建筑拙劣卑鄙,极不自然、没有什么艺术效果、让人无法从中获得愉悦和享受,这些建筑都是对于神圣的亵渎。它们的起源都是对神大不敬的异教徒文化,重新被应用到今天也没有什么神圣性质可言;运用这些建筑语言的人,也是趾高气扬、毫不谦逊;这些建筑形式因为过于陈旧也近乎瘫痪……这种建筑艺术风格从它出现的时候开始,就让它的建筑师成了抄袭专家,让建造工匠们成了奴隶,让使用居住在里面的人变成穷奢极欲的腐化分子。这种建筑艺术让人的思想活动处于空转状态,发明创新变为不可能,但是它却能够让各种奢华得到满足,让各种傲慢得到强化。我们对这类建筑唯一需要做的事情就是把它们清除干净,然后把沾在脚上的灰尘抖落干净,从此再也看不见它们。我们无法再忍受看见任何与所谓的五柱式相关的东西,或者与其中任何一个柱式有关的东西;无法忍受再看见什么多立克柱式、爱奥尼柱式、科林斯柱式、复合式柱式;无法忍受再看见任何与希腊、罗马文化相关的建筑元素;那些背叛维特鲁威的精神或者符合帕拉第奥设计原则的所有建筑,我们都无法再继续忍受下去了。"[2]

上面的这些艺术批评所透露出来的愤怒和不满,显然与我们在前面所讨论过的各种理论有所不同,它透露出一种前所未有的、带有强烈宗教神学色彩的仇恨(odium theologicum)显然已经介入到艺术辩论中来了。它的出现完全是对崇尚科学技术那一派艺术理论的挑战。这种理论所反对的对象变了,这也标志着他们的理论基础已经有所改变:"这种建筑形式中的道德本质却是堕落的。"这是在对古典艺术的指控中最新增加的一条罪状。这并不是说包括缺乏灵活多变的特点、缺乏生气,建筑形式不自然、缺乏理性等在内的旧罪名获得免除。在新的法庭审判中,这些指控仍然照旧被重新提出来。在浪漫诗人和崇拜技术的学者教授针对艺术想象力的损害、对人类理性智慧的荒唐亵渎等指控之外,现在又有人出面宣布一项新的罪名:古典建筑是对良知的敌视,是对人类灵魂的挑衅。

从充满了各种偏见的建筑艺术理论中,我们已经清楚地看到一种新的影响力量出现了,它并不会单独地出现,成为一种独立于其他的艺术批评理论,但是它的巨大影响力则是来自

---

[1]《威尼斯的石头》,第三卷,第二章,第4节
[2]《威尼斯的石头》,第三卷,第四章,第35节

indeed always existing separately in criticism, but deriving their persuasive force from a separate motive of assent. The ideals of romanticism and the logic of a mechanical theory are not the sole irrelevancies which falsify our direct perception of architectural form. We see it *ethically*.

How did the ethical judgment come to be accepted as relevant to architectural taste? How far on its own principles did it establish a case against Renaissance architecture? And can those principles find place at all in a rational aesthetic? These are the questions which now require solution, if we are to guard against, or do justice to, a still powerful factor in contemporary taste. For although few serious students of architecture would now confess themselves Ruskinian, and none would endorse those grand anathemas without reserve, the phrases of Ruskin's currency are not extinct. In milder language, certainly, but with even less sense that such ideas require argument or proof, the axioms are reiterated: architecture is still the 'distinctively political' art, its virtue, to 'reflect a national aspiration,' and all the faults and merits of a class or nation are seen reflected in the architecture that serves their use.[1]

The ethical critic of architecture has three different forms of arrow in his quiver, all of which are sent flying at the Renaissance style—an unperturbed Sebastian—in the two passages we have quoted. First, the now blunted shafts of theology: Renaissance architecture is 'impious.' Next, a prick to the social

---

[1] It is significant of the now axiomatic character of such ideas that we find them included by courtesy in the works of writers *whose actual bias and method are utterly opposed to the ethical*. Thus Professor Moore, in one of the few volumes which have been devoted to considering Renaissance architecture as a distinctive art, bases his whole treatment quite consistently upon a *mechanical* ideal of architecture: an ideal in which a most scholarly study of Gothic has no doubt confirmed him. Fitness of construction is his sole and invariable test of value. Not one word occurs throughout with regard to any single building about the kind of human character it indicates or promotes. Yet he prefaces this scientific work, not by any declaration of mechanical faith, but by a rapid liturgical recitation of all the ethical formulae. 'The fine arts,' he says. '*derive their whole character*' from 'the historical antecedents, moral conditions... and religious beliefs of the peoples and epochs to which they belong' (the aesthetic sense of a people apparently contributes nothing to the character of their work): 'Into the service of this luxurious and immoral life,' he continues (speaking of the Renaissance), 'the fine arts were now called; and of the motives which animate such a life they become largely the expression.' They 'minister to sensuous pleasure and mundane pride,' and the architect sets himself to his task 'in a corresponding spirit.' The point of interest here is not simply that the principle implied is false or misleading though it will presently be shown that it is both—but that it is neither demonstrated nor even applied. *It no longer forms part of a conscious system of thought,* but of a general atmosphere of prejudice. The mechanical case derives no authority or support from the ethical case; the ethical case is not illustrated by the mechanical. The ethical formulae have no function in the argument of the book; they are even opposed to it: but they are so familiar that they can be automatically stated and automatically received. A better example could hardly be desired of that unanalysed confusion in architectural criticism which is the reason of this study. — Charles Moore, *The Character of Renaissance Architecture*.

于一种不同于任何其他的价值基础。我们在前面讨论过的浪漫主义和崇尚技术的逻辑性都不能独立地把我们对于建筑艺术形式的判断引向歧途的,还要借助于*道德的说教*。

道德判断是通过怎样的方式成为建筑艺术的审美标准呢?这种道德判断依据自身的原则,在批判文艺复兴时期建筑艺术的道路上到底能走多远?这些原则在理性的美学领域里能否站得住脚?如果我们打算对这个在今天依然具有极大影响力的思想观念进行驳斥或者予以肯定,我们就必须对这些问题进行回答。尽管在今天几乎没有什么严肃的建筑学生再来宣称自己信奉拉斯金的理论,也不会有人再毫无保留地认同拉斯金对于古典主义建筑发出的恶毒诅咒,但是,拉斯金理论中的某些说教却还没有消失。它以一个新面孔出现,语言也柔和许多,论述方式也不易使人察觉出其中某些判断或者论据其实仍需要加以证明,但却被轻轻地放过了。这个重新出笼的定理中所使用的词语是这样的:建筑艺术仍然属于一种"特别的、表达着政治理念的"艺术,它的本质反映出一个"全民的期望",一个阶层,甚至一个国家的优点和缺点都明显地表现在为它们服务的建筑艺术上面的。[1]

在我们引用的两段对建筑艺术的道德进行评论的文字中,我们看到了理论家们射向古典主义建筑艺术的三支箭:第一支,文艺复兴时期的建筑是亵渎神圣的,这支箭现在已经有些秃钝,不如从前那样锋利了。第二支箭,直刺社会大众的良知:文艺复兴时期的建筑导致

---

[1] 有一些理论家,*尽管他们的真实观点与论证方法恰恰是与用道德标准作为建筑艺术评判标准的做法背道而驰的*,但是,他们还是在自己的著作中夹杂着这种已经成为公认准则一样的论调。这个现象的意义非同小可。穆尔教授(二十世纪初哈佛大学艺术系教授)专门著书论述过文艺复兴时期建筑艺术,而且是把它当作艺术作品来看待的,这一点更加难得。他的全部论述,从头至尾非常严格地遵守着一个基本思想,那就是建筑艺术反映了*建造技术*这一基本理念。大多数人在研究过哥特建筑艺术过后,很容易就会接受他的这个观点。恰当的建造技术是穆尔教授唯一的、也是贯彻始终的检验衡量建筑艺术的标准。在这个著作中,从头到尾只字不提任何建筑与人体特征的关系。然而在这部充满科学精神的著作里,穆尔教授也没有告诉读者有关自己坚持以建造技术为衡量准则的基本思想,却在前言部分,口若悬河地大谈各种道德说教的公式。他说:"纯粹的美术作为艺术,它是从自己的前辈和当时的道德现状中,是从自己所属的时代和当时的宗教信仰中,*获得自己全部特征的*……"(在他看来,艺术家的审美爱好显然对于自己的艺术作品没有任何影响)他继续说道:"在这个时期(指文艺复兴时期),美术被拿来为奢侈、不道德的生活方式服务;而美术作品把表现这种生活方式当作自己的创作动力。"这些作品"尽情地满足感官的快乐和世俗的炫耀"。而建筑师正是用同样的精神把满足感官的快乐和世俗的炫耀当成了自己的全部任务。在这里,问题的关键不是说这种论述所隐含的原则是错误的,或者说在误导读者,而是说作者根本就没有向我们展示他自己的论据,也没有说明这个原则是如何运用到建筑实践中去的。后来的历史证明,作者在这里所下的结论不正确,是在误导读者。采用这一类的论述根本不能够成为任何缜密论证的组成部分,这样的说法仅仅是当时带有强烈偏见的社会风气的一种表现。通过技术方面的论证根本不能说明道德的观念。而这些道德说教的公式在这本书里对于作者陈述自己的论点也没有发挥出什么作用,反而是与单纯的技术观点发生抵触;但是,这类道德说教在当时是司空见惯的口头禅,作者有口无心地在说,读者心不在焉地在听。建筑艺术理论中充斥着这类不加分析的混乱言论,恐怕找不到比这本书更有代表性的例子。这也是我们自己这本书之所以出现的主要原因。——见查尔斯·穆尔(Charles Moore)《文艺复兴时期建筑艺术的特征》。

conscience: Renaissance architecture entails conditions, and is demanded by desires that are oppressive and unjust; it 'makes slaves of its workmen and sybarites of its inhabitants.' Last, most poisoned, and the only menace to the martyr's vital part: Renaissance architecture is bad in itself, inherently, because it is insincere (for instance) or ostentatious; because the 'moral nature of it is corrupt.' These darts, if the fury of intolerance which first rained them has abated, still stand conspicuous in the body of the saint.

The attack is met on the other side by a contumacious brevity of argument, appropriate indeed to martyrdom, but hardly convincing to the mind. 'The spheres of ethics and aesthetics are totally distinct: ethical criticism is irrelevant to art.' This, together with some manner of diatribe against Ruskin, is all that is vouchsafed in reply when, as now, fashion veers for a moment, and with more ardour than understanding, in the direction of our Georgian manners. 'Ethical criticism is irrelevant to art.' No proposition could well be less obvious. None, We shall see reason to admit, could be less true. But one confusion begets another, and this axiom, too, now adds its darkness to the dim region where the controversies of architecture are sorrowfully conducted.

The ethical case deserves a closer study and a less summary retort.

First, then, for the origins of our habit. The ethical tendency in criticism is consequent upon the two we have already discussed. The Romantic Fallacy paved the way for it. The Mechanical Fallacy provoked it.

The essential fallacy of romanticism was, we saw, that it treated architectural form as primarily symbolic. Now there is evidently no reason why an art of form. if it be regarded as significative at all, should have its meaning limited to an *aesthetic* reference. Romanticism, it is true, was concerned with the imaginative or poetic associations of style. But when once this habit of criticism was established—when once it seemed more natural to attend to what architecture indirectly signified than to what it immediately presented—nothing was required but a slight alteration in the predominant temper of men's minds, an increased urgency of interest outside the field of art, to make them seek in architecture for a *moral* reference. Romanticism had made architecture speak a language not its own—a language that could only communicate to the spectator the thoughts he himself might bring. Architecture had become a mirror to literary preferences and literary distastes. Now, therefore, when the preoccupations inevitable to a time of social change and theological dispute

了一种社会不公、令人压抑的后果,这种建筑"让工匠成为奴隶,让建筑主人成为腐化堕落的暴君"。最后一支箭,也是最毒的一支,它包含让文艺复兴时期的建筑艺术最为恐惧的指控:这种建筑自身本来就不是好东西,它从遗传基因上就继承了不诚实与追求浮华排场的血统,"它的道德本性就是堕落的"。面对这三支箭,文艺复兴时期的建筑如同圣徒塞巴斯蒂安(an unperturbed Sebastian)那样从容镇定、面不改色。如果说当初发射那些如同雨点般的毒箭是出于无法克制的愤怒,那么到现在经过这么长的时间以后,这种愤怒或许已经得到了释放与发泄,但是,那些已经被射出的毒箭现在还依然明晃晃地挂在圣徒受伤的身上。

对于这种攻击的反抗却是很简单无力的,只有一句"道德的范围与美学的范围是两个根本不同的东西:道德批评与艺术欣赏之间是毫无关系的"。这样的言辞当然符合为自己信仰而献身的圣徒的性格,但是对于澄清普通人的思想却没有什么帮助。这种冷静地指出道德与艺术无关的言语,加上另外一些出于对拉斯金的厌恶而破口大骂的吼叫,二者加到一起成为今天反对拉斯金理论的主要力量。这种力量到目前又得到进一步强化,因为艺术潮流发生了变化,当今成为主流的是所谓的乔治时期的建筑风格。但是,人们对于这种乔治时期建筑艺术的热衷也是出于一种狂热,而不是出于理解。"从道德标准出发的批评理论与艺术毫无关系。"这个判断再明显不过了,没有什么其他说法要比它来得更简单明了。经过以下的分析之后,我们有充分的理由来证明这个判断是一个千真万确的判断。但是,对于一个问题的误解导致了对另外一个问题的误解。在建筑艺术的辩论中,各种主张众说纷纭,人们悲观地参与讨论。在这样的氛围里,上面的这个判断则把那微微透露出来的一点光亮又给掩盖住了,变得更加昏暗了。

这个从道德标准出发而引出的艺术批评理论需要我们好好地深入探讨一下,而不是简单地概括总结,然后急着予以反击。

首先,来看看我们目前思维习惯的来源。在艺术批评理论中掺进道德标准是在两件事情发生之后才出现的。关于这两件事情,我们在本书的最开始做了讨论:第一件事情是浪漫主义艺术理论的谬误,它为道德标准的出现铺平了道路;第二件事是技术决定论的谬误,它直接导致了道德标准进入艺术批评理论。

我们看到,浪漫主义艺术理论的最根本谬误在于,它认为建筑艺术形式的最重要作用是它对于象征意义的体现。既然一件艺术品形式的最重要的职能是它的象征作用,那么艺术形式也就显然不能仅仅局限于美学的考量。浪漫主义艺术的确是非常注重一个艺术形式或者艺术风格所具有的形象象征意义与诗意的联想。当这种艺术理论被广泛接受以后,也就是说,人们更关注建筑艺术所暗示的意义,而非建筑本身直接呈现的内容,理论家们不需要再特别地做什么,只要把自己脑子里固有的想法与偏见略微偏转或扩大一点,也就很自然地延伸到了*道德*的层面。浪漫主义已经迫使建筑艺术必须用原本不属于建筑艺术自身的语言去讲话,而这种语言却只能是观察者自己设想的语言,自己熟悉的语言。结果是,建筑艺术成了反映文学艺术爱好的一面镜子,理论家的喜好与厌恶都反映在这里。因此,当一个时代的社会变革与宗教辩论恰恰都集中在道德标准的时候,艺术语言也就不可避免地反映这种社会现象,也就充满了这类的道德说教。建筑的艺术风格成了理论家口中的拿来象征那些参与其中的工匠、艺术家、投资人,甚至当时民众的性格特征的东西。再根据这些人的性格特征中道德成分的多少,根据理论家认为这些道德成分是否值得称赞或是谴责,以及称赞或

had become predominantly moral, the language of art, reflecting them, was rife with ethical distinctions. The styles of architecture came to symbolise those states of human character in the craftsman, the patron or the public which they could be argued to imply. They were praised or blamed in proportion as those states were morally approved.

But this was something more than romanticism. No doubt, when all the imagery of nature is employed to heighten the contrast between the rugged integrity of the mediaeval builders and the servile wordiness of the modern; then, indeed, the ethical criticism is a form of the romantic. The moral appeal becomes imaginative and the religious appeal poetic. Nevertheless, the arguments which could dismiss the Romantic Fallacy will not suffice to meet the ethical case. The difference between the two seems fundamental. It is, as we saw, unreasonable to condemn an architectural purpose because it fails to satisfy a poetic predilection, for the standards of poetry and of architecture are separate in their provinces and equal in their authority. But, *prima facie*, it is not in the same sense unreasonable to condemn an architectural purpose because it offends a moral judgment; for the moral judgment claims an authority superior to the aesthetic, and applies to all purpose and action whatsoever. Hence. architecture falls within its province. If, then, it can be shown that moral values exist at all in architectural style, these, it may be pleaded, must form our ultimate criterion; these will determine what we *ought* to like, and a criticism which ignored their existence would be frivolous and partial. It would not, that is to say, be a final criticism; for to the moral judgment belongs the verdict upon every preference. Why, then, should the criticism of architecture stop short of the last word? And if, from this plain course, the seeming opposition between aesthetic and moral values should deter us, might not aesthetic good prove, on a due analysis, reducible to terms of moral good? This reduction, in effect, the ethical criticism of architecture attempted to achieve. Nor was there anything absurd in the attempt.

The ethical criticism, then, though it claims a different sanction and raises a wider issue, arose from the romantic. It arose, also, as a protest against the mechanical theory. Its motive was to assert the human reference of art against the empty cult of abstract technique. We have already seen that the extreme constructional ideal of architecture was no more than a phase of nineteenth century materialism. It ignored feeling. It neglected alike the aesthetic conscience and the moral. It appealed solely to an intellect which recognised no law but the mechanical. It was an episode in the dehumanisation of thought: a process which, carried to its logical conclusion, renders all values unmeaning. Such a process, however powerful its impulse, could not but provoke in many minds an immediate

者谴责的程度,然后进行建筑艺术批评。

这已经远远超越了浪漫主义的界限。当自然界的景象被拿来作为类比对象,借此强化了中世纪粗犷、毛糙的地域性的建筑与现代强调通用性之间的对比,在这个时候,用道德的标准来批评建筑艺术无疑就是浪漫主义的一种具体体现。道德说教成为一种想象中具有形象的东西,而宗教的教条也就变成诗歌艺术的同义词。但是尽管如此,我们在前面批驳浪漫主义的那些论述,在这里还不足以用来批驳这种从道德标准出发的艺术理论。这两者之间有着根本的差别。我们知道,如果一个建筑中的某种用途不符合诗意的期待,但是,任何人都无法接受以诗意的名义来指责它,因为大家都知道诗意和建筑毕竟是截然不同的两种东西。两种艺术各有各的范围,各有各的判断标准。可是采用道德标准对建筑的目的进行一番评估和判断,乍看起来(prima facie)就不再是没有道理的了。因为道德的权威性要高于美学的权威性,无论任何的目的、任何的行为,都应该接受道德的检验。建筑艺术也就因此落入了道德的掌控范围之内。每一种建筑艺术风格便都具有了各自的道德价值,理论家们便据此建成了一个终极道德评判标准,这些标准反过来决定了我们应该喜欢什么。任何艺术理论如果缺少了这些内容,那么它一定是微不足道的,是不完整的理论。也就是说,根据这种理论所作出的任何判断都不是最终的审判,因为各种喜好与偏爱,到最后,都属于道德判断的内容。建筑艺术理论怎么可能是个例外呢?怎么可以躲避道德的审判呢?按照这个逻辑,如果美学与道德两种价值发生冲突,让我们无所适从,那么,我们能不能经过分析,把美学上认为是好的东西转换成能够被高尚的道德标准可以接受的东西呢?这正是按照道德标准所建立的建筑艺术理论所期望达到的目的。这种想法本身并没有什么不妥。

尽管这种理论宣称自己的评判标准与以往的理论不同,它涉及的内容也相当广泛。但是这种按照道德标准作为衡量建筑艺术的理论,从本质上讲,仍然是浪漫主义的东西,它是为了抗衡技术决定论这一理论才得以出现的。这种按照道德标准评判建筑艺术的理论,它的动机是打算在抽象的技术判断之上加进人文的因素。我们在前面的分析中看到,建筑艺术的终极技术理想不过是十九世纪唯物主义理论的一个片断,它否认人的感受。同时它也把美学的认识与道德的良知都忽略了。这种技术决定论只认同理性,而这种理性也只是机械的推导和演算,并不是普遍的规律。它在思想发展过程中否定了人的存在价值,主张非人化的片面理性。按照技术决定论的逻辑推导下去,自然就会得出任何价值都是没有意义的这么一个荒诞的结论。这样的理论无论多么有说服力,也会立刻遭到很多人的反对。而这种反对意见大多是来自道德与宗教方面的,因为这些领域里的东西原本是唯物主义观点最

resistance. But it was a resistance in the field of ethics and theology. For here were the interests which materialism seemed most obviously and immediately to challenge: here, at any rate, were the interests which it was all-important to safeguard. Aesthetic values are a luxury; they are readily forgotten when more vital conflicts become acute. Thus, the necessary counter-attack to the movement of science was consequently ethical in temper. Its concern was with conduct and not primarily with art. It was, in effect, a Puritan revival. The intellectual alternative was strict: either a truculent materialism (with consequences for architecture already analysed) or a moralistic ardour more severe than any that had been dominant since the seventeenth century.

Here were two sinister antagonists. The amiable provinces of art, which lay forgotten and unguarded at their side, soon trembled with the conflict. Architecture became a rallying point; for while the constructive basis of the art exposed it obviously to the scientific attack, its ecclesiastical tradition invited for it, no less, a religious defence. In this region, where the air was dense with ancient sentiment, the moral losses suffered in the territory of metaphysics might, even by a shaken army, be made good. It was a Puritan revival, but with this difference: the fervour of Puritanism was now active in vindicating the value of art. It insisted that architecture was something more than a mechanical problem. It gave it a human reference. But, unluckily, this Puritan attack, far from clearing the path of criticism, did but encumber it with fresh confusions no less misleading than the logic of inhuman science. Art was remembered, but the standards of art remained forgotten. The old Puritanism of the seventeenth century had weighed the influence on life of art as a whole. It had condemned it and driven it forth from its Republic with all the firmness, and something less than the courtesy, which Plato extended to the poets. But the Puritanism of the nineteenth century attempted, while retaining art and extolling its dignity, to govern its manifestations. It sought to guide the errant steps of the creative instinct. It sought also to explain its history. And it did so, as was natural to it, by moral laws and divine authority. At Oxford even the Chair of Poetry was disputed between the creeds. And, in architecture, once granted the theological prejudice, aesthetic dogmas are not likely to be lacking to prove that all the vices which were supposed to have accompanied the return of the Roman style in Europe must be inherent also in the Roman architecture itself. These dogmas survive the sectarian quarrel which gave them birth. The charge outlives its motive; and Renaissance architecture is still for many a critic the architecture of ostentation and insincerity once attributed to a 'Jesuit' art.

明显、最直接的攻击对象。这些受到唯物主义理论攻击的思想是不容侵犯的，必须要加以保护。美学价值不是最基本的东西，它们是奢侈品，当冲突对立十分尖锐的时候，这些美学的东西就很容易被忽略。因此，对于科学运动的反击也就自然地成为道德的责任，而且需要其充满激情地全身心地投入。道德所关心的是人们的行为活动，不仅仅局限于艺术。道德价值的重新提倡所直接导致的结果就是清教徒思想的复兴。这时人们的思想倾向只有两个截然不同的选择：要么是坚持残暴的唯物主义（它所产生的建筑艺术我们已经分析过了），要么是坚持严格的道德标准，要知道，这时的道德底线要比十七世纪以来任何时期的道德标准都要来得更加严苛。

这是两个均非善类的冤家对头。原本平静和睦的艺术领域，很久以来没有什么人来这里打扰，艺术领域自己也没有采取任何防御措施。当这两个冤家跑到这里来厮杀的时候，整个艺术领域跟着一起颤抖。建筑艺术则是整个艺术领域里两个冤家交手最激烈的地方。当建筑艺术中的技术部分暴露在科学的攻击之下的时候，艺术中的宗教传统则站出来试图保护建筑艺术不受科学的攻击。建筑艺术领域里的气氛本来就是很沉重的，里面充满了各种古老的思想感情与文化积淀。形而上学领域蒙受唯物主义和科学的冲击之后，其结果就是让其中的道德价值观明显丧失，这时哪怕是一支已经几乎溃不成军的队伍前来解围，但对于道德价值观来说也是一个不小的救星。前来救驾的这支军队是重新复活起来的清教徒，但是这次的清教徒已经不是以往的那个清教徒了，现在是以艺术价值保护者的身份出现的。他们坚持相信一条：建筑艺术绝对是超越单纯技术问题的一门艺术，一定要与人的价值相关。不幸的是，清教徒在交战时的表现，不但没有把艺术理论道路上的障碍清除，反而是给艺术理论带来新的思想混乱。在误导人们对于艺术的认识方面，这种新的思想混乱的影响绝对不亚于不讲人性的科学技术决定论所产生的影响。艺术忽然间又被人想起来了，但是艺术标准却还是被遗忘的。十七世纪的清教徒们曾经把艺术对于人生的负面影响，强调到了无以复加的地步，并对所有属于艺术范畴的内容进行过无情的谴责，坚决地主张把艺术从人们的生活中清除干净。其严厉程度绝对不亚于柏拉图从自己的《理想国》中把诗人驱赶出去的做法。二者不同的地方在于，相比之下，柏拉图略微还保持住一点风度和礼貌。但是到了十九世纪，这时的清教徒一方面在呼吁保留并赞美艺术的美誉尊严，另一方面却试图控制艺术的表现内容。清教徒要给艺术家规定出一条道路，让他们不至于在创作的时候迷失自己，最后偏离正确的方向。清教徒也试着对艺术历史进行解释，当然是按照道德的规范与神学的权威进行解释，这也是可想而知的。牛津大学诗学教授（Chair of Poetry）也曾经被卷入到这些教条的辩论中。在建筑艺术领域接受这些神学宗教立场之后，当然地会认定，古罗马建筑风格在欧洲的兴起，必定会把古罗马建筑艺术风格中所固有的那些罪恶也再次带回到我们的生活当中来，建筑美学理论中的说教也就自然地不会缺少关于这方面的内容。这些说教也就很自然地会超越当初参与辩论的宗派而在社会上散布开来。当提出这些教条的那些宗派消失之后，这个观念却依然存在。指控的直接动机与起因，或许已经过去了，没有人会在乎它是怎么起来的，但是被指控的罪名却仍然留在那里。对于很多建筑艺术理论家来说，文艺复兴时期的建筑就等同于浮华张扬的艺术，是不诚实的艺术，这种艺术风格甚至

The sectarian import of style, though somewhat capriciously determined, might provide an amusing study. The Roman architecture stood for the Church of Rome. The association was natural, and had not the Papacy identified itself with the Renaissance almost at the same time and in the same spirit as it had provoked the rise of Protestantism? Thus the classical forms, although a generation earlier they had echoed in many a Georaian church to strictly Evangelical admonitions, were now arbitrarily associated with the Pope, or—should their severity be in any way mitigated—with the Jesuits. The Gothic, on the other hand—Pugin notwithstanding—was commonly regarded as the pledge of a Protestant or, at the worst, of an unworldly faith. And it is easy to understand that in the days ot Bradlaugh and of Newman, these rectitudes of architectural doctrine were of greater moment than aesthetic laws.

The soil was therefore prepared. The sects had ploughed upon it their insistent furrows. And now the winds of architectural doctrine blew loudly, beating strange seed. The harvest which resulted is historic. *The Seven Lamps* appeared and *The Stones of Venice*. The method of the new criticism was impressive and amazing. For here, side by side with plans and sections, mouldings, and all the circumstance of technical detail, the purposes of the universe were clearly, and perhaps accurately, set forth, with a profusion as generous as, in this subject, it had previously been rare. The prophets Samuel and Jeremiah usurp the authority of Vitruvius. They certainly exceed his rigour. Dangers no less desperate than unexpected are seen to attend the carving of a capital or the building of a door; and the destruction of Gomorrah is frequently recalled to indicate the just, if not the probable, consequences of an error in these undertakings.

But the new criticism did not limit itself to denunciation. A moral code, at once eloquent and exact, was furnished for the architect's guidance and defence, and determined the 'universal and easily applicable law of right' for buttress and capital, aperture, archline and shaft. An immense store of leaning and research, of reason also, and sensitive analysis, far superior to that which Ruskin brought to painting, lay imbedded in these splendid admonitions, and seemed to confirm the moral thesis. And it no doubt added greatly to the plausibility of the case that the principles which he presented with the thunder and pageantry of an Apocalypse had been carried out, from foundation to cornice, in almost meticulous detail. Impressive principles of right! They could be fitted to every case, and as we read we cannot but suspect that they are able to establish any conclusion.

The moralistic criticism of the arts is more ancient, more profound, and might be more convincing, than the particular expression which Ruskin gave to it. It is not

一度被广泛地认定为完全是天主教中"耶稣会"一派所倡导的艺术。

这种由某个宗教派别根据自己对于艺术风格的理解所作的解释,几乎可以断定是随心即兴式的,但是他们的见解也的确是很有点意思的。古罗马的建筑当然就是代表了罗马的教廷,这种联想是再自然不过的了,难道不是教皇本人认同文艺复兴式艺术与精神吗?与此同时,难道不是因为罗马教会所认同的精神才导致宗教改革与新教的出现吗?因此,古典主义的建筑形式在这时被认定就是代表了教皇,或者代表了天主教会中的耶稣会这一宗派。他们忘记了就在他们发表这个见解之前的二十年左右,英国新教中的福音神学一派曾经严格地按照教义规定,在乔治时期建造了不少采用古典主义建筑语言的教堂,但是眼下,这种古典主义的建筑语言则成了教皇和耶稣会的专利。而与此相对立的哥特建筑风格则是代表了新教的信仰,最起码它是代表了超越当时主流宗教的信仰。对于哥特建筑的认知,普金(Pugin)是绝对不会认同的。在无神论者布拉德劳(Bradlaugh)和红衣主教纽曼(Newman)生活的那个时代,把古典主义的建筑与教皇与耶稣会等同,把哥特风格的建筑和新教画上等号,这类教条式的建筑理论要比抽象的美学理论更有市场。

土地已经平整完毕,宗教组织也已经在这块土地上深耕播种。现在,建筑艺术清规戒律的暖风又在紧吹,暖风撒播了一些奇怪的种子在这片土地上。这自然导致了具有历史意义的收获。《建筑艺术中的七盏灯》与《威尼斯的石头》两部著作相继问世。两部书所展示的艺术理论方法令人印象深刻,甚至有点难以置信。在这两本书里,建筑平面图、剖面图、线脚的细部等插图,一个接一个,各种具有代表性的技术性详图也不少。整个宇宙的目的性也是清晰详细地展示在世人的面前,而且是经过十分准确的描述。建筑艺术中广泛的题材都前所未见地罗列在书中,丰富多彩。先知撒母耳(Samuel)和耶利米(Jeremiah)的说教取代了维特鲁威的建筑艺术权威。先知的说教当然要比维特鲁威的更严厉。注重柱头上的雕刻,注重一扇门的制作,这些都是很危险的事情。这种危险与其说让人觉得有些意外,倒不如说是在哗众取宠。蛾摩拉城(Gomorrah)的毁灭在这两本书里时常被提起,其目的就是为了证明,注重那些华丽雕刻与装饰的行为,必然会导致自己的毁灭;这绝不是为了吓唬你们在这里说说而已的。

这种新的理论并不满足于对于艺术仅仅作一番点评或者指责;它还要建立一套道德行为规范,准确、周密地规定建筑师的具体手段,规定哪些可以做,哪些不可以做,为建筑师制定了一套"放之四海而皆准"的正确法则,而且是简单易行的,例如壁柱、柱头、开洞、拱券、立柱等正确做法。这两部著作同时也是两个巨大的宝藏,里面有各种学问、研究课题,也有很多理性分析、感性认识。拉斯金带给建筑艺术的内容要比他带给任何其他艺术领域的东西要多得多。这一切都存于他在书中给人的谆谆教诲之中,同时他也借助这些内容再次阐述它们所包含的道德准则。拉斯金所大声疾呼、大肆渲染的这些道德准则如同《天启录》一样庄重;这些准则在一些建筑实例上获得应用,从基础到屋檐线脚,再到各种细微处的节点大样,这些都无疑更强化了它们的说教效果。这是多么让人印象深刻的正确原则啊!它们可以被应用到任何场合。当我们阅读他的著作的时候,我们甚至不知道这些理论能建立起什么样的结论。

这种以道德准则为核心的艺术理论实际上是非常古老的传统,要比拉斯金所讲述的内容具有更加悠久的历史和更加深远的意义,也更加令人信服。而且也并不是基督教所特有

specifically Christian. It dominates the fourth book of Plato's *Republic* no less than the gospel of Savonarola. It is one of the recurrent phases of men's thought: a latent tendency which it was Ruskin's mission rather to re-awaken than create. The ethical criticism of architecture is likely therefore to survive the decay of the individual influence which brought it back to force. The dictator's authority has long since, by his own extravagance, been destroyed. The casuistries of *The Stones of Venice* are forgotten; its inconsistencies quite irrelevant to the case. They are the unchecked perversities of genius, which an ethical criticism is not bound to defend, and which it would be idle, therefore, to attack. We are concerned, not with the eccentricities of the leader, but with the possible value and permanent danger of the movement which he led. And it is more necessary at this date to emphasise the service which he rendered than to decry the logic of his onslaught.

In the first place, Ruskin undoubtedly raised the dignity of his subject, no less than he widened its appeal. He made architecture seem important, as no other critic had succeeded in doing. The sound and the fury, not unduly charged with significance; the colour of his periods; the eloquence which casts suspicion on the soundest argument and reconciles us to the weakest; the flaming prophecies and the passionate unreason, had that effect at least. They were intensely dynamic.

In the second place, it is fair to remember that Ruskin asserted the *psychological* reference of architecture. No ingenuity of technique would satisfy him, nor any abstract accuracy of scholarship, however mediaeval. Mere legalism, mere mechanism, mere convention, and everything which, outside the spirit of man, might exercise lordship over the arts he combated. No doubt his psychology was false. No doubt he utterly misinterpreted the motive of the craftsman and dogmatised too easily on the feelings of the spectator. Probably he took too slight account of the love of beauty as an emotion independent of our other desires. But still in some sense, however illusory, and by some semblance of method, however capricious, the principle was maintained: that the arts must be justified by the way they make men feel; and that, apart from this, no canon of forms, academic, archaeological or scientific, could claim any authority whatsoever over taste. This was a great advance upon the mechanical criticism; it was an advance, in principle, upon the hieratic teaching of the schools.

But the psychological basis which Ruskin sought to establish for architecture was exclusively moral, and it was moral in the narrowest sense. He searched the Scriptures; and although the opinion of the prophets on Vitruvian building might seem to be more eloquent than precise, he succeeded in enlisting in favour of his prejudices an amazing body of inspired support. But it is easy to see that an equal expenditure of ingenuity might have produced as many oracles in defence of

的观念。柏拉图《理想国》第四卷的主要内容就是阐述道德标准,古希腊时期的教条所透露出来的严厉规定,绝对不亚于十五世纪出现的萨佛纳罗拉(Savonarola)福音所宣扬的道德标准。这是一种再次重复出现的人类思想观念,也是一种不易被人察觉到的潜在规律,拉斯金就是试图唤起人们这种已有的观念,而不是标新立异地创造出新理论。而建筑艺术中的道德理论也将会在鼓吹这些教条的理论家们去世很久以后仍然继续流传。蛮横的独裁者由于自己的过分嚣张也让他们的权威早已丧失。《威尼斯的石头》一书中所宣扬的根据伦理道德作为艺术判断依据的理论早已为人们所忘记,至于书中那些自相矛盾的地方也没有人再去计较。这些著作是刚愎自用的天才理论家自以为是的教训别人的言辞,他的道德说教几乎是没有任何作用的。它不可能保护自己,也更不可能对敌人实行攻击。对于这种理论的倡导者所表现出来的那种自大怪癖,我们根本不以为意,我们只想关注一下他所提倡的这个运动到底有多少价值,或者认清这种理论能否产生无穷的贻害。眼下最关键的是看看这种理论所带给我们的实际好处与实际效果是什么,而不是纠缠这种攻击理论所犯的错误。

首先,拉斯金把建筑艺术提高到一个崭新的高度,不但扩大了它的接触范围,也提高了它的尊严。这一点是没有任何疑问的。他让广大民众都相信建筑艺术是很重要的一门艺术,任何一位其他的理论家都没有像拉斯金这样成功过。拉斯金大声的疾呼和愤怒的咆哮,其中也包含着一定的道理,不是没有它的意义。他的理论具有强烈的时代色彩。他滔滔不绝的言辞让我们对自己最有力的论据不禁表示出怀疑,在他面前我们毫无还口的机会。他的说教当中那种火一样的语言有如先知们的教导,他言辞中透露出的激动与豪情让人忘记其中的非理性,这些都在感染着我们。拉斯金的艺术理论充满了无限的感人力量。

其次,我们应该要了解,是拉斯金强调了建筑艺术中必须考虑到人们的*心理因素*。单单依靠精巧的技术还不能够令他满意,单纯的学术研究也太抽象,无论内容是多么的中世纪,也无法让他满意。墨守成规、迷信技术、食古不化的做法,或许在艺术实践中有各自的天地;但是,任何作品如果没有人的灵性,拉斯金都坚决反对。我们不怀疑,拉斯金对于心理学在艺术领域里的作用有他错误的理解;我们不怀疑他明显地误解了艺术工匠的动机,草率地得出结论,并且将自己的结论变为教条,然后据此判断艺术对人产生的作用。或许他很轻易地忽略了我们对于美的热爱实际上是独立于其他七情六欲的一种感受。但是,无论那里面存在有多少错觉,无论他的方法是多么的不够完整、不够理性,拉斯金的艺术理论在总体上还是坚持了一个最基本的原则:艺术必须按照它带给人们的真实感受作为评价判断的依据。离开了这一点,任何形式的准则,无论是学术的也好、考古的也好,还是科学的也好,不管这些准则所宣扬的艺术风格、爱好是什么,统统都没有任何说服力。这一点,要比技术决定论高明许多,是一个巨大的进步。从原则上讲,这是一种超越了以往各种僧侣诵经式学派的巨大进步。

但是,拉斯金带给建筑艺术的心理学基础却是除了道德说教以外不包含任何其他内容的心理学,而他在讨论道德问题的时候,采取的是最为狭隘意义上的道德概念。他仔细察看了《圣经》的经文,那些传达神明旨意的先知们,在自己所书写的经文中,涉及古代维特鲁威建筑艺术的那些内容都是一笔带过,而不是详细描述;但是,拉斯金非常了不起地从中整理出一大堆经文条目,来支持他个人的偏见。问题是,我们也可以从《圣经》经文中找出同样多的神谕,来证明复兴时期的帕拉第奥理论也自有它的道理。当学者们试图在英国两大

Palladio as it showed grounds for his perdition. The time is gone by when scholars, passing to their innocent tasks through the courts of Hawkesmoor or of Wren, were startled to recognise the Abomination of Desolation standing, previously unnoticed, in the place where it ought not. And a criticism which would be willing—were they propitious—to prove a point of theory by citing the measurements of the Ark, must now seem obsolete enough. But if the theological argument has ceased to be effective, its interest for the history of taste remains immense. And the fact that, a hundred years after Voltaire, one of the foremost men of letters in Europe should have looked for architectural guidance in the *Book of Lamentations* is one which may well continue to delight the curiosity of anthropologists when the problems of aesthetic have been rejected as unfruitful, or abandoned as solved.

Ⅱ. More persuasive than the theological prejudice, and more permanent, is the political. If, as we have said, the romantic fallacy reduced taste to a mere echo of contemporary idealism, it it encouraged men to look in art always for a reflection of their existing dreams, what must be the verdict on Renaissance architecture of an age whose idealism was political and whose political ideal was democracy? For here was an architecture rooted in aristocracy, dependent on the very organisation against which society was now reacting. It had grown up along with the abuses which were henceforth to be expelled from the moral ordering of life. And these abuses—to use the question-begging phrase of modern criticism—'it expressed.'[1] It had exalted princes and ministered to popes. It stood for the subordination of the detail to the design, of the craftsman to the architect, of conscience to authority, of whim to civilisation, of the individual will to an organised control. These things were hateful to the philosophy of revolution. They were hateful no less to the philosophy of *laissez faire*. The architecture of the Renaissance shared inevitably their condemnation. Moreover, the minds alike of the good citizen who gloried in industrialism, and of the thinker who shrank from it, were turned to the future rather than the past. Even the mediaeval daydreams of Morris were a propaganda and essentially prophetic. Now the neo-Gothic experiment and the architecture of steel, whatever their initial failures, could claim to be still untried; from them might still spring the undreamt-of pinnacles which should crown the Utopias of the capitalist and the reformer. But the Renaissance style represented inertia, and the hypocrisy of a dead convention. It promised nothing, and in the commercial monotony of the

---

[1] Abuses in the organisation of society may sometimes, as in the French eighteenth century, be a precondition of certain achievements in the arts. But the artistic achievements do not on that account 'express' the social conditions, though the one may recall the other to our mind. It would be as true to say that the view from a mountain 'expressed' the fatigue of getting to the top. Whether the mountain is worth climbing is another question.

建筑师豪克斯摩尔（Hawkesmoor）和伦恩（Wren）的作品中寻找自己判断依据的时候，却惊奇地发现，问题的关键根本就不在这里——古代的废墟令人意外地出现在本来被认为是不该出现的地方，这样的事情早已成为过去了。这种艺术理论就好比是想借助于诺亚方舟的尺寸大小来证明自己的论点一样，到头来却发现毫无用处。如果说，这种理论中所引用的神学论述不足为凭，但是其中的历史因素对于人们喜好的影响却是不容忽视的。伏尔泰（Voltaire）之后已经一百年了，欧洲最有影响力的一位文化人仍然在《耶利米哀歌》（Book of Lamentations）中寻找建筑艺术的发展方向，而美学的问题则完全被忽略，或者认为美学问题已经是完全解决了的问题而被搁置。这一点真的是值得人类学家好好研究的一个现象。

二、与神学理论中那种不容他人置喙的顽固相比，从政治角度来论述艺术要更加容易被人接受一些，影响也相对持久。正如我们在前面说过的那样，浪漫主义的谬误在于把人们对艺术的喜爱简化为是对一种当代理想的回音，它总是鼓动人们在艺术作品中寻找自己现有的梦想。假如是这样的话，那么对于文艺复兴时期的建筑艺术该怎样来做一个概括呢？要知道，文艺复兴时期的理想正是它的政治理想，这个政治理想就是追求民主、平等。但是这时的建筑艺术完全是贵族寡头政治的体现，而贵族政治完全是当时全社会反对的对象。建筑艺术的发展正是在这种不协调政治环境中形成的，而这种贵族政治则是逐渐被社会生活所排斥，不为新的道德标准所接受的。用现代艺术理论家的话来说，这种不协调的政治环境正是"建筑艺术所表现的内容"。[1]这时的建筑艺术让王公贵族十分满意、十分开心，也能够很好地满足教皇的需求。这个时期的建筑艺术代表了建筑细部服从于整体设计构思；代表了工匠服务于建筑师；代表了良知需要服从于权威；代表了异想天开的怪异念头服从于文明传统；代表了个人服从于组织。这一切都正是富有革命性的理想所痛恨的东西，同时那些主张"自由放任"（laissez faire）思想的人对它的痛恨一点儿也不少。建筑艺术必定要分担大家对于文艺复兴时期艺术的各种谴责。这时的社会上存在着两大类人群，一群是讴歌并积极投身工业化革命的先进市民，另一群是在工业化革命面前畏缩发抖的旁观者。有趣的是，这两大类人群都不约而同地寄希望于未来来实现各自的理想，而不是回到过去来寻找自己的安慰。甚至像莫里斯（Morris）这样的憧憬中世纪艺术的梦想家也成了未来的鼓吹者与先知。各种按照哥特艺术风格制造出来的假古董和伴随工业革命出现的钢铁建筑都宣称自己代表着未来，无论开始的时候二者是多么的步履蹒跚、不尽如人意，但是都坚持要试一试，绝不放弃。他们都觉得自己会成就前所未有的大事业，一个幻想着自己会登上工业革命乌托邦的最高峰，一个幻想着成为社会改革的胜利者。至于文艺复兴时期的艺术风格，那不过是代表着因循守旧的旧势力，一种虚伪的、没有生命力的传统。在那里，人们看不到未来和希

---

[1] 在个别情况下，当我们的社会组织结构在受到破坏的时候，却成就了某些艺术创作。这种社会组织的破坏成为能够取得这种成就的前提条件，例如在十八世纪的法国就是这样。但是，艺术的成就绝对不在于它"表现了"当时的社会现状，尽管不断有人这样宣称。这就好像在说，一座高山的景色"表现"了爬到山顶上的那种疲惫不堪。至于这座山到底值不值得去爬则是另外一个问题。

time the joy that had been in it had died out. 'The base Renaissance architects of Venice,' remarks Ruskin bitterly, 'liked masquing and fiddling, so they covered their work with comic masks and musical instruments. Even that was better than our English way of liking nothing and professing to like triglyphs.'[1] A gloomy style, then; a veritable Bastille of oppressive memories; a style to be cast down and the dust of it shaken 'from our feet for ever.'

On its constructive side the new criticism was no less flattering to a democratic sentiment. It set out to establish, and delighted its public by providing, a 'universal and conclusive law of right' that should be 'easily applicable to all possible architectural inventions of the human mind'; and this in the 'full belief' that in these matters '*men are intended without excessive difficulty* to know good things from bad.' Good and bad, in fact, were to be as gaily distinguishable in architecture as they notoriously are in conduct. And the same criterion should do service for both. Because a knowledge of the Orders, which was the basis of architectural training, is not, of itself, a passport either to architectural taste or practice, it was argued that training as such was corrupting. The exactitudes of taste, the trained and organised discrimination which, in the collapse of the old order, men had indubitably lost, were declared to be of less service in framing a right judgment of architecture than the moral delicacy they conceived themselves to have acquired. From the fact that the sculptures of a village church have, or once had, an intelligible interest for the peasant, it is argued that all architecture should address itself to the level of his understanding; and this paradox is so garnished with noble phrases that we have well-nigh come to overlook its eccentricity. This prejudice against a trained discernment is significantly universal among writers of the ethical school. They describe it as 'pride,' as 'pedantry,' as 'affection';[2] a habit of speech which would be inexplicable since, after all, training is not a very obvious vice or fatal disqualification, did we not relate it to the combination of romanticism and democracy in which this view of architecture takes its rise. But their habit makes it easy to understand that the ethical criticism was certain to gain ground. It appealed to a sincere desire for beauty in a society that had cast off, along with the traditions of the past, the means by which a general grasp of architectural beauty had in fact been maintained. It offered the privileges of culture without demanding its patience. A new public had been called into being. Works on architecture could never again be addressed: 'To all Joiners, Masons, Plasterers, etc., and their Noble

---

[1] *The Stones of Venice*, vol. I. chap. ii. p. 13.
[2] e. g. *The Stones of Venice*, vol. III. chap. ii. § 38.

望,随着它长时间地到处散播,过去它曾经带给人们的快乐到现在也已消失殆尽。拉斯金毫不留情地斥责说:"文艺复兴时期威尼斯的那些低下的建筑师们喜欢化装舞会的面具和吹拉弹唱,所以他们就在建筑物上装饰了一些戏剧面具和乐器之类的东西。即便是这样,它们也要比我们今天英国建筑上什么都没有强出不知多少倍。英国的建筑师什么爱好都没有,只好在建筑上装饰一排排的三槽板。"[1]这么说来,文艺复兴时期的建筑简直就是一种阴沉沉的风格,是现实生活中令人压抑的巴士底狱(Bastille),是一种必须要把它彻底推翻、打倒在地的建筑风格,"然后再把沾在脚上的尘土彻底地抖落干净"。

这种艺术理论同时具有积极建设性的一面,它极大地迎合了崇尚民主思想的热忱。这种理论也不是仅仅停留在说教而已,它还积极倡导实践,为广大民众提供了一套"放之四海而皆准且无可辩驳的正确准则",这些准则是"很容易地被应用到各种场合的,只要是人们的头脑能够想到的情形,这种设计原则都是适用的";而且这种理论坚信,如果按照这个理论来进行建筑设计,"人们可以毫不费力地识别出什么是好的建筑,什么是坏的建筑。"建筑艺术中的好与坏可以像人们的行为那样很容易地判断出来,这两个领域其实是可以使用同一套准则作为评判的尺子。作为建筑艺术训练的基础,古典柱式本身并不能让一个人知道什么是好与坏,也不能让人知道在面对一项建筑设计任务的时候该如何去做。这种训练方法在某些人眼里就是误人子弟,简直是堕落。随着旧有的训练方式逐渐被淘汰,人们自然就永远地失去了古典建筑艺术的熏陶,失去了甄别建筑艺术的训练与具体方法。但是,这些损失在鼓吹以道德标准作为建筑艺术衡量标准的理论家看来根本不值得大惊小怪,因为他们认为自己建立起来的道德准绳要比古典柱式方面的训练高明得多。当他们注意到某一个村子里的教堂上曾经有一个雕刻是当地农民们可以理解、看明白的内容的时候,他们便得出结论说,一切建筑艺术都必须做到让农民看得懂才行。这个似是而非的说法充满了高尚的思想和感人的词语,结果是我们大家都没有注意过它的偏差。这种仇恨专业训练的偏见言论,在鼓吹以道德标准来判断建筑艺术的学者中间是极为普遍的现象。他们把专业训练称为"高高在上""酸腐的学究""矫情做作"[2];这种陈腔滥调并不能让人明白它们的用意,因为归根到底,专业训练本身并不是什么罪恶,不能说因为训练的缘故让人们反而变得愚笨了。只有当我们把它同浪漫主义思想与民主思想联系到一起的时候,我们才能够看明白他们这些说法的真正含义。这一系列说法就是很容易让人们接受他们的把道德准则作为艺术批评的理论。它的目的就是,要在被遗忘的社会阶层里,唤起他们对于美的追求。这个阶层的人们已经把自己过去的传统,连同过去曾经有过的对于建筑艺术进行欣赏的工具统统抛弃了。这种新理论就是让这些人一下子重新有了文化的特权,同时又不必经过耐心艰苦的努力。一个全新的公共大众群体就这样形成了。建筑理论著作再也不需要和与之相关的专业人士进行沟通了,什么"木匠、石匠、泥瓦匠,以及尊敬的赞助人等等"统统可以不要。一场巨大的艺术民主运动可以通过投票的方式来决定什么是美,来决定人们的喜好。拉斯金是第一位唤起民众对于建筑艺术产生兴趣的理论家;被他唤醒的民众很自然地就把他的信条当成自己的信条。至于说他的信条中有哪些是对建筑艺术的实际创造与投资赞助等工作是具有实

---

[1]《威尼斯的石头》,第一卷,第二章,第13页
[2]《威尼斯的石头》,第三卷,第二章,第38节

Patrons.' A vast democracy was henceforth to exercise its veto upon taste. Ruskin was the first to capture its attention for the art of building, and it was natural that a public which he had enfranchised should accept from him its creed. It had no effective experience either in the creation or in the patronage of architecture by which that creed might be corrected. Architecture supposedly 'Ruskinian'—though not always to the master's taste—triumphed henceforth in every competition. Architecture in modern theory was a book for all to read. Democracy, looking to the memorials of a world it had destroyed for some image of its own desires, saw in the writing on the wall a propitious index of its own destiny. The orders of Palladio which had dignified the palaces of the *ancien régime* were easily deciphered: Mene, Mene, Tekel, Upharsin. Thus the history of architecture was made a pledge of social justice, and the political currents, strongly running, destroyed all understanding of the Renaissance.

The political prejudice in taste justifies itself by an appeal to moral values; but it does not, like the theological prejudice, indulge in oracles from revelation. It is ethical, but it is ethical in a utilitarian sense. It judges the styles of architecture, not intrinsically, but by their supposed effects. The critic is sometimes thinking of the consequences of a work upon the craftsman; sometimes of the ends which the work is set to serve, and of its consequences upon the public. But in all cases his mind moves straight to the attendant conditions and ultimate results of building in one way rather than another. The importance of the matter is a social importance; the life of society is thought of as an essentially indivisible whole, and that fragment of it which is the life of architecture cannot—it is suggested—be really good, if it is good at the expense of society; and to a properly sensitive conscience it cannot even be agreeable. Purchased at that price it becomes, in every sense, or in the most important sense, bad architecture. The architectural doctrines of such a man as Morris—a picturesque fusion of artistic with democratic propaganda—are for the most part of this type. The underlying argument is simple. Ethics—or politics—claim, of necessity, precisely the same control over aesthetic value that architecture, in its turn, exercises by right over the subordinate functions of sculpture and the minor arts; and Renaissance architecture is rejected from their scheme.

Even so, it is clear that criticism will still have two factors to consider: the aesthetic quality of architecture and its social result. To confuse the social consequences with the aesthetic value would be an ordinary instance of the Romantic Fallacy. Those were not necessarily the worst poets whom Plato urbanely ushered out of his Republic; for the practical results of an art are distinct from its essential quality. Even for our practice we require a theory of aesthetic value as well as a theory of ethical value, if only in order to give it its place within the ethical

际意义的,人们并不关心。一时间,在拉斯金理论影响下出现的建筑艺术作品席卷了大大小小的建筑设计竞赛方案;虽然有些参加评审的大师们并不看好这些做法,但是没关系,设计师乐此不疲。现代建筑艺术理论已经认定,建筑艺术就是一本人人都可以阅读理解的书籍。民主的思想摧毁了世界上一大批建筑艺术纪念碑,现在又回到其中去寻找自己所需要的形式,这实际上就是一种警讯;但是它并没有明白这种警讯的含义,却误以为这是自己的光荣使命。那些曾经让王公贵族的建筑荣耀一时的帕拉第奥建筑理论和柱式太容易理解了,那不过是伯沙撒写在墙上的几个字吗? Mene, Mene, Tekel, Upharsin(希伯来语,被音译成拉丁字母的拼写,意思是:算啊,算啊,权衡,分解,暗示了巴比伦伯沙撒的王国将被分解,但是,神的警告并没有引起伯沙撒的注意,结果,他的王国分裂了)。对建筑艺术历史的理解变成了对社会正义的向往,而政治上的洪流毁掉了人们对于文艺复兴艺术的全面理解。

用政治上的偏见来判断建筑艺术只能依靠道德的价值观才能得以进行,但是,它不同于神学观念上的偏见,它并不完全依赖于神明的启示和宣谕。它也讲求伦理道德,但是,它所讲的道德完全是从实用角度来看的。这种理论在判断建筑艺术风格的时候,不是从艺术风格的内在本质入手,而是假想这种风格会给社会带来什么样的效果,再根据这些效果来进行评估。理论家们有时会想到艺术作品会带给工匠的影响是什么,有的时候会考虑这个建筑物的目的与功能是什么,在另外的场合会考虑这个建筑会对社会民众产生什么样的作用,等等。但是,在任何情况下,理论家总是在考虑那些跟随建筑出现的次要情况,就是不考虑建筑本身,而且总是攻击一点,不及其余。一件艺术作品的重要与否完全取决于它的社会效果。全体社会的生活状态在这里是一个不可分割的整体,建筑艺术所占据的那一小块儿如果不顾全社会的效果而只关注自己,那么它的结果肯定也不会好到哪里。对于一个具有良知的人来说,他也绝对不会认同不顾社会效果的艺术作品的。一件建筑艺术作品如果真的是以牺牲了社会效果作为代价来成全自己的话,那么对任何人来说,这样的建筑作品就无疑是一座坏建筑。以莫里斯为代表的建筑艺术理论教条就是坚持这类观念的说教,都是一些追求图画般构图的艺术理念,外加一些民主思想的宣传口号,如此而已。这种理论背后的主线其实很简单:道德原则(或者说政治上的原则)必须左右建筑艺术的美学原则,正如同建筑艺术原则必须左右其中的功能与雕刻装饰等部分一样。以此作为依据,文艺复兴时期的建筑自然地被排斥在外。

即便是这样,艺术理论还是不可避免地要考虑两条因素:建筑艺术的美学价值和建筑艺术的社会效果。错把社会效果当成美学价值正是浪漫主义艺术理论的一种最常见的谬误所在。柏拉图并不是挑选最差的诗人才把他们驱逐出理想国的。一件艺术作品的成品是否制作精良与它的根本性质是两个截然不同的东西。即使是从实用角度来看,我们在需要道德价值理论的同时,也还是需要一个美学价值理论,因为这样我们才能在满足道德原则的前提下对于美学的高下做一个衡量。因此,问题的次序应该是这样的:从美学角度来看一个艺术风格,它的优点是什么? 这个艺术风格的社会价值是什么? 这个艺术风格的优点和社会价

scheme. The order of thought should be: what are the aesthetic merits of a style; what is their social value; how far are these outweighed by their attendant social disadvantages?

But the critics of architecture who assail the Renaissance style are far from proceeding in this sequence; nor do they establish their social facts. We may well doubt whether the inspired Gothic craftsman of that socialist Utopia ever existed in the Middle Ages. No historical proof of his existence is advanced. If we base our judgment on the *Chronicle of Fra Salimbene* rather than on the *Dream of John Ball*, which has the disadvantage of having been dreamt five hundred years later, we shall conclude that the Gothic craftsman was more probably a man not unlike his successors, who over-estimated his own skill, grumbled at his wages, and took things, on the whole, as they came. Some stress is not untruly laid upon his 'liberty': a Gothic capital was, now and then, left to his individual imagination. But how minute, after all, is this element in the whole picture. The stress laid upon its springs from that disproportionate interest in *sculpture* as opposed to architecture, the causes of which have already been traced to Romanticism and the cult of Nature. But just as sculpture is not the aesthetic end of architecture, so, too, sculpture is but a small part of its practical concern. The foundations are to be laid, the walls and piers erected, the arches and the vaultings set. In all this labour there was nothing to choose between the Mediaeval and the Renaissance style: neither more nor less liberty, neither more nor less joy in the work. The Renaissance, too, had its painting and its minor arts —its goldsmiths, carvers and embroiderers—destined in due course to enrich what had been built. Here, if we trust the pages of Vasari and Cellini, was no lack of life and individual stir.

The Renaissance 'slave' toiling at his ungrateful and mechanical task is, no less, a myth. Such persons as may have formed any intimacy with his successor, the Italian mason, on his native ground, will realise that he is capable of taking as vital a pride and as lively a satistaction in the carving of his Ionic capital as the mediaeval worker may be supposed to have derived from the manufacture of a gargoyle; that he by no means repeats himself in servile iteration but finds means to render the products of his labour '*tutti variati*'; and that so far from slavishly surrendering to the superior will of his architect, he permits himself the widest liberty *perchè crede di far meglio*,—whereby, indeed, now as in the past, many excellent designs have been frustrated.

But the mediaeval labourer, in this Elysian picture, has his toil lightened by religious aspiration. No doubt he took pleasure in his cult and got comfort from his gods. But how was it with the Renaissance workman at the lowest point of his 'slavery and degradation,' the dull tool whose soulless life is revealed in the

值因为它所产生的副作用而降低了多少呢?

对文艺复兴时期艺术提出批评指责的建筑理论家们对于上面这几个问题根本不予理睬,他们也不愿意实事求是地面对社会的现实问题。我们真的非常怀疑那些中世纪满怀社会理想的哥特艺术工匠们到底是否确实存在过。到现在我们还没有看见这方面令人信服且深入细致的研究成果。如果我们在十三世纪的意大利教士萨利比尼(Fra Salimbene)所写的《编年史》(*Chronicle of Fra Salimbene*)和十九世纪莫里斯所写的《英国教士约翰·鲍尔梦想录》(*Dream of John Ball*)之间作一个选择的话,因为后者是五百多年以后凭想象写出来的东西,前者要可靠得多,我们会选择相信前者。如果是这样,那么,我们就会从《编年史》中了解到,中世纪哥特艺术的工匠们和后来每一个时代里的工匠们实际上并没有什么两样,他们喜欢吹嘘自己的技艺,计较工钱的多少,从总体上来说也是得过且过。这里要说明一点,这些中世纪的工匠的确会"自由地"发挥一下自己的想象,一座建筑中的一些柱头上的雕刻偶尔会留给这些工匠一些空间,让他们根据自己的喜好来发挥一下。但是,同整个建筑比较起来,这点儿发挥的空间实在是微乎其微。把这种简直不成比例的雕刻部分拿来大书特书,反而对建筑整体避而不谈,这种做法实际上就是浪漫主义艺术理论与崇拜自然的那些理论的结果。但是,雕刻并不是建筑艺术的美学目的,雕刻不过是建筑在具体建造制作过程中的很小一部分工作而已。建筑的基础结构必须要做,墙体和柱子必须要竖立起来,拱券和拱顶必须坚固地搭起来,这一切费时费工的工作对于中世纪和文艺复兴的人们来说,没有任何差别:谁也不会比别人有更多的发挥空间,谁都不会觉得这些艰苦的工作是一种快乐享受。文艺复兴时期也有隶属于建筑艺术的绘画、装饰、五金等手工艺,各种艺人、匠人都在各自努力,把建筑打造得更漂亮一些。如果我们觉得瓦萨利(Vasari)和塞利尼(Cellini)所记载的艺术家传记是真实的记录的话,那么这些记录告诉我们,这个时期的艺术家个个都是充满了独特个性、有血有肉活生生的人。

文艺复兴时期的那些所谓的"奴工"从事着没有尊严的繁重体力劳动,这又是人们编造出来的神话故事。这个时期的工人与前一个时代里的工人一样,都有自己亲切可爱的地方,有属于自己的生活。意大利的石匠在自己生长的土地上,也对自己打造出来的华丽的爱奥尼(Ionic)风格的柱式感到骄傲,这种感情与中世纪的石匠在成功地打造出大教堂上面雨水排放口处的怪异兽头造型(gargoyle)后所产生的那种自豪感根本没有任何的差别。他绝对不是悲惨地重复着单调的苦工,而是在自己的制作中随时发挥着自己的才智;他根本不需要绝对地服从建筑师的指示和要求,而是坚信自己要比建筑师更高明,更懂行(perchè crede di far meglio)——结果和过去一样,让建筑师的精美设计构思不得不留下一些遗憾。

然而中世纪的工匠们有着神话传说般的工作环境,因为时时感受着宗教的关爱和恩泽,所以艰苦的劳动变成了轻松愉快的经历。我们不怀疑他们可以从自己的信仰中获得安慰。但是,文艺复兴时期那些"丧失尊严的奴隶们",在最黑暗的时刻,又是怎样用没有生命的工具和材料创造出巴洛克艺术的呢?下面我引述一段利奥波德·冯·兰克(Ranke)描写的在

baroque? This is Ranke's description of the raising of the great obelisk before the front of St. Peter's, which Domenico Fontana undertook for Sixtus v. :

'It was a work of the utmost diffculty—to raise it from its base near the sacristy of the old church of St. Peter, to remove it entire, and to fix it on a new site. All engaged in it seemed inspired with the feeling that they were undertaking a work which would be renowned through all the ages. The workmen, nine hundred in number, began by hearing Mass, confessing, and receiving the Communion. They then entered the space which had been marked out for the scene of their labours by a fence or railing. The master placed himself on an elevated seat. The obelisk was covered with matting and boards, bound round it with strong iron hoops; thirty-five windlasses were to set in motion the monstrous machine which was to raise it with strong ropes; each windlass was worked by two horses and ten men. At length a trumpet gave the signal. The very first turn took excellent effect; the obelisk was heaved from the base on which it had rested for fifteen hundred years; at the twelfth, it was raised two palms and a quarter, and remained steady; the master saw the huge mass, weighing, with its casings, above a million of Roman pounds, in his power. It was carefully noted that this took place on the 30th April 1586, about the twentieth hour (about three in the afternoon). A signal was fired from Fort St. Angelo, all the bells in the city rang, and the workmen carried their master in triumph around the inclosure, with incessant shouts and acclamations.

'Seven days afterwards the obelisk was let down in the same skilful manner, upon rollers, on which it was then conveyed to its new destination. It was not till after the termination of the hot months that they ventured to proceed to its re-erection.

'The Pope chose for this undertaking the 10th of September, a Wednesday, which he had always found to be a fortunate day, and the last before the feast of the Elevation of the Cross, to which the obelisk was to be dedicated. On this occasion, as before, the workmen began by recommending themselves to God; they fell on their knees as soon as they entered the inclosure. Fontana had not omitted to profit by the suggestions contained in a description by Ammianus Marcellinus of the last raising of an obelisk, and had likewise provided the power of one hundred and forty horses. It was esteemed a peculiar good fortune that the sky was covered on that day. Everything went well: the obelisk was moved by three great efforts, and an hour before sunset it sank upon its pedestal on the backs of the four bronze lions which appear to support it. The exultation of the people was indescribable and the satisfaction of the Pope complete. He remarked in his diary that he had succeeded in the most difficult enterprise which the mind of man could imagine. He

FIVE

罗马圣彼得大教堂前面竖立起那座大方尖碑的过程,来看看那个时期的工匠是怎样工作的。这是在建筑师多米尼哥·方坦纳(Domenico Fontana)的指挥下,为教皇西克斯图斯五世(思道五世)(Sixtus V.)改造教堂前面广场的情形:

"具体的工作是十分困难的一件事,就是把一座巨大的方尖碑从过去旧的圣彼得教堂存放圣器的仓库搬到现在新建大教堂前面的广场上来,并把它竖立起来。所有参与其中的人都觉得自己很幸运,有机会参加一场让他们可以名留青史的壮举。参加的工人达到九百人之众。他们先是聆听弥撒,接着是忏悔,然后是接受神的保佑,与神结成共患难的兄弟。仪式过后,大家便分头走向各自的位置。每个人的位置与工作都是事先安排好,用栏杆或者绳索划分好了的。总指挥站在一个高高的位置上发号施令。那个巨大的方尖碑已经用木板和垫子包裹捆扎好,再用铁箍把这些保护层固定住。三十五组滑轮带动数不清的粗大绳索构成了一个庞大的起重设备,每一组滑轮配备两匹马和十个人。一切准备就绪以后,号角一响,旗开得胜,那个在仓库角落沉睡了一千五百年的巨大方尖碑被拉了出来,到了中午十二点的时候,这个方尖碑已经被抬起,距离地面大约二尺三寸,而且是非常平稳的。总指挥看着这个庞然大物,估算了一下它的重量,包括那些保护层在内,总重量绝对不下一百万罗马磅。他当然很得意地看到这些完全在自己的控制之内。这个重大事件发生时间被清楚地记录下来:一五八六年四月三十日,下午三点(当时的日晷刻度是指向二十时)。这时,从圣安琪罗古堡(Fort St. Angelo)传来一声炮响,全市的大大小小教堂钟声齐鸣。人们兴高采烈地把总指挥抬起来,在场地上载歌载舞,欢庆这一伟大的壮举,欢呼声和笑声响彻云霄。

七天以后,人们用同样有组织、有秩序的工作,把方尖碑放到一个巨大的滑轮平板车上,顺利地把方尖碑运送到新的地点。运到广场上以后,大家决定,等过了炎热的夏天以后,再把它竖立起来。

教皇选定的日子是在九月十号,星期三,因为他相信这是一个幸运之日,是圣十字架树立纪念日前的最后一天。而竖立这座大方尖碑的目的就是为了这个纪念日。这一天,和上一次一样,工人们同样先向上帝祈求力量,祈求他的保佑。他们到了工地以后的第一件事,便是双膝跪到,向上帝祷告。方坦纳当然不会忘记阿米阿努斯·马尔切利努斯(Ammianus Marcellinus)所写的古罗马历史书中记载的那些技艺和工艺,上一次成功地把方尖碑抬出来并运送到这里就是参考了古书中记载的方法。这次也不例外,他总共配备了一百四十四马。那一天,天上云彩很厚,遮住了日头,大家觉得这是一个好兆头。一切进展十分顺利,前后分三个步骤,把巨大的方尖碑竖立起来,在太阳落山前一小时的时候,方尖碑成功地被安放在广场上新建成的基座上面。基座下面有四个青铜狮子,好像在用力托起这个方尖碑似的。人们欢呼雀跃,喜悦的场面无法用文字描述,教皇本人当然也十分满意。他在日记中谈到这件事的时候说,人们能想象出来的世上最为困难的一件事,他今天做到了。 教皇请人制作奖章来纪念这件事,他也收到了世界各国送过来的赞美诗歌,他也向世界各地的君主们发出正式的告示,宣布这个方尖碑的落成。他也勒石刻碑纪念此事,并撰写奇怪碑文一篇,吹嘘这

caused medals commemorating it to be struck, received congratulatory poems in every language, and sent formal announcements of it to all potentates. He affixed a strange inscription, boasting that he had wrested this monument from the emperors Augustus and Tiberius, and consecrated it to the Holy Cross; in sign of which he caused a cross to be placed upon it, in which was unclosed a supposed piece of the true Cross.'[1]

The modern labourer has lost these joys; but he has not lost them on account of his Palladian occupations. Whether he be set to build the Foreign Office in the Italian manner, or the Law Courts in the mediaeval manner, or a model settlement in the democratic manner, his pagan pleasure and his piety are equally to seek. Here, indeed, is the fallacy of the writers of this school: an idealised mediaevalism is contrasted with a sharply realistic picture of Renaissance architecture in modern life: the historical Renaissance, the historical Gothic, they are at no pains to reconstruct. Conducted without impartiality, arguments such as these are but the romance of criticism; they can intensify and decorate our prejudices, but cannot render them convincing. Even so, and did they prove their case, the superior worth of a society might justify the choice, but would not prove the merit of the style of architecture which that society imposed. The aesthetic value of style would still remain to be discussed. Or is that, too, upon a due analysis, within the province of an ethical perception? That is the question which still remains.

III. The last phase of ethical criticism has at least this merit, that it strikes at architecture, not its setting. It takes the kernel from its shell before pronouncing upon taste.

There are those who claim a *direct* perception in architectural forms of moral flavours. They say, for example, of the baroque (for although such hostile judgments are passed upon the whole Renaissance, it is the seventeenth century style which most often and most acutely provokes them) that it is slovenly, ostentatious, and false. And nothing, they insist, but a bluntness of perception in regard to these qualities, nothing, consequently, but a moral insensibility, can enable us to accept it, being this, in place of an architecture which should be—as architecture can be—patiently finished and true. Baroque conceptions bear with them their own proof that they sprang from a diseased character; and his character must be equally diseased who can at any subsequent time take pleasure in them or think them beautiful.

---

[1] Ranke's *History of the Popes*, trans. S. Austin, vol. I. book iv. § 8. I have quoted the passage at length because, besides indicating the religious enthusiasm of the workmen, and their delight in the work (two supposed monopolies of the Gothic builders), it illustrates the superb spirit of the baroque Pope, who gave Rome, for the second time, an imperial architecture.

个方尖碑是他从古代罗马奥古斯都（Augustus）和提贝里乌斯（Tiberius）两位皇帝手里夺过来的，云云，并把它奉献给神圣的十字架。作为表达这样一份思想感情的标志，他命人在方尖碑的顶上放上一个十字架，据说制成十字架的其中一块材料还是从耶稣受难的那个十字架上取来的呢。"[1]

　　现代社会里的劳工已经不会再有这样的热情了，但这绝对不是因为在建造帕拉第奥风格建筑的时候才失去的。无论是在建造意大利风格的外交部办公大楼，还是建造中世纪哥特风格的最高法院，甚至是具有现代民主思想的居住区等项目的时候，劳工们的世俗快乐与宗教虔诚实际上都很缺乏。而鼓吹用道德标准作为建筑艺术评判标准的这些理论家所犯的谬误在于：他们用理想化了的中世纪艺术与现实生活中出现的文艺复兴时期艺术进行对比，然后歌颂前者，贬低后者。至于历史上存在过的那个真实的中世纪和真实的文艺复兴时期到底是怎样的，他们并不关心，也不会去帮助读者作深入的了解。在对待两个不同时期的态度上不能保持公平、中立，在论述语言上不能恰当、准确地描写，而是用带有强烈感情色彩的浪漫语言进行叙述，这样的理论可以强化或者修饰人们的偏见，但是他们的结论却无法令人信服。即便是他们从道德角度出发作出的判断是对的，即便是社会的崇高价值可以确认他们推崇的建筑艺术风格是正确的，但是，这种理论到最后还是没有说明被社会认可的建筑艺术风格本身到底好在哪里。艺术风格自身的美学价值还有待进一步阐述。或许经过仔细的分析以后，有可能得出结论说，艺术风格的美学价值最终仍然是属于道德判断的范畴的吗？但是，这个问题也同样需要进一步回答。

　　三、前面所说的用道德标准来衡量建筑艺术的理论至少有一点还算是优点：它直接面对的是建筑本身，不是它周边那些次要的东西。它是在对建筑艺术的好坏作出判断之前，把建筑的外壳砸碎，把核心从中间拿出来进行评估。
　　有一些人宣称他们可以*直接*地从建筑形式上感受到它所具有的道德倾向。例如，这些人会说，巴洛克艺术就是反映出它的草率、浮夸和虚假。（这种措辞其实是可以拓展到对整个文艺复兴时期艺术的批评，但是这些评论主要还是针对十七世纪的那些建筑艺术风格而作出的。）他们会坚持认为，巴洛克艺术除了对于美感的感觉迟钝之外，没有什么其他可说的；这样一来，除了对于道德问题毫无知觉之外，别的任何理由都无法让我们接受这样的建筑艺术，因为真正的建筑艺术只能是由耐心加上诚实创造出来的。巴洛克艺术的产生明显地表明这种艺术形式完全是产生于一种病态的性格，而后来凡是能够欣赏这种艺术并从中感受到乐趣与美感的人们也必定是具有同样病态的性格的。这些艺术肯定是从一种道德败坏与堕落的社会环境中生长起来的，是为卑鄙的目的服务的。这个事实刚好证明了我们的一个

---

[1] 蓝柯（Ranke）所著《教皇的历史》，S. 奥斯汀翻译，第一卷，第四册，第 8 节。我把这一段几乎全部都引用过来，除了想表现一下当时的劳工们满腔的宗教热情之外，还想借此表现一下他们的工作热情（关于这两点，很多人固执地偏信是只有哥特时期的工匠才具有的两种特征）。这一大段文字精彩地描绘了这位巴洛克时期的教皇那高昂的激情，他重新把帝国的风范带回到罗马这座城市。（第一次指的是罗马帝国时期。）

They may have grown up in a corrupt society and served ignoble uses. That fact would but confirm our judgment: it does not furnish its ground. Its ground is in the work itself; and this is not bad because it is ugly; it is ugly because, being false, ostentatious, slovenly and gross, it is obviously and literally bad.

This contention is supported by admitted facts. The detail of the baroque style is rough. It is not finished with the loving care of the *quattrocento,* or even of the somewhat clumsy Gothic. It often makes no effort to represent anything in particular, or even to commit itself to any definite form. It makes shift with tumbled draperies which have no serious relation to the human structure; it delights in vague volutes that have no serious relation to the architectural structure. It is rapid and inexact. It reveals, therefore, a slovenly character and can only please a slovenly attention.

The facts are true, but the deduction is false. If the baroque builders had wished to save themselves trouble it would have been easy to refrain from decoration altogether, and acquire, it may be, moral approbation for 'severity.' But they had a definite purpose in view, and the purpose was exact, though it required 'inexact' architecture for its fulfilment. They wished to communicate, through architecture, a sense of exultant vigour and overflowing strength. So far, presumably, their purpose was not ignoble. An unequalled knowledge of the aesthetics of architecture determined the means which they adopted. First, for strength, the building must be realised as *a mass*, a thing welded together, not parcelled, distributed and joined. Hence, the composition (the aesthetic unity of parts) must be imposing; and no one has yet suggested that the baroque architects lacked composition—either the zeal for it or the power. Next, again for the effect of mass, the parts should appear to flow together, merge into one another, spring from one another, and form, as it were, a fused gigantic organism through which currents of continuous vigour might be conceived to run. A lack of individual distinctness in the parts—a lack of the intellectual differentiation which Bramante, for example, might have given them—was thus not a negative neglect, but a positive demand. Their 'inexactness' was a necessary invention. Further—again for the suggestion of strength—the *scale* should be large; and hence, since a rough texture maintains a larger scale than a smooth, an inexact finish was preferred to one more perfect. Last, for the quality of exultation: for vigour not latent but in action; for vigour, so to speak, at play. To communicate this the baroque architects conceived of Movement, tossing and returning; movement unrestrained, yet not destructive of that essential repose which comes from composition, nor exhaustive of that reserve of energy implied in masses, when, as here, they are truly and significantly massed. But since the architecture itself does not move, and the movement is in our attention,

判断：这种艺术是没有任何道理可言的。它的道理就是作品自己。我们认定它是不好的艺术不是因为它难看，它之所以难看那是因为它是虚假的、虚张声势地浮夸；因为它在敷衍应付，而且效果是令人作呕的，因此它才成为不折不扣的坏艺术。

这个论点还有一些公认的事实作为依据。巴洛克建筑艺术的细节都很粗糙，它没有十五世纪里艺术作品所具有的那种精致与细腻，明显地让人感到这时的艺术家并不十分热爱艺术，甚至连中世纪时期的那些生硬的雕刻也要比它们精致一些。这时的艺术家似乎没有兴趣要表现任何具体的形象，有时甚至连一个完整的造型都不是，更不用说表现什么东西了。巴洛克的雕塑有时候只是一些零乱的挂帘或者布饰，与人体的结构毫无关系；有的时候这些装饰构成了模模糊糊的漩涡状的形态，与建筑结构也没有什么直接关系。一切都在快速地变化着，一切都不确定。正因为如此，巴洛克建筑艺术表现出一种草率、粗糙的特征，只能让人们粗略地看一眼，不能细瞅。

这里所说的事实的确是真的，但是最后的推论却是错的。如果巴洛克的建筑师和工匠们是为了偷懒，那他们会干脆什么装饰都不做，而且还可以博得一个"纯净简洁"、道德高尚的好名声。然而，这些艺术家有自己明确的目标，这个"明确"的目标却需要"不明确"的建筑艺术手段来达成。他们希望借助于建筑艺术来表达一种兴高采烈的氛围和一种容器盛装不下最后溢出来的巨大能量。暂时先让我们假设这些艺术创作的动机并不是龌龊和卑贱的勾当吧。这些艺术家由于出色地掌握了建筑艺术美学的理论与技巧，采取了一系列的具体艺术手段来达到自己的创作目的。首先，为了表达力量与能量，建筑不能是零散的小块，然后再通过连接体把它们拼接起来，建筑必须要以整体的形式出现。因此建筑体量在构图上必须气势逼人才行。到目前为止，我们还没有听到任何指控，说巴洛克缺乏构图知识。无论是在追求体量的热情上，还是在体量造型的把握上，巴洛克建筑艺术都有过人之处。其次，为了体现这种整体体量的效果，每一个建筑局部都服从于整体的走势，互相之间总是你中有我，我中有你，我从你中来，你从我中去，整个形式融合到一起，形成一个庞大的有机体，一种连续不断的生命力在其中无休止地流动。每一个局部没有自己明确的界限绝对不是被动地被忽略，而是一种积极主动地创造。早年伯拉蒙特（Bramante）的那种条理分明地处理各个局部的手法与观念，在这里显然是被有意识地加以排除了。现在出现的这些"飘浮不定"的效果是一种必不可少的创新。也同样是出于强调力量的目的，整个建筑的尺度感也必须够大才行；因此，艺术家选择了粗糙的表面取代以往常见的光滑细腻的表面质感，用不明确的分界来取代过去精准的交接。最后一点，为了表现兴高采烈的氛围，每一个细节都要把生命力表现出来，而不是含蓄地隐藏在其中，要让生命的活力参与到建筑空间与人相互作用的互动中来。为了表现这些特征，巴洛克建筑艺术家选择了动感作为自己努力的目标，让建筑中的元素在这里被抛出去，从那里又旋转回来。动感是不受限制的，但是又不能破坏整个构图的均衡与稳重，同时也不能把整个建筑体量所蕴藏的全部能量释放干净，那样会让人感到这个巨大体量缺乏后续的力量。这样才能真正表现出建筑的厚实。由于建筑物本身肯定是不会动的；那么我们观察到的动感都是借助于设计手段形成的，把我们的注意力一下子带到这里，一下子又带到那里，有时候会让我们的注意力停留在某一个地方，有时候又会把注意力引到别处去。一切的视觉效果都取决于设计手段对于我们注意力的控制。这种被建筑设计

drawn here and there by the design, held and liberated by its stress and accent, everything must depend upon the kind of attention the design invites. An attention that is restrained, however worthily, at the several points of the design; an attention at close focus and supplied by what it sees with a satisfying interest; an attention which is not *led on*, would yield no paramount sense of movement. Strength there might be, but not *overflowing* strength; there would be no sense of strength 'at play.' For this reason there exist in baroque architecture rhythm and direction and stress, but no repose—discord, even—till the eye comes to rest in the broad unity of the scheme, and the movements of the attention are resolved on its controlling lines. In proportion as the movement is tempestuous, these lines are emphatic; in proportion as it is bold, these are strong. Hence, sometimes, the necessity—a necessity of aesthetic, if not of constructive logic—for that worst insolence and outrage upon academic taste, the triple pediment with its thrice-repeated lines, placed, like the chords in the last bars of a symphony, to close the tumult and to restore the eye its calm.

In this sense alone is baroque architecture—in the hands of its greatest masters—slovenly or ostentatious, and for these reasons. But we do not complain of a cataract that it is slovenly, nor find ostentation in the shout of an army. The moral judgment of the critic was here unsound because the purpose of the architect was misconceived; and that was attributed to coarseness of character which was, in fact, a fine penetration of the mind. The methods of the baroque, granted its end, are justified. Other architectures, by other means, have conveyed strength in repose. These styles may be yet grander, and of an interest more satisfying and profound. But the laughter of strength is expressed in one style only: the Italian baroque architecture of the seventeenth century.

This brings us to the last charge. Real strength, the critic can reply, may be suffered to be exultant, though it is nobler in restraint. But the strength of the baroque is a deceit. It 'protests too much,' and for the usual reason: that its boast is insecure. Its mass is all too probably less huge, its vistas less prolonged, its richness less precious, than it pretends. The charge of false construction, as construction, has, it is true, been dealt with; the argument from science fell, as we saw, to nothing. But this is an argument of moral taste. Can we approve a style thus saturated with deceit: a style of false facades, false perspectives, false masonry and false gold? For all these, it must be agreed, are found in the baroque as they are found in no other style of architecture. It is an art, not indeed always, but far too often, of 'deceit.'

与建筑艺术所左右的注意力是按照收放有章法那样被控制住的,不是随意泛滥地,无论是多么高贵的立意,也都在具体设计中有所控制;这种注意力也是有选择地集中在一些点上,那里有充分的内容来满足这种注意力的要求,当眼睛停留在那里的时候,感官可以得到极大的满足。注意力如果不是被有意识地引导到某些焦点的话,那么在整个过程中就不会产生动感,也就不会有高潮迭起的效果。仅仅表现出力量还不够,一定要让这种力量有盛装不下溢出来的感觉才行,否则就达不到"力量正在发挥着作用"的效果。基于这样的认识,巴洛克建筑艺术中借助于动感节奏、韵律、方向感,以及重点强调的地方,来引导人们的视线不断地移动,直到获得了整个设计的和谐感觉以后才会停顿下来。这时,细节上的厚重、沉稳,对于建筑视觉的动感没有任何帮助,这时的细节有时甚至会有种不协调的感觉。对于动感的注意因为那些流淌着线条而让我们达到了目的。与此相对应的,因为这里要表现的是很剧烈的动感,所以建筑中的很多线条为了加强这种感觉而可以做得有些夸张;因为建筑追求的是奔放的效果,所以线条也很粗犷。这些出于美学考虑的夸张手段,因为不完全是出于建造技术逻辑上的需要,因此会冒犯一些学者们的感受与审美趣味。例如重复渲染三次的山花上面有三组线条就好比一首交响乐结束时的最后一个重音和弦,作为一段喧闹的乐曲的终结,让我们的眼睛又重新回归原先的安静。

在文艺复兴艺术大师的手中,巴洛克建筑艺术就是这样产生出某些人认为的匆忙草率与浮华炫耀的效果的,但是这种效果都是有其逻辑与原因的。我们不会对着奔腾的河流抱怨说它流淌得太匆忙了,也不会对军队的呐喊声抱怨说它太过于喧嚣。用道德标准衡量艺术的那些批评家们所作的评论之所以不成立,那是因为他们并没有理解巴洛克艺术家的创作意图,而被他们指控为粗制滥造的那些细节恰恰是这些艺术家绞尽脑汁力图达到的效果。如果对于巴洛克艺术的意图没有疑问的话,那么,它的艺术手段则完全成立。别的建筑艺术风格采用其他的艺术手段获得了庄重的效果,它们或许更加气势磅礴,给人的感受或许更隽永,意义更深长。但是,在兴高采烈的欢乐声中表现这种宏大的力量只有一种艺术风格:十七世纪里的意大利巴洛克建筑。

说到这里,我们注意到这些批评声音中的最后一个。批评家们会这样说,真正的力量会在兴高采烈的情绪中受到损害,有节制的艺术表现更高贵。巴洛克艺术所表现出来的力量实际上只是一种欺骗行为。这种艺术"总是在大声喧哗,好像在表达不满的情绪",这只能说明一个明显的事实,它的虚张声势是在掩盖内心的缺乏安全感。巴洛克建筑的体量并不真的如同它给人的感觉那样巨大,它的视觉空间没有看上去那样深远,它丰富的造型不过是装腔作势的姿态,根本没有什么价值可言。从建造技术方面来说,这些造型上的元素根本没有必要,是虚假的东西。关于这个指控我们已经讨论过了,的确是没有技术上的需要,这里不再赘言。来自科学技术方面的论点最终是站不住脚的,这一点我们也已经讨论过了。现在再次提起这个技术话题完全是因为道德上虚假与否的缘故。外观造型都是虚假的,空间形态与透视关系都是假的,砖石材料都是假的,金碧辉煌的装饰都是假的,我们能够接受并且肯定一种充满了虚假与欺骗的艺术风格吗?我们刚刚列举的所有这些,只有在巴洛克建筑艺术中才有,其他任何时期的建筑艺术都没有这些。巴洛克是一种艺术,虽然我们不能说它从头到尾都是欺骗,但是,它里面所包含的"欺骗"的东西实在是太多了。

This is probably the commonest of all the prejudices against the Renaissance style in its full development. But here, too, the facts are sounder than the conclusions.

The harmfulness of deceit lies, it must be supposed, either as a quality in the will of the deceiver, or in the damage inflicted by the deceit. If, in discharge of a debt, a man were to give me instead of a sovereign a gilded farthing, he would fail, no doubt, of his promise, which was to give me the value of twenty shillings. To deceive me was essential to his plan and the desire to do so implied in his attempt. But if, when I have lent him nothing, he were to give me a gilt farthing because I wanted something bright, and because he could not afford the sovereign and must give me the bright farthing or nothing bright at all, then, though the coin might be a false sovereign, there is evidently neither evil will nor injury. There is no failure of promise because no promise has been made. There is a false coin which, incidentally, may 'deceive' me; but there is no damage and no implied determination to deceive, because what I required in this case was not a sovereign but the visible effect of a sovereign, and that he proposed to give—and gave.[1]

I am probably *not* persuaded into believing that the false window of a Renaissance front is a real one, and the more familiar I am with Renaissance architecture, the less likely am I to believe it; but neither do I wish to believe it, nor does it matter to me if, by chance, I am persuaded. I want the window for the sake of the balance which it can give to the design. If the window, in regard to its utilitarian properties, had been wanted at that point, presumably it would have been made. But, on the contrary, it was—very likely—definitely *not* wanted. But its aesthetic properties—a patch of its colour, shape and position—*were* required in the design, and these I have been given. Had it been otherwise there would have been artistic disappointment; as it is, there is no disappointment either practical or artistic. And there is no deceit, for, as the architect is aware, the facts, should I choose to

---

[1] This may seem obvious enough, and too obvious; but, as Wordsworth wrote in a famous preface: 'If it shall appear to some that my labour is unnecessary and that I am like a man fighting a battle without enemies, such persons may be reminded that whatever be the language outwardly holden by men, a practical faith in the opinion which I am wishing to establish is almost unknown. If my conclusions be admitted, and carried as far as they must be carried if admitted at all, our judgments... will be far different from what they are at present, both when we praise and when we censure.' It is, in fact, for lack of stating the case at length and rendering it obvious, that the attack on the inherent falsity of the baroque is repeated in every history of architecture which appears in this country or in France. The attack varies in severity, and in extent. Either the whole Renaissance style is made 'intolerable' by deceit, or it becomes intolerable at its seventeenth century climax; or, if not intolerable, it is a very serious blemish and to be apologised for. But no critic desires or, desiring, has the courage to justify the Renaissance method, *qua* method, root and branch, and to insist that the baroque style was the first to grasp the psychological basis, and consequent liberties of architectural art. Yet such is the fact.

这个批评大概是所有对整个文艺复兴时期艺术提出的众多偏见中最常见的一种。但是,事实仍然是要比结论来得更强。

人们这样痛恨虚假一定是因为他们坚信:虚假的害处要么是弄虚作假的人在故意欺骗害人,要么上当的人会受到伤害。如果为了还清债务,一个人本应还给我一个英镑的金币,但他却给了我一个镀过金的最小面值的一分硬币,不用说,他欺骗了我,他给我的价值根本不是他承诺的二十先令。欺骗我显然是他的目的,这也表现在他的所作所为上。但是,如果当初我并没有借给他钱,当他知道我喜爱金光闪闪的东西,便说他要送给我一个镀金的一分硬币;因为他不富裕,没有一个英镑的金币送给我,只好送我一个镀金的一分硬币,否则就什么都不能送了。这时,尽管这个镀金的硬币不是真正的金币,但是,他并没有存心要骗我,我也没有因为这个硬币不是很真的金币而受到任何伤害。整个事件没有任何欺骗和食言行为,因为从一开始就没有什么骗人的企图。如果我不留意,或许会误以为这个镀金的硬币是真的,但是,这不会给我带来任何伤害,我也不会认为是被骗,因为从一开始我并没有期待一个真正的金币,我只是要一个看上去金光闪闪的东西,而他答应给我一个那样的东西,他最后也实现诺言,给了我这样的东西。[1]

无论怎么说,在面对文艺复兴时期建筑作品的时候,我恐怕不大会误以为那上面的假窗户是一个真窗户,我们对这个时期的建筑作品了解得越多,就越不会产生这样的错觉。我本来就不会相信那是真窗户,而且它到底是不是真窗户,我根本就不在乎。为了建筑立面的平衡,我只是认为那里应该出现一个窗户。至于这个窗户到底有没有实际采光功能并不那么重要,因为如果需要采光通风,它一定会那样去做,这没有任何困难。如果我们看到的窗子是没有实际功能的假窗户,那一定是它不需要,而且是明显地不需要。它的出现完全是出于美学方面的考虑,需要在那里出现一定的色彩、形状和大小,是设计上的需要,而一个假窗户刚好满足了这样一种需要。如果没有它,视觉上就会不完美。现在把它摆在那里,建筑师明显地知道,任何人都会看到这是一个假窗户,他没有打算欺骗任何人;任何人如果想发现

---

[1] 这一点看起来很明显,也许过于明显;但是,正如诗人华兹华斯在一篇著名的序言里写道的那样:"如果在某些人看来,我在这里所辛辛苦苦从事的工作根本没有必要,认为我是在进行一场没有敌人的战争,我想在这里提醒他们注意,不管人们叽叽喳喳地说什么,我下面要讲的话中所包含的一个实用性观念,可以说是你们从来没有听说过的。如果我的结论能够被接受,同时这个结论也能够传播得很久远的话,那么,我们在进行判断的时候……所得到的结果,肯定要比今天我们所看到的结论强出无数倍,而且无论你是持赞同的意见,还是持反对的意见,结果都一定会比现在的情形要好无数倍。"实际上,由于论述得不充分,论证得不够清楚,因此,在英国和法国,随便翻开任何一本建筑艺术历史书籍,我们都会看到不断反复重复地针对巴洛克固有缺陷所进行的批判。批判的火力大小不同,火力扫射的范围也有所不同。无论是整个文艺复兴时期的艺术均被认为是"无法忍受"的艺术,还是在十七世纪当巴洛克艺术达到高峰时期是"无法忍受"的艺术,或者说,即使这些还不是"无法忍受"的内容,那它们也是具有致命的瑕疵的,也要对它们的不幸表示一下怜悯和同情。但是,没有一位理论家有勇气愿意站出来,来替文艺复兴时期的艺术方法说一句公道话,以证人的身份证明:这种艺术方法是一种彻底的艺术方法,巴洛克风格的艺术是第一个以心理学理论为基础的艺术实践,也是后来建筑艺术呈现自由开放局面的滥觞。然而,事实的确就是如此。

know them, are readily discoverable. True, if I find the apparent stonework of the window is false, there is an element of genuine aesthetic disappointment, for the quality of the material has its own aesthetic beauty. But the baroque architects did not prefer paint to stone. Ruskin was not more disappointed than Palladio that the palaces of Vicenza are of stucco. Few generations realised more clearly the aesthetic quality of rich material; as the bronze and *lapis lazuli* of the altar of S. Ignazio in the Roman *Gesù* may suffice to show. But these architects placed aesthetic values in the scale of their importance, and where economic or other barriers stood in their way, preferred at least, and foremost, to indicate *design*. And, since, in the rich material, part only of the charm resides in the imaginative value of its preciousness—its rarity, the distance it has come, the labours and sacrifices it has cost—and a far greater part in the material beauty, for the sake of which those sacrifices are made, those labours undertaken, the baroque architects, seeing this, sought to secure the last by brilliant imitation, even when, of necessity, they forewent the first. Nor was the imitation, like many that are modern, sordid and commercial—a meticulous forgery. It was a brave impressionism, fit to satisfy the eye. The mind was deluded, if at all, then merrily, and for a moment.

An impartial spectator who found so much contrived—and so ingenuously—for his delight would, on taking thought, no more complain of all these substitutions—these false perspectives and painted shadows—than grow indignant because, in the Greek cornice, he is shown false eggs and darts. For this is no mere flippancy. Imitation runs through art; and Plato was more logical who rejected art, on this account, altogether, than are those critics who draw a line at the baroque. When we have imitated in one way long enough, our convention is accepted as such. The egg and dart moulding is a convention. The baroque habit is a convention also. It is objected that it is a convention which actually deceives and disappoints. But when we are familiar with it, and have ceased desiring to be shocked, this is no longer the case. Its critics, in fact, complain of the baroque that in it they encounter deceit too often; the cause of the complaint is that they have not encountered it enough.

Morally, then, Renaissance 'deceit' is justified. It does not follow on that account that *aesthetically* it is always equally to be admitted. If 'deceit' is carried beyond a certain point, we cease to get architecture and find stage decoration. There is nothing wrong about stage decoration; in its place there is not even anything aesthetically undesirable. It has a sole defect: that it fails—and must fail inevitably—to give us a high sense of permanence and strength. But these are qualities which are appropriate, above all others, in a monumental art; qualities,

真相的话，也不是什么困难的事儿，走到窗户跟前，看一眼立刻就知道了。没错，当我发现原先以为是石头的窗框其实是用抹灰做出来的，的确不免有几分失望，因为石头有它自己的美感和品质。巴洛克的建筑师也不是放着真实的石头不用，怪癖地一定要使用抹灰和涂料来仿造石头不可。帕拉第奥在维琴察（Vicenza）设计的王公府邸使用了抹灰以替代天然石材，我想帕拉第奥本人对这一点恐怕要比拉斯金更加遗憾。罗马的耶稣会圣伊纳爵教堂（S. Ignazio）里面的祭坛上使用了青铜和青金石（lapis lazuli）作为装饰，说明这个时期的建筑师和艺术家比其他任何时期都更重视使用名贵材料的美学价值，这一点是毫无疑问的。但是，这些艺术家和建筑师会根据建筑各个部分的重要性来决定那里的美学价值，当财力和物力受到制约的时候，他们会尽可能地保证设计构图的完整而采用替代的材料。名贵的材料本身因为珍贵而具有了其他材料所没有的想象空间和价值感，因为少见、因为来自遥远的地方、因为需要花费大量的人工才能寻找得到等等因素都增加了高级材料的价值。这些因素实际上远远超过材料本身的物理性质所带给人们的美感。考虑到需要花费大量的人力物力才能获得某些名贵的材料，巴洛克时期的建筑师们觉得有必要通过自己精湛的技艺来模仿制作出这些名贵材料的效果，即使他们以前用过这些真实的材料，现在也愿意采用模仿的手段。他们的这些模仿手段和我们今天以贪财营利为目的伪造仿冒商品不同，他们不是想欺骗我们；而只是为了创造出那样的一个效果而已，创造那样一个印象，让我们的眼睛获得一种满足。我们的意识便可以暂时从这种愉快的视觉效果中把它当作是真实的材料。

　　一个不带任何偏见的人，当他看到这么多巧妙的手段，有些甚至是天真幼稚的手段，例如近大远小的透视效果是做出来的，阴影是画上去的，等等；而这样做的目的完全是为了取悦于自己，他大概绝不会因为那些材料是模仿出来的而一个劲儿地抱怨。对于虚假的材料，如果他产生任何愤怒情绪的话，其程度大概也不会超过古希腊檐口线脚中那些蛋形装饰与枪头装饰所带给他的那种虚假感觉多少，因为那些蛋形的物体和枪头都不是真实的。我们在这里看到的所谓的虚假做法并不是故意睁着眼睛说瞎话的恶意举动。艺术中的模仿手段从来就存在，因为这一点，柏拉图倒是比这些理论家更为彻底，也更符合逻辑，他排斥一切艺术形式，而不是单单把巴洛克的艺术挑出来加以指责。当一种模仿形式持续了足够长的时间以后，它也就被人们所接受，成为一种习俗。古希腊檐口上的蛋形装饰与枪头装饰就是如此。巴洛克的这种采用仿造的材料来达到设计效果的做法也是这一类的行为，是一种习惯做法。然而巴洛克的习惯做法却被指控为欺骗、效果卑劣的坏习俗。但是，当我们熟悉了这种做法，同时也不再期待从中会有什么惊人的发现以后，在心平气和的时候，我们就会发现那些指控其实根本就不是那么回事儿。这些提出批评的人士抱怨说，在巴洛克艺术中总有被欺骗的感觉，那是因为他们还没有充分地接触这种艺术风格，对它了解得不够的缘故。

　　通过上面的论述，我们可以确认，文艺复兴时期建筑艺术中出现的这种所谓"欺骗行为"，从道德角度来讲是没有问题的，是站得住脚的。但是，我们还不能因为它在道德方面没有问题，便断定它在美学方面也就肯定没有问题了。如果这种"欺骗行为"超过一定的限度，那它就不再是建筑艺术，而变成了舞台布景。舞台布景本身也没有什么不对，舞台布景在美学追求上也不比其他艺术形式低人一等，它唯一的一点不足就是缺乏一种永久性的特质。这一点是无法避免的，它不能给予我们永久性与坚实的力量。永久性和坚实的力量正是我们在建筑艺术中所期待的特征和品质，这两点要比其他任何特征都突出。这么说来，某

therefore, which we have a right to expect in architecture. Here, then, is some justification for the theory that the *degree* of pretence is important. True, it is important aesthetically, and not morally, but it is important. But then the baroque style had the most penetrating sense of this importance. It recognised that the liberty to pretend—which the Renaissance had claimed from the beginning—though unlimited in principle, must be subject in practice to the conditions of each particular problem that the architect might undertake. It was a question of psychology. The scope of architecture, in a period as keenly creative as the seventeenth century, was a wide one; its influence was felt through everything that was made. The gaiety of life, no less than its solemn permanency, sought architectural expression. And the baroque style—the pre-eminent style of the pleasure-house, of the garden—was able to minister to this gaiety. The aesthetic pleasure of surprise may be a low one in the scale; but it is genuine, and not necessarily ignoble. And the same is true of the mere perception of dexterity. To obtain these, on their appropriate occasions, the thousand devices of baroque deceit were invaluable. Humorous or trifling in themselves, they gained an aesthetic interest and dignity because the unity of baroque style allied them to a general scheme.

Besides these ingenuities of the casino, the grotto, and the garden, there were architectural opportunities of a frankly temporary sort. There was the architecture of the *festa*, of the pageant, of the theatre. There was no reason why this should not be serious, supremely imaginative, or curiously beautiful. But it was not required to be, or seem, permanent. There was here no peril of that disappointment, which pretence involves, to the just expectations we form of monumental art. And these occasions, for which the baroque style remains unequalled, were an endless opportunity for architectural experiment. They were the school in which its psychologic skill was trained.

Last, there was monumental architecture. The resources learnt in the theatre must here be subject to restraint. Here we must hold secure our sense of permanence and strength. No falsities, no illusions, can here be tolerated that, when the eye discovers them, will lower our confidence in these qualities. But deceptions which pass unnoticed, and those which have no reference to stability and mass—deceptions of which the psychologic effect is negligible—may even here be admitted. The Parthenon deceives us in a hundred ways, with its curved pediment and stylobate, its inclined and thickened columns. Yet the sense of stability which it gains from these devices survives our discovery of the facts of its construction. The Italian mastery of optics was less subtle than the Greek, but it was put to wider uses. Perhaps the most familiar instance of its employment is in the galleries which connect St. Peter's with the colonnade of Bernini. Here the supposedly parallel

种程度上的摆出有气势的架子在建筑艺术中还是有必要的,也是很重要的一点。当然我们知道,这只是从美学和视觉来讲的重要性,不是道德上的,但是,它毕竟还是相当重要的。而巴洛克艺术风格对于这种重要性有着最深刻、最透彻的认识。它知道这种手段的极限在哪里,每一个具体作品都是建筑师根据实际情况酌情处理的结果。从文艺复兴时期开始,建筑师和艺术家们就清醒地认识到这一点;尽管从原则上讲,没有任何人会限制该如何如何做,但是,巴洛克的建筑师都恰到好处地把握住了它的程度。这是一种心理学方面的问题。在十七世纪这样一个极其富于创造力的时代,建筑艺术所包含的范围十分广阔,它所涉及的内容都必将受到巴洛克艺术的影响。生活的愉悦与永恒的性质都在建筑上面有所表现。巴洛克的艺术,包括建筑与园林景观,都反映了这种愉快的生活。对于意外惊喜的热衷可能在美学欣赏的标尺上处于一个很低下的位置;但是,这的确是当时的一种发自内心的真实喜爱,况且这种爱好也不是什么卑下的行为。在巴洛克艺术中应用的那些内容广泛的巧妙手段也是一样,没有什么不妥。获得这些技艺,并在适当的时机加以运用,这也是一门学问,这些数以千计的不同技巧是巴洛克艺术风格的宝贵财富。它们反映出来的幽默也好,轻浮也好,总之形成了一股强大的美学势力,并获得世人的尊重。它们的共同兴趣成就了巴洛克艺术风格的统一。

除了赌场、游乐场和公园以外,巴洛克建筑的新颖设计也出现在很多临时的设计方面,如欢庆场面的布置、比赛会场的布置、剧场设计等。这些场合也是十分严肃的艺术创作,需要极大的想象力,结果需要绝对地吸引人们的注意力。但是,它们都是过后便被拆除的临时设施,不需要对永久性加以考虑。没有人对它反映出来的那种永久性的东西提出任何批评。那些看上去很坚固的东西显然是虚假的,因为很快就会被拆除,但是它符合人们对于纪念性建筑物特征的期待。在这方面,巴洛克的建筑风格是其他风格所没有办法比拟的,这方面的需求也为巴洛克艺术提供了各种尝试的机会。这些临时性建筑成为心理学技能得到训练的大学校。

最后,我们这个社会需要大型的纪念性建筑物。从剧场布景中学习来的那些技艺在运用到大型建筑上面的时候需要格外小心,因为我们必须准确地把握住建筑中必须具备的永久性和巨大的力量。在这里,任何虚假的东西和利用视错觉产生的效果都是不能被接受的。而一旦其中的虚假成分和视错觉成分被发现,那么,我们对这个纪念性建筑的信心就会动摇。但是,某些手段在确定不会被察觉的时候,或者与坚固、耐久、体量等性质无关的内容,也就是那些对心理没有什么影响的内容,还是可以使用的。古希腊的帕台农神庙(Parthenon)通过上百种手段来影响我们的视觉效果,例如凸起的台阶、弯曲的屋檐山花、倾斜的柱子、加粗的柱子等。但是,即便是在我们发现这些艺术处理手段之后,我们从这座建筑所获得的整体感觉并没有受到任何影响,我们对这座神庙的坚固性没有任何怀疑。意大利巴洛克时期的这些艺术大师,在视觉调整的手法方面远不如古希腊的手法来得含蓄,但是,他们要比古代希腊在更多的场合使用这种视觉调整手段。最著名的例子要数圣彼得大教堂前面的两旁侧廊,它们连接着大教堂主体和伯尔尼尼设计的椭圆形大柱廊。两旁的侧

lines converge on plan and lengthen the perspective. This, indeed, is by no means a remarkably successful expedient, since what is gained for the eastward perspective is lost in that towards the west. But there is no loss of monumentality. The important point, realised by the architects of this period, is that, even in monumental architecture, the question of 'deceit' is one rather of degree than principle, rather of experiment than law. A design that is in the main substantial, and of which the serious interest is manifest, can 'carry' a certain measure of evident illusion and, needless to say, an indefinite amount of illusion which escapes all detection save that of the plumb-line and measure. An entire facade of false windows may be theatrical. A single such window, especially where its practical necessity is for any reason obvious, lowers in no sense our confidence in the design. Between these extremes the justifiable limits of licence are discoverable only—and were discovered—by experiment.

We have dwelt merely on a few conspicuous examples of the moral judgment in architecture, selecting for defence the worst excesses of the most 'immoral' of the styles. The main principle in all these matters is clear: the aesthetic purpose of the work determines the means to be employed. That purpose might conceivably give a clue to the nature of the artist—to his fundamental tendencies of choice. But we must understand it rightly. The moral judgment, deceived by a false analogy with conduct, tends to intervene before the aesthetic purpose has been impartially discerned. An artist may fail in what he has set before him, his failure may be a moral one, a recognisable negligence, but it is manifested, none the less, in an aesthetic failure, and is only to be discovered for what it is by a knowledge of the aesthetic purpose. It follows that we cannot look to the morality of the artist in his work as a criterion of the aesthetic value of the style.

Ⅳ. Thus far it may seem that whenever the criticism of architecture has taken moral preference as its conscious principle, it has forthwith led to confusion. Whether its method has been theological or utilitarian or intuitive, it has come to the same end: it has raised a prejudice and destroyed a taste without cause, logic, or advantage.

Are we then to say, with the critics on the other side, that moral issues are utterly different from aesthetic issues, and expel the moral criticism of architecture, its vocabulary and its associations, altogether from our thought? For this, we saw, has been the favourite retort, and this is the method which those critics who have an exacter sense of architectural technique have tended to adopt.

But among the consequences of the moral criticism of architecture, not the least disastrous has been its influence on its opponents.

We have, in fact, at this moment two traditions of criticism. On the one hand

廊在一般人的期待中是互相平行的,但是在平面布局的时候,设计师有意识地让它们形成一个角度,这样从大教堂望出去,广场的位置会让人感觉更加深远。但是,这个例子也不是一个十分成功的案例,因为向东望去所形成的透视效果,在向西望去则被抵消。但是,无论这种手段如何应用,一点也不影响整个建筑群的纪念性效果。这个时期的建筑师认识到非常重要的一点,那就是:即便是在纪念碑性质的建筑上,所谓的"虚假手段"也只是程度上的问题,绝对不是原则上的问题;是具体问题具体分析的实验性问题,而不是普遍适用的定律。具有重要性质的公共建筑也是可以利用一定的视错觉的手段的,更不用说,有数不清的这类细节其实早已逃过我们的眼睛和其他感官,有些东西只有借助于水平尺或者铅锤才能发现其中的偏差。假如整个建筑一面墙上的所有窗子都是假的,那么,这面墙就不是建筑,而是舞台道具了。假如只是其中一个窗子是假的,特别是当那个窗子的功能明显不需要的时候,那么无论如何也不会让我们对这个建筑失去信心的。在这两种极端情况之间的那些情况该如何拿捏,则完全视具体情况而定,只能在具体实践中找到答案。

我们只是有选择地就建筑艺术中几个突出的道德问题作了一番论述,为那些被指控为"最不道德的"艺术风格进行了一番辩护。在这些讨论中,我们遵循了一个显而易见的基本原则:一件艺术作品的美学目的决定了相应使用的手段。而这里的美学目的自然为艺术家提供了思路和线索,决定了他的艺术创作最基本的选择和取向。但是我们必须清醒地认识到,道德的判断是因为受到错误的类比关系,把艺术手段与人们的社会行为等同起来,结果在美学目的达到之前便从道德角度介入干扰,因此导致美学的目的得不到公平的对待。一个艺术家可能在他面前的障碍上摔倒,他摔的这一跤或许是道德上的判断失误,也可能是明显的意外;但是,不管怎么讲,最后还是要落实到美学的判断上来,他的失误最后要根据作品的美学目的来判断它的作品是否在美学上仍然是失败的。我们不能把艺术家的道德价值观作为评判标准来检验他的艺术作品,不能用道德标准来取代艺术风格的美学价值。

四、到目前为止,我们看到一个现象:只要是在建筑艺术的理论中掺进来一些道德的因素,并且以此作为一项原则,那么这种理论就必然会导致思想的混乱。不管这种道德因素是以何种方式介入到建筑艺术的,神学的也好、技术的也好,还是凭借直觉,最后的结果都是一样的:它直接导致某一种偏见的产生,把审美的过程给彻底地破坏了,让人们无法正常进行美学的判断。而这种破坏又是毫无道理、毫无逻辑,也没有产生任何好处的。

那么,我们是不是可以说,道德问题根本不同于美学问题呢?是不是应当把道德问题的相关语汇、与之相关联的种种因素都从我们的思想中清除出去呢?我们知道我们那些道德理论家肯定是不会同意这个看法的。这种近乎拌嘴式的反问听起来很熟悉,这正是技术决定论那一派具有严密科学分析方法的理论家所极力倡导的观念,也是他们的理论所坚持的基本点。

以道德标准作为审美标准的建筑艺术理论的后果之一就是让它的反对者们受到灾难性的打击。

到目前为止,我们基本上是面对着两种理论传统。一种就是我们在这一章里着重讨论

there is a tradition in which the errors examined in this chapter find their soil; a tradition of criticism constantly unjust, sometimes unctuous, often ignorant; a tradition, nevertheless, of great literary power. Into this channel all the currents of the Romantic Fallacy, all the currents of the Ethical, flow together. It is the Criticism of Sentiment.

On the other hand is a body of criticism sharply opposed to this. It has two forms: the '*dilettante*' —in the older and better sense of that word—and the technical: two forms, different indeed in many respects, but alike in this—that both are *specialised*, both are learned and exact and in some sense cynical. They derive their bias and their present character from an obvious cause: a sharp reaction, namely, against the Criticism of Sentiment. The amateur, the pedant, the mechanic, have always existed; but, until the Criticism of Sentiment arose, their exclusiveness was a matter of temperament and not of creed. On the contrary, the older 'pedants,' with Vitruvius at their head, claimed every kind of moral interest for their art, and were fond of arguing that it involved, and required, a veritable rule of life. But the exacter criticism of our own time, in natural disdain for the false feeling and false conclusions of the opposite school, restricts the scope of architecture to a technical routine, and reduces its criticism to connoisseurship. This, then, is the second tradition: the Criticism of Fact.

The consequences, for the criticism of sentiment, of its lack of exact knowledge and disinterested experience in the art of architecture, have already been set out. But what are the results, for the critics of 'Fact,' of their aversion —historically so justified—to the methods of 'Sentiment'? The results are clear. The appreciation of beauty, cut off from the rest of life, neither illuminates experience, nor draws from experience any profundities of its own. It loses the power to interest others, to influence creation or control taste: it becomes small and desiccated in itself. And another result is equally apparent. Appreciation, thus isolated, discriminates the nice distinctions of *species*, but loses sight of the great distinction of *genus*: the distinction between the profound and the accomplished. An accurate and even interest studies Francois Boucher with Bellini; an equable curiosity extends itself indifferently to the plans of Bramante and the furniture of Chippendale. For, in the last resort, great art will be distinguished from that which is merely aesthetically clever by a nobility that, in its final analysis, is moral; or, rather, the nobility which in life we call 'moral' is itself aesthetic. But since *it interests us in life as well as in art*, we cannot—or should not—meet it in art without a sense of its imaginative reaches into life. And to separate architecture, the imaginative reach of which has this vital scope—architecture that is profound— from architecture which, though equally accomplished, is nevertheless vitally

的这种错误理论所赖以生存的土壤。这种传统从来就是蛮横不公正的,有时候看上去虚情假意,多数情况下傲慢无知;但是这种传统有着巨大的文学艺术上的力量。这种传统中包含了浪漫主义的谬误,包含了各种各样的道德决定论理论,这些流派汇集到一起形成了一股洪流潮水。我们可以把这一传统概括为"强调感性的艺术批评理论(Criticism of Sentiment)"。

与上面的各种理论形成鲜明对比的理论组成了另外一种传统。这类传统表现为两种形式:一种是艺术爱好者(dilettante)所提出的理论,一种是过于注重技术的理论。这两种理论在表面上看起来有很多的不同,但是二者有一点非常相近:它们都过于专门化,两种理论都很有艺术以外的学问,都讲究严谨精确,在某种程度上,二者都有几分瞧不起别人的态度。他们的偏见和性格特点基本上来自一个地方:它们都强烈地反对那些强调感性的艺术理论。业余爱好者、专业的学究、技术专家,这些人从古到今从来没有断过,但是,直到感性的艺术理论兴起之前,他们的孤傲还只是个人的秉性、脾气偶然发作而已,还没有结成一个宗派。而且几乎相反,不但没有结成宗派,而且过去的学究还愿意将各种道德的说教囊括在自己的理论著作中,维特鲁威应该算是这类学究的祖师爷。他们喜欢在自己的理论中说明艺术与生活中的规律是一致的,也需要生活中的规律对艺术加以规范。但是,我们今天的这些严谨理性的理论家们因为出于对另外一种理论传统的敌视和厌恶,把自己的理论严格限定在技术的范围之内,把艺术批评理论变成了一种类似技术鉴定的规则。这一类就是第二种传统,我们称之为"强调事实的艺术批评理论(Criticism of Fact)"。

至于"强调感性的艺术批评理论",因为它们缺乏严谨的理性知识,缺乏对建筑艺术的公正认识,它们所导致的后果我们已经详细地论述过了。至于"强调事实的艺术批评理论"的传统,它憎恨感性的结果又是怎样的呢?结果也是十分明显的。对于美的欣赏和理解,如果把它与生活的其他方面面割裂开来,那么这种欣赏和理解也就不能充分理解人们对于美的体验和感受,也不能从这些亲身体验与感受中领悟出美的深刻含义。这种离开感性体验而发现的美缺乏感染别人的力量,不能激发别人的创造力,也无法影响世人的审美观念与审美爱好。这样的理论只是让自己变得越来越渺小,最后变得脱水、干枯。另外还有一个结果也是显而易见的。这种与周围其他因素割裂的审美体验只是注意到*物种*之间的差别,却忘记了这些物种之上还有不同的*属性*:也就是说,有的艺术作品是具有深远影响的作品,有的只是完成制作得不错而已。如果用精确的分析方法,平等地对待十八世纪的弗朗索瓦·布歇(Francois Boucher)与十五世纪的齐奥瓦尼·贝利尼(Bellini)的艺术作品,那么我们就只会看到两种作品各自的特点;但是把它们联系到一起的时候,我们就能够看出二者其实是属于不同的属性的。如果是只见树木不见森林地研究,那么伯拉蒙特的建筑设计与十八世纪英国的古董家具大概没有什么差别了。但是,归根到底,一件伟大的艺术作品与那些手艺精湛的产品之间还是有很大差别的,它们之间的不同就在于伟大的艺术作品最根本的特征是其具有崇高的品质,到最后这又是道德的判断问题。或者换句话说,生活中那些被我们称为"道德高尚"的崇高品质,其本身同时也是美的。由于*一件伟大的艺术作品从艺术角度以及从生活角度对我们都产生作用,都在影响着我们*,所以,我们不可能,也不应该,在研究艺术作品的时候脱离生活;在面对艺术作品的时候不会联想到它的社会

trivial, is a necessary function even of aesthetic criticism.

There is, in fact, a true, not a false, analogy between ethical and aesthetic values: the correspondence between them may even amount to an identity. The 'dignity' of architecture is the same 'dignity' that we recognise in character. Thus, when once we have discerned it aesthetically in architecture, there may arise in the mind its moral echo. But the echo is dependent on the evoking sound; and the sound in this case is the original voice of architecture, whose language is Mass, Space, Line, and Coherence. These are qualities in architecture which require a gift for their understanding and a trained gift for their understanding aright: qualities in which men were *not* 'intended without excessive difficulty to know good things from bad,' and by no means to be estimated by the self-confident scrutiny of an ethical conscience; qualities, nevertheless, so closely allied to certain values we attach to life, that when once the aesthetic judgment has perceived them rightly, the vital conscience must approve, and by approving can enrich. To refuse this enrichment, or moral echo, of aesthetic values is one fallacy; the fallacy of the critics of Fact. To imagine that because the 'conscience' can enrich those values it has, on that account, the slightest power, with its own eyes, to see them, is the contrary, the Ethical Fallacy of taste.

Morality deepens the content of architectural experience. But architecture in its turn can extend the scope of our morality. This sop, which that Cerberus unchastised shows little disposition to accept, may now be proffered in conclusion.

Values (whether in life or art.) are obviously not all compatible at their intensest points. Delicate grace and massive strength, calm and adventure, dignity and humour, can only co-exist by large concessions on both sides. Great architecture, like great character, has been achieved not by a too inclusive grasp at all values, but by a supreme realisation of a few. In art, as in life, the chief problem is a right choice in sacrifices. Civilisation is the organisation of values. In life, and in the arts, civilisation blends a group of compatible values into some kind of sustained and satisfying pattern, for the sake of which it requires great rejections. Civilisation weaves this pattern alike in life and in the arts; but with a difference in the results. The pattern that is realised in conduct is dissipated with each new experiment; the pattern that is realised in art endures.

Our present experiment in democratic ethics may be the best which the facts of life afford: or it may not be the best, and yet be necessary. But, in either case, though morality in action may stand committed to a compromise, the imagination of morality need have no such restrictions. It should have some sense of the values it is forced to subordinate or to reject. Of those values the arts, enduring from the past,

意义，这是不可能的。美学理论最基本的作用之一就是要在意义深远的建筑艺术作品与平庸的作品之间作一个区分。

在道德价值和美学价值之间，事实上的确存在着一种对应关系，这种关系是正确的类比，不是错误的，这种关系甚至将二者看成是相互对等的：那就是两种价值体系中都具有的"尊严"。建筑艺术作品里的尊严与我们在生活中所看到的尊严实际上是一回事儿。因此，当我们在建筑艺术中感受到那种尊严存在的时候，我们会在自己的意识中很自然地联想到人的品德中所包含着的同样的特质。但是这种联想取决于建筑艺术作品的冲击力，意识中的联想是视觉中艺术作品的回音。建筑艺术作品发出的声音是借助于建筑艺术固有的语言的，亦即体量、空间、线条、和谐的关系等。建筑艺术中有某些品质是需要具备一定的能力才能够理解的，而专业训练正是获得这种能力的途径：这种能力绝对不是像有人说的那样，"可以毫不费力地识别出什么是好的，什么是坏的"，更不可能想当然地凭着自己的主观愿望，按照道德的标准去衡量建筑艺术就可以做到的。这种品质与我们生活中的某些价值观念紧密联系在一起，只要是在美学判断上认为它是可以接受的时候，我们会自然而然地在道德良知上确认它的正当性。而这种从道德良知方面对它的肯定也丰富加深了我们的美学体验。否认这种美学体验因为道德的回应而得到深化，这就是美学价值判断中的一种谬误，是片面强调事实的艺术批评理论的谬误。同时如果因为道德良知可以丰富加深我们的体验而夸大道德的作用，借此贬低我们通过眼睛来亲自感受建筑艺术的作用，认为眼睛看到的东西微不足道，这也是一种谬误，是道德决定论的谬误。

道德价值观念加深了建筑艺术体验的内容，建筑艺术也反过来扩大了道德观念的范围。建筑艺术理论中的各种谬误就好比是看守地狱大门的那个凶猛多头犬（Cerberus），它对我们刚刚得到的这个结论似乎没有兴趣；我们不指望能够打动它。但是，这个结论可以用来给本篇做一个概括。

各种价值观（艺术中的和生活里的都一样）不可能在各自最为膨胀的时候会与其他的价值观兼容。精巧纤细与粗犷坚实、冷静安详与冲动冒险、威武庄重与滑稽幽默等等，要想与对方同时存在的话，那么两方面都要做出一定的妥协才行。伟大的建筑与伟大的人物是一样的，并不是在囊括一切的情况下来成就自己的，而是在有限的几个方面做到了极致。所以，艺术也和人生一样，最重要的课题是懂得正确地放弃和牺牲一些东西。一种文明就是一个充满了各种价值观念的有机体。人类生活中的一种文明把各种彼此兼容的价值观混合在一起，形成了一种可以延续下去的生活模式；为了保持这样的生活模式，很多东西是必须加以排斥的。艺术也是如此。文明在生活中形成了某种可以让社会得以持续的模式，文明也在艺术中形成了同样的模式，但是两种模式的结果却大不相同。在社会生活中形成的行为模式每一次都是全新的尝试，而在艺术中形成的模式则会保持不变。

在今天，民主的道德价值观念是正在进行中的一种尝试。它或许是我们从生活中所能够设想出来的最好的一种模式；也许它不是最完美的，但是，它是必不可少的。无论是哪一种情况，道德观念在实际生活中是随时准备接受妥协、可以打折扣的，但是在思想中的道德需求是不会这样做的。理想中的道德价值观念强迫我们必须在服从与反对之间作一个抉择。艺术从过去传承下来的那些价值观就保持着这样一种特征，我们要么继续延续它，要么

retain the impress.

Without the architecture—together with the poetry and other arts—of the Greeks, we should have a poorer conception, even morally, of the possible scope and value of balance and restraint; without the architecture of the eighteenth century, a poorer sense, even morally, of the possible scope and value of coherence—of a fastidious standard consistently imposed; without the architecture of the Renaissance, a far poorer sense of the humanist conviction: the conviction that every value is ideally a good to be utterly explored, and not indolently misprized—the conviction which spurred the Renaissance builders, as it spurred their painters and their thinkers, to attempt, in a sudden and ardent sequence, the extremest poles of opposite design, and in each attempt to discern for a brief instant the supreme and perfect type: a humanist passion which made of architecture the counterpart of all the moods of the spirit, and while, Cortez-like, it laid open the round horizon of possible achievement, never disowned allegiance to a past which it deemed greater than itself.

坚决反对它。

　　假如没有古代希腊的建筑艺术，连同古代希腊的诗歌艺术等其他艺术形式，那么，我们对于什么是平衡与取舍以及它们所涉及的范围与价值的认识就会很缺乏。假如没有十八世纪的建筑艺术，那么，我们对于什么是整体和谐以及它们所涉及的范围与价值的认识就会很缺乏；而和谐是一项十分难达到的标准，却又无时无刻不在提醒着我们。假如没有文艺复兴时期的建筑艺术，那么，我们就会更缺乏对于人文主义精神的认识以及对人文主义的信念；这种信念让我们执著地认为，各种价值从理想上来讲都是值得我们怀着热情去亲自探索的，而不是以旁观者的姿态，躲在那里懒惰地蔑视它。正是这种人文主义的信念让文艺复兴时期的建筑师们如同那些杰出的画家、思想家一样，满腔热忱地投入到各种尝试中去。各种设计的倾向都被尝试过，甚至非常极端的设计手段也不放过。在每一种尝试中都争取做到当时最出色、最完美：这就是人文主义的热情，它把当时的各种思想与精神都反映在建筑艺术上。这个时期的建筑艺术，如同西班牙殖民主义者科尔蒂斯（Cortez）那样，开疆辟土，力图在各个方向都最大限度地扩大自己的殖民地，但是同时也从来不忘记对过去历史的尊重，因为他们相信，历史上的伟人们，在从前已经创造出比自己更辉煌的业绩。

SIX
The Biological Fallacy

Of all the currents that have lapped the feet of architecture, since architecture fell to its present ruin, the philosophy of evolution must be held to have been the most powerful in its impulse, the most penetrating in its reach. The tide of that philosophy, white with distant promises, is darkened, no less, by the wreckage of nearer things destroyed. Have these waters, then, effaced the characters which, upon the walls of architecture, Romance overlaid with others of its own, Science disfigured, and Ethics sought falsely to restore?

So long as the sequence of Renaissance styles continued unbroken, the standards by which architecture was judged grew and developed with architecture itself. A formative force took possession of critical taste, while it controlled creative power. The large outline of tradition stood fast; but, as within it shape succeeded shape, reason—with due conservative cries and proper protests—yet followed, understood and sanctioned. Style dictated its own criterion; taste accepted it. The past died because the present was alive. Style itself, and not *the succession* of styles, engrossed men's thought. The sequence, *as a sequence*, was not studied. But when, in the nineteenth century, the sequence was cut short and a period of 'revivals' was initiated, the standards of taste were multiplied and confused; past things became contemporary with present. Sequence—the historical relation of style to style—now was studied, when sequence itself had ceased to be. If the different stages of a historical evolution are brought simultaneously to life—if only to the life of chattering spectres—style no longer can affirm its nights unquestioned. Claims that once were owned must then be adjusted, challenged and compared. When architecture, once a clear directing voice, is heard to speak 'with tongues' forgotten and confused, men must hearken for interpretation, and find it, then, in the sound of every passing gust of thought.

第六章
基于生物进化论的艺术理论之谬误

自从建筑艺术变成今天这种废墟状态以来,已经有很多不同的大潮冲刷过它的脚面。从冲击力的强度和渗透程度来看,进化论的观念应该是产生过最强烈效果的一股大潮。当这个进化论观念衍生出来的艺术理论形成的时候,那股潮水从远处涌来,泛着白色的光亮,给人们带来了无限的希望和期待,但当它冲到我们眼前的时候,展现在我们面前的是它沿途所摧毁那些物体的残骸,黑压压的一片。黑暗的浪头在气势上一点也不亚于潮水刚形成时候那泛着白色光亮的大浪。浪漫主义艺术理论曾经把自己的观念强行罩在建筑艺术的外表面上;崇尚科学的艺术理论则扭曲了我们对建筑艺术的认识;把道德标准作为艺术准则的理论则要在建筑艺术中恢复到它们对历史的错误理解上来。这一切的因素,当它们作用在建筑艺术上的时候,必然改变了建筑艺术的面貌。那么,经过这些大大小小潮水的冲刷和洗涤以后,我们自然要问,建筑面貌上的这些被改造过的痕迹已经被冲刷掉了吗?

只要文艺复兴时期各种建筑艺术风格是不间断地在演变着,那么建筑艺术的评判标准也就会随着建筑本身的演变而不断成长,会变得越来越成熟。最开始的形成过程决定了审美观念和美学理论的倾向性,左右了艺术家的艺术创造力。艺术传统在整体上保持着稳定的状态,但在这个稳定轮廓之下,各种形式则不断地交替出现。伴随着这些不断交替出现的各种形态,保守的声音和抗议的声音也就跟着出现。但是,理性的分析暂时还没能够跟上步伐,出现的理论更没有被理解、被接受。建筑艺术风格在没有理论标准的情况下,风格本身就是它自己的评判标准,建筑风格是否受到人们的喜爱成了审美爱好的唯一衡量标准。过去的旧风格被淘汰完全是因为有最新的东西出现了。这些彼此不同的建筑风格自身是人们所关注的全部内容,这时的人们不会去考虑这些不同风格的连续性与变化规律。演变过程作为过程没有得到足够的重视和研究。但是到了十九世纪,这种风格的转变过程被人为地大大缩短了,历史上有过的各种"建筑风格"这时不断地被人们重新"复兴"回来;人们的审美习惯与爱好开始被搞混乱了,各种审美标准也跟着成倍的出现,自然更加重了这种混乱。本来明明是过去的旧东西,现在却和当代的新东西同时出现在建筑物上。一个风格与另一个风格之间的历史关系,也就是它们之间的次序和演变过程开始受到关注的时候,建筑风格演变过程本身却已经不复存在了。在历史演变过程中曾经出现过的不同时期、不同阶段的风格,现在却如同鬼怪显灵一样,同时粉墨登场,存在于某些现实生活中。风格本身不再具有自己的权威性,人们开始对有些建筑风格的正当性产生怀疑。过去与某种艺术风格相关的结论现在则必须要加以调整、被质疑、被拿来与其他风格的结论相比较。在过去的某一个时期里,它的建筑艺术是不言自明地只有一个方向,一种声音;但现在则是南腔北调都在发出各自的声音,有些还是早已被人遗忘的声音,而这些不同的腔调所带来的信息也是含糊不清,令人费解的。人们无法听清楚这些声音,只好依赖于旁人对它的翻译解说,结果发现那

Three such sounds in the wind were those we have examined, each of them borne from a source remote from architecture itself. Poetical enthusiasm, the zeal and curiosity of science, the awakened stir of a social conscience, are voices in the criticism of architecture still to be discerned. But the philosophy of evolution—vast in its sweep, universal in its seeming efficacy, and now less an instrument of science than a natural process of the unconscious mind —was a steadier wind more strong than these. What has been its bearing upon the appreciation of Renaissance architecture? Has it assisted us, or not, to see its value as an art and to judge it for that third condition of well-building—its 'delight'? It is the gain and loss which 'evolution' has brought to taste that now must be computed.

In one sense the gain has been obvious. Of the evolutionary influence on criticism the most evident result has been a wide enlargement of our sympathy.

A sharply-defined circle formed the limit of eighteenth-century vision; within it, all was precisely seen, brilliantly illuminated; beyond it, outer darkness. That sympathetic traveller, the Président de Brosses, has nothing to say of the paintings of Giotto save that they are '*fort mauvaises*' ; Goethe, even, at Assisi, does not remark on them at all; nor on the two churches of St. Francis: the vestiges of the classic temple engage all his attention. The architectural histories of the time, after citing a few historic landmarks like the Tower of Babel, hasten on to the business in hand—the 'better manner' of their own day. Step back from '*le grand siècle*' and you are in '*le méchant temps.*' And when the obligations of devotion compelled these fastidious amateurs to pass an hour beneath a Gothic groin, they took care, at least, that a festive chandelier should hang from it to provide a haven for the outraged eye, and that richly scrolled and classic woodwork should accommodate the physical requirements of their piety. Secure in the merits of 'the better manner' they neither sought, nor were able, to do justice to the past.

The release from this contracted curiosity was brought about by two main causes. It was brought about, aesthetically, by the Romantic Movement. It was brought about, intellectually, by the philosophy of evolution. The Romantic Movement placed a poetic value, for its own sake, on the remote. The philosophy of evolution, with its impartial interest in all things, placed a scientific emphasis, for its own sake, upon *sequence*. Both these were enlargements of our curiosity.

每一阵阵吹过来的大风中,夹杂着各式各样的思想。

　　大风中夹杂着的三种声音我们已经在前面讨论过了,每一个声音都是从距离建筑艺术十分遥远的异域土地上,从空中传播过来的。对诗歌艺术的热情讴歌,对于科学的崇拜和好奇,对于社会良知的觉醒,直到今天,这三种声音对我们建筑艺术理论的影响仍然是显而易见的。但是,进化论的思想是最强劲并且最持久的大风,它要比上面提到的这三种声音都更加猛烈。这股进化论狂风席卷地球各个角落,而且具有放之四海而皆准的架势,谁也不能置身事外。但是在今天看来,这种思想已经不再是那么强有力的科学工具,更多的是它已经成为人们潜意识里的思维模式与思考过程。这种进化论的思想对于我们理解和欣赏文艺复兴时期的建筑艺术有什么样的作用呢?关于这个时期的建筑作为艺术有什么样的价值的问题,进化论的思想是否曾经对我们有所帮助呢,还是根本就没有任何作用?也就是说,对于"优秀建筑"必须具备的条件中第三个条件,亦即建筑使人感到愉悦这个条件是否有所帮助呢?进化论思想对艺术审美习惯的影响,是收益还是折损呢?因此我们必须首先要厘清建筑艺术审美习惯的现状。

　　从一个角度来看,进化论思想对建筑艺术审美习惯有所帮助,这一点是很明显的。在它对艺术理论的各种影响中,最明显的一个结果就是它极大地扩大了我们对艺术理论的兴趣与认同。

　　围绕着十八世纪建筑艺术界的整体情况,我们面前所看到的是一个界限分明的圆圈。圈子里面灯火通明,光亮耀眼,一切都看得清清楚楚;圈子以外则是漆黑一片,什么也看不见。那位对于自己所见所闻都报以同情心的作家查尔斯·德·布罗西斯(de Brosses),人称布罗西斯议长(他曾出任家乡小镇议会的议长),在看过乔托(Giotto)的绘画作品之后,却说那些作品除了"极其糟糕"之外,没有别的什么好说的;歌德在造访过阿西西(Assisi)之后,也就是圣方济各教派的发源地,他也没说过什么。甚至连圣方济各本人兴建的那两座小教堂也没能引起他的兴趣,但是古希腊、古罗马时期的神庙遗迹则让他流连忘返,每一个细节都不放过。这个时期出现的建筑历史书籍在罗列几个历史古迹,例如巴比伦通天塔(Tower of Babel)之后,便很快转向讨论作者们最关心的自己所处时代的"更好的艺术手法"。当你从这个被法国人标榜为"伟大的世纪"的十七世纪(le grand siècle)后退一步,就进入了"邪恶的时代"(le méchant temps)。当宗教信仰让这些普通的信徒不得不来到哥特风格的教堂祈祷,被迫花上一个钟头与上帝交流思想的时候,他们至少会在哥特教堂的拱券上悬挂起一个华丽的大吊灯;这样他们受到冒犯的眼睛终于可以停留在一个能够集中注意力的地方,那些雕梁画栋的木制装修可以为他们的虔诚提供一些物质上的烘托。理论家们对于自己时代的"更好的手法"太有信心了,他们根本不会去关注过去,不会去探究其中的缘由,实际上他们也没有能力去做这件事情。

　　改变这种对一切都失去好奇心、失去兴趣的状态,并且从中挣脱出来的原因主要来自两个方面:在美学方面,浪漫主义艺术运动改变了这种死气沉沉的状态;在理性方面,进化论的思想让人们重新认识了艺术规律。浪漫主义艺术为了自己的利益,在建筑艺术领域里大力推行诗歌所代表的遥远时空的意境;而进化论思想也是为了自己的利益,着重强调结果之前的一系列*演变过程*,但是它一视同仁地在所有领域里提倡科学的观念。这两种思想方法扩大了我们对建筑艺术的兴趣与看问题的视野。

But the Romantic enlargement fails because, although it finds an aesthetic value in the past, the value it finds is too capricious and has no objective basis. And the evolutionary enlargement fails because it is not interested in 'value' at all. It does not deny that values exist, but it is of the essence of its method that it takes no sides—that it discounts value and disregards it. The intellectual gain is effectively a loss for art.

The object of 'evolutionary' criticism is, *prima facie*, not to appreciate but to explain. To account for the facts, not to estimate them, is its function. And the light which it brings comes from one great principle: that things are intelligible through a knowledge of their antecedents. *Ex nihilo nihil fit*; the nature of things is latent in their past. The myriad forms of architecture fall, by the compulsion of this principle, into necessary order. The interest of the study shifts from the terms of the sequence to the sequence itself. In such a view there is no place for praise or blame. The most odious characteristics of an art become convenient evidences of heredity and environment, by means of which every object can be duly set in a grand and luminous perspective. This tendency of the mind was a needed corrective to the Ethical Criticism; and the clear light of philosophic calm replaces, in these expositions, the tragic splendour of denunciatory wraths. Nevertheless, the direction of the tendency is unmistakable. It is a *levelling* tendency. The less successful moments of the architectural sequence have an equal place with the greatest. More than this, the minor periods, the transitional and tentative phases, acquire, when our interest is centred in the sequence, a *superior* interest to the outstanding landmarks of achieved style. For the intellectual problem is, precisely, to connect these landmarks with one another and with their obscure origins. Hence not in architecture alone, but in many other fields of study—in religion, for instance, and mythology—a sharp prominence is given to what is primitive and submerged, at the expense, inevitably, of the classic points of climax. When there is prominence there is soon prestige. The coldest scrutiny must recognise one value—namely, intellectual *interest*; and interest takes by degrees the place of worth. Thus the ennobled cult becomes for us the bloody sacrifice, civilised usage a savage rite, and the Doric temple justifies its claim on our attention by reminding us that it was once the wooden hut. The question is no longer what a thing ought to be, no longer even what it *is*; but with what it is connected.

But Renaissance architecture is a very unfortunate field for the exercise of this kind of criticism, for the reason, already established, that it was an architecture of taste; an architecture, that is to say, which was not left to develop itself at the

浪漫主义的参与虽然带给我们对于过去历史的兴趣,但是它的价值观却让它看问题的重点集中在异想天开、喜怒无常的怪异现象上,缺乏具体真实的客观基础与目的,因此它在建筑艺术领域里的影响最后失败了。进化论的思想则是因为强调过程,而对于艺术本身就没有表现出任何兴趣。它并不否认价值的存在,但是由于它的思想重点在于演变过程,强调对于前后需要不偏不倚,不能对任何一方有所偏爱,这也就决定了它的方法不能抓住重点。因此这种进化论的思想是降低了艺术价值的意义,到最后终于抛弃了艺术本身。抽象的理性似乎取得了胜利,但这种胜利却让艺术蒙受惨痛的损失。因此,进化论思想在建筑艺术领域里的影响最后也以失败告终。

由"进化论思想"衍生出来的艺术理论给我们的第一印象是,它所关心的重点不是欣赏,而是着重进行解释。它的作用在于历数过程中的各种事实,而不是在于评估每一个事实的价值。进化论思想带给建筑艺术的一道光明就在于它的一个最伟大的基本原则:任何事物都是可以为人们所掌握的,掌握的途径就是了解它的前身。事出必有因(Ex nihilo nihil fit),每一件事情的本性都隐藏在它自己的过去。根据这一个基本原则,建筑艺术中的无数风格样式就必然构成了一种秩序,对于这些风格样式的研究逐渐被忽略,取而代之的则是对这些风格样式之间演变过程的研究。根据这样一种进化论的观念,没有一种建筑风格值得特别的赞誉,也没有另外一种会受到责难。这样一来,艺术中最丑恶的特征变成了从上一代遗传下来的结果,也成了今后的生长环境,然而每一种艺术特征都有理由迈向一个宏伟与光明的未来。这种思想倾向对于克服用道德标准来衡量建筑艺术的艺术理论是必不可少的,它富有哲理的基本原则在不慌不忙的论述中带给人们一片明亮的天空,完全不同于道德理论中那种充满了愤世嫉俗的责难与悲愤的语气。不管怎样,这种理论所揭示的演变趋势则是明确的。那是一个把一切都抹得平平整整的趋势,建筑艺术那些很平庸的作品与最辉煌的作品,在整个演变过程中都具有相同的价值与作用,都有自己的位置。不仅如此,由于在进化论的影响下,人们考虑问题的重点放在了演变的过程;因此那些次要的阶段,也就是那些具有过渡性质的、试验性质的阶段获得了极大的重视,而那些取得了辉煌成就的时刻则被有意无意地忽略了。因为理性关注的焦点在这时恰恰就是如何把这些成就辉煌的时刻联系起来,让彼此找到相互关系,进而找到它们被隐藏的起源。不仅仅在建筑艺术领域里是这样,很多很多领域,例如在宗教界、在神话研究领域,都是如此。十分突出的一点就是大家把研究的重点放在了最原始的状态,放在了那些不易被人注意的环节,其代价是那些达到成就高峰的阶段则被忽略。当某一阶段成为了研究的重点,它也就成了被关注的对象,因此便具有了一定的优越感,名声也变得显赫起来。最冷酷无情的深入追究,最终一定会找到一种有价值的结果,这就是理性追求的目标所在,目的在一定程度上取代了价值判断。这种研究方式的结果是:高尚的宗教在我们的眼里变成了血淋淋的牺牲;文明的行为方式变成了野蛮的祭奠仪式;多立克神庙庄严的造型之所以能够令我们心驰神往是因为它起源于早期的木头棚子。问题的焦点不再是事物本身应该是怎样的,甚至不关心它们*目前的状态*是怎样的,一切的一切都集中在它们过去是与什么相关联的。

但是,采用这种艺术理论来检验建筑艺术,文艺复兴时期的建筑艺术是一个非常不幸的领域,因为这个时期的建筑艺术是根据人们的审美爱好人为地兴起的,就是说,它并不是盲目被动地遵循进化论规律演变而来的。它完全不顾在它之前那个时代的遗产,文艺复兴时

blind suasion of an evolutionary law. It cast off its immediate past and, by an act of will, chose—and chose rightly—its own parentage. It scorned heredity; and, if it sometimes reflected its environment, it also did much to create it. It could change its course in mid-career; it was summoned hither and thither at the bidding of individual wills. Brunelleschi, at its birth, searching with Donatello among the ruins of Rome, could undermine tradition. Michaelangelo, independent of the law as Prometheus of Zeus, controlled its progress more surely than did any principle of sequence. And the forces which he set loose, a later will—Palladio's—could stem, and the eighteenth century revoke. Here was no procession of ordered causes, but a pageant of adventures, a fantastic masque of taste.

With what result for criticism? Because Renaissance architecture fits ill into the evolutionary scheme, it is on every side upbraided. Because its will was consciously selfguided, it is called capricious. Because it fails to illustrate the usual lessons of architectural development, it is called unmeaning. Because there is no sequence; because the terms are 'unrelated'—or related not strictly, as in the older styles, by 'evolution'—the terms are *ipso facto* valueless and false. A certain kind of intellectual interest is frustrated: *therefore* aesthetic interest is void. This is the evolutionary fallacy in taste.

At its hands, as at the hands of the Romantic Fallacy, Renaissance architecture suffers by neglect and it suffers by misinterpretation. It suffers by neglect: the historian, committed to his formulas of sequence, is constrained to pass hurriedly by a style which fits them so ill and illustrates them so little. But it suffers also by misinterpretation, for that slight account of the Renaissance style which is vouchsafed is given, as best may be, in the formulas of the rest. It is drilled, with the most falsifying results, into the lowest common terms of an architectural evolution. The prejudice to taste is not merely that facts are studied rather than values; it is not merely that the least worthy facts are studied most, and that the stress falls rather on what is historically illuminating than on what is beautiful. The prejudice is more profound. For evolution was schooled in the study of biology; and historical criticism, when it deals in values at all, tends unconsciously to impose on architecture the values of biology. Renaissance architecture is blamed, in the general, because it is self-guided and 'arbitrary'; yet it is condemned, in the particular, by the unjust dooms of 'necessary' law. Let us take a typical

期的建筑艺术把它之前留下来的东西统统抛弃,而且是有意识地、主动地这样做的,然后根据自己的审美爱好选择了自己的父母和血缘。文艺复兴时期的建筑艺术,对于血统的遗传持一种嘲讽蔑视的态度。如果说这时的建筑艺术作品有时候反映了当时的一些实际情况的话,那些被反映出来的实际情况也基本上是这些人自己创造出来的。文艺复兴时期的建筑艺术在创造过程中,不会自始至终贯彻一个想法、一个概念,很有可能在中间的某一个时刻,因为某一个人的突发奇想而改变原来的设想,前一个时刻还可能是如此这般,转眼之间就可能变成了另外一种形式。在文艺复兴时期的初期,布鲁乃列斯基与多纳特罗（Donatello）一起在罗马城里从古代遗留下来的残垣断壁中寻找灵感与艺术语言,这种举措已经开始让当时的传统习惯做法遭到一定的打击。米开朗基罗则是如同遭到宇宙大神宙斯（Zeus）惩罚的普罗米修斯一样（Prometheus）,打破当时所有的传统规律,在自己的创作中根据自己的意志,任意地改变原有的逻辑与演变规律。自米开朗基罗开始的,这种根据艺术家自己的意志,在设计中自由地发挥自己想象力的艺术实践在后人,比如帕拉第奥,那里得到进一步的发扬光大,直到十八世纪的时候才告终止。在这个过程中,我们看不到前后演变的必然逻辑关系;我们看到的是大家在争奇斗艳般地发挥着各自的想象力,是各种审美爱好不受约束地展现着自己的幻想。

　　面对这样的艺术场面,艺术理论所得出的结论又是什么呢？因为文艺复兴时期的建筑艺术实践根本不符合进化论的规律,因此,这一时期的艺术在各个方面遭到责难。由于这一时期的建筑艺术完全是按照人的意志来决定自己的方向的,因此,这种建筑艺术被说成是胡乱编造的怪想;由于它不能与其他的建筑艺术演变规律合拍,因此它被说成是毫无意义的;由于它与此前的艺术完全脱节,没有延续它之前的传统,不符合"进化论"的规律,因此它被说成是没有任何价值的东西,是虚假、错误的东西。我们从这里看出,某些特定的理性逻辑思维对文艺复兴时期的建筑艺术很恼火,因此对它的美学价值就忽略了。这正是进化论思想在审美规律上的一种谬误。

　　用以进化论思想为核心的艺术理论作为衡量的尺子,文艺复兴时期的建筑艺术一方面遭到忽视,同时另一方面也遭到误解。这和浪漫主义艺术理论对待这一时期的建筑艺术如出一辙。说它遭到忽视,那是因为具有进化论思想的理论家,在事先已经形成了一套演变过程的公式,当他把这个公式套在建筑艺术身上的时候,发现文艺复兴时期的建筑艺术实践并不完全吻合这个公式,更不能证明他自己的这个公式的准确性,因此,他只好对于这一时期的艺术实践匆匆带过。说它遭到误解,那是因为文艺复兴时期建筑艺术从这种进化论艺术理论中获得的那一点点关注,完全类似于其他的艺术风格所得到的关注,把它等同于整个建筑艺术进化演变过程中的其他阶段,对它的描述也与对其他风格的描述没有什么两样,都是最基本层次的进化论语言。这种在审美观念上的偏见,不仅仅在于进化论艺术理论所关注的只是事实的罗列,而不是评估这些事实的价值;不仅仅在于进化论艺术理论所关注的更多的是次要的事实,而不是那些成就突出的主要事实;问题还在于这种理论只关注哪些事实与现象能够帮助说明历史发展规律,而不在于什么是美的东西。这种偏见所具有的影响极其深远。由于进化论思想的产生是基于对生物的研究结果,因此,当这种以历史过程为重点的艺术理论无法避免地要对艺术作品的价值作出判断的时候,它在毫无察觉的情况下,把生物过程中的价值规律强加在建筑艺术上面。所以,从大的原则上来说,文艺复兴时期的建筑艺

presentation of the style, and see how this occurs.

The architecture of the Renaissance, we are told, and rightly, falls into three fairly distinct periods. There is the period of the Florentine Renaissance—the period of the *quattrocento*—tentative, experimental, hesitating, with a certain naïve quality that makes for charm but hardly for accomplishment: the period of which Brunelleschi is the outstanding figure. Of this manner of building the Pazzi Chapel is the earliest pure example, and the 'Carceri' Church of Giuliano da Sangallo, at Prato, one of the latest. This is the period of immaturity.

The second period is that of Bramante and of Raphael. It is much more sure of itself; its aim is clearly defined and supremely achieved. The tentative Brunelleschian charm has vanished, and a more assured and authoritative manner has taken its place. Here, as at no other time, is struck a complete equipoise between majesty and refinement. The architecture of Bramante and Raphael and Peruzzi is as free from the childish and uncertain prettiness of the work which precedes it as from the 'grossness and carelessness, of that which followed. It shares the faultless ease of the painting of its period. Raphael's ruined villa 'Madama,' Peruzzi's palace of the Massimi, the Farnesina, which these two names dispute, a score of other Roman houses, with that at Florence of the Pandolfini, all have this greatness, this distinction of design. Behind them is discerned the image of the grandest: Bramante's vision of St. Peter's, ill-starred, unrealised.

It is a short period—a single generation well-nigh covers it. But it is the climax of the Renaissance and its prime. It synchronises with the climax of painting and civilisation. It is the architecture of Leo X . and of Leonardo: the architecture of a time that could see its prototype in the assembled genius of the 'School of Athens.' This is the second period of Renaissance architecture: its supreme efflorescence.

And now begins the decline; the perfect equipoise could not be sustained. The inevitable decay sets in. It takes two complementary shapes: exaggeration and vacuity. The noble disposition of architectural forms gives place to restlessness: dignity is puffed into display. The sense of grandeur becomes the greed for size. It is the period of the Baroque: the period of decadence. The problem of style once solved—Bramante's school had solved it—nothing can remain but an abuse of power, and architecture feels the strain of too much liberty. As the architecture of Bramante stood linked to the art of Leonardo, so this of the baroque shares in the general corruption of the time: a time when 'gods without honour, men

术是必须受到谴责的，因为它是"毫无根据地"在自己创造自己；从具体的细节上来说，它也是应该受到批判的，因为它受到"必然规律"的惩罚，在劫难逃。下面让我们看看进化论理论对于文艺复兴时期建筑艺术风格的代表性论述方式是怎样的，然后再看看为什么会出现对它的这种责难。

进化论艺术理论是这样说的：文艺复兴时期的建筑艺术分为三个明显不同的阶段。这个判断是没有错。第一个阶段是佛罗伦萨文艺复兴阶段，也就是十五世纪时期。这个阶段的特征是尝试性的，一切都在实验阶段，里面充满了犹豫和不确定，同时具有一定的天真成分在其中，这让这一时期的建筑艺术获得了一些讨人喜欢的特点，但是，这一时期的建筑艺术没有什么突出的成就。这一时期的代表人物是布鲁乃列斯基，最早期的代表性建筑作品是他的巴齐礼拜堂；晚期的代表性建筑作品是位于意大利普拉托（Prato）市的人称"监狱"教堂（'Carceri' Church），它的建筑师为朱利亚诺·达·桑伽罗。这是一个不成熟时期。

第二个阶段是以伯拉蒙特和拉菲尔（Raphael）为代表的时期。这个时期的建筑师对自己更加有信心，目标更加明确，同时取得的成就也杰出辉煌。布鲁乃列斯基那一时期所表现出来的尝试性特点被一扫而光，那种讨人喜爱的特点也没有了，取而代之的是一种坚定、刚毅、充满了权威的艺术形象。文艺复兴时期的建筑艺术和以往其他时期不同，它同时具有宏伟的气势和精湛的技艺。以伯拉蒙特、拉菲尔、裴鲁齐（Peruzzi）作品为代表的建筑艺术不再带有任何前一阶段不成熟的可爱与孩子气，同时也看不到后继者那种"令人作呕的胡搞和肆无忌惮"。他们的建筑艺术与当时的美术作品一样精致、完美。拉菲尔设计的别墅"玛达玛（Madama）"（已毁）、裴鲁齐设计的玛希米豪宅（Massimi）和法内西纳豪宅（Farnesina）（也有人认为这个豪宅是拉菲尔设计的），在罗马地区为数不少的其他住宅，以及他在佛罗伦萨设计的庞多菲尼豪宅（Pandolfini），等建筑作品，都具有一种豪迈的气魄，在建筑艺术上都是杰出的作品。在这些作品背后，我们看到的是最伟大建筑艺术作品的身影：伯拉蒙特主持下设计的罗马圣彼得大教堂，可惜他的设计命运不济，没有能够实现。

这一阶段持续时间不长，也就只有这一代人。但是它是整个文艺复兴时期的高峰，是最重要的阶段。建筑艺术的高峰与绘画艺术、社会文明是同步的。这是以教皇利奥十世（Leo X）和达·芬奇（Leonardo）为标志的时代，这个时期的建筑典范可以从拉菲尔的绘画作品"雅典学派"的大型壁画中看见。这就是文艺复兴时期建筑艺术的第二个阶段，那是一个无比辉煌的全盛阶段。

从这时开始进入衰退阶段，宏伟的气魄与精湛的技艺无以为继。败落不可避免。建筑艺术的败落表现在两个互为表里的现象上：歇斯底里的夸张和颓废迷惘的空虚。高贵的建筑艺术形式变成了急促的不安定感，尊严和高贵被炫耀加噱头所取代。对宏伟气魄的崇尚变成了对于傻乎乎巨大建筑物的贪求。这就是我们所说的巴洛克风格：这是一个堕落的时代。这个时代的建筑艺术风格在伯拉蒙特学派的手中已经解决，但是到了这时，除了对这个风格进行毫无节制地滥用之外，没有别的。这时的建筑艺术充满了随心所欲的自由发挥空间。正如伯拉蒙特的建筑对应着达·芬奇的艺术，巴洛克建筑艺术对应着的恰好是一个腐败的时代：在这个时代里，"上帝失去了荣耀，人类失去了人性，少女不再天真，山人也不再质朴，他们像一群白痴聚集在一起，出现在被污染了的画布上，虚假的舞台布景一样的建筑物

without humanity, nymphs without innocence, satyrs without rusticity gathered into idiot groups on the polluted canvas and scenic affectations encumbered the streets.' Scenic affectations, broken cornices, triple and quadruple pediments, curved facades, theatrical plans, gesticulating sculpture: everything is irrational, exaggerated, abused. These are the dreams of a collapsing mind; this is the violence of a senile art: a sort of architectural delirium foretelling the approach of death. But senility, if sometimes it is violent, is at other moments apathetic; and the approach of dissolution, if it is heralded by delirium, is foreshadowed also in coma. Thus the third period of the Renaissance is marked sometimes by an opposite mood to its extravagance. The exquisite proportions of Raphael are hardened, in this decline, into academic formulas; architecture, when it is not ostentatious, becomes stiff, rigid, and inert. Simplicity becomes barren, and a restrained taste, vacant. And as the end draws near this vacancy is set in all finality on architecture's features by the Empire style. The Renaissance dies, its thoughts held fixed, by a kind of wandering memory, upon the classic past whence it arose, and which, in its last delusion, it believes itself to have become.

Such is the theme which, in their several manners, our historic repeat. But is it not too good, a little, to be true? Is it not a little like those stories of Herodotus that reveal too plainly the propensity of myth? This perfect image of the life of man—why should we look to find it in the history of architecture? This sequence of three terms—growth, maturity, decay—is the sequence of life as we see it in the organic world, and as we know it in ourselves. To read the events of history and the problems of inanimate fact in the terms of our own life, is a natural habit as old as thought itself. These are obvious metaphors, and literature, which has employed them from the beginning, will not forego their use. It is by words like these that the changes of the word will always be described. But, at least, it might be well to make certain that the description fits the facts. The criticism of architecture, with the solemn terminology of evolution, now too often forces the facts to fit this preconceived description. It is true that of late years a slightly more worthy appreciation of the baroque style—it would be truer to call it a mitigation of abuse than an appreciation—has crept from German into English criticism. But the new, less vivid, colours are still woven on the old pattern. Immaturity prime, and decay follow one another in predestined sequence. Architecture is still presented to us as an organism with a life of its own, subject to the clockwork of inevitable fate. After Brunelleschi the herald, and Bramante the achiever, must come Bernini and the fall.

Let us retrace the biologic myth. The period of Brunelleschi is tentative and

排列在街道两旁。"装模作样的布景式建筑手法，加上随意打破的檐口线脚，三个、甚至四个重复叠加在一起的山花，波浪式的建筑正立面造型，追求戏剧效果的平面布局，动作夸张的雕塑，等等。在这种建筑风格中，一切的一切都是毫无理性可言，任何一处都是无节制地夸张和胡搞。这正是人们在失去理智，神志开始有些不清的时候所出现的梦幻状态，是艺术在变得老朽、衰败时呈现出来的暴力倾向。这是一种类似于死亡即将来临前的挣扎，在建筑艺术上表现为一种精神错乱式的疯狂。如果衰败灭亡之前有时候表现为疯狂挣扎，那么在另外的时候则表现为麻木不仁。疯狂是最终衰亡的前兆，在死亡来临之前它表现为昏迷不醒。因此，文艺复兴时期建筑艺术中的第三个阶段，表现为两种极端对立的状态有时会交替地出现。以拉斐尔艺术为代表的那种精巧微妙的比例关系在这个时期变得死板，成了学院派的公式。这个时期的建筑艺术，当它忘记表现自己的时候，它又变得僵硬、笨拙、迟钝。简洁变成了贫瘠，对于审美追求的节制变成了空洞无物。当死亡终于来临的时候，这种没有实质内容的空洞则表现为各种建筑细节手法的堆砌，五花八门，无所不用其极，形成了所谓的帝国风格（Empire style）。文艺复兴死亡了；它的思想凝固了，借助于飘忽不定的松散记忆，它凝固在古典主义的艺术风格上面，而古典主义恰恰又是它的发源之处。在它最后消失的时候，它确信自己已经变成了古典主义的一部分。

　　这就是艺术历史不断地告诉我们的文艺复兴艺术的发展过程主线，当然具体的说法有很多种版本，不过主要线索都是如此。但是，这个描述是不是有一点过于圆满而让人有点不敢相信其真实性呢？是不是有一点类似古希腊第一位历史学家希罗多德（Herodotus）那样，把内容丰富的希腊神话用平实、通俗的白话加以简单的概括呢？上面的描述是人的一生的典型经历，我们为什么一定要在建筑历史中寻找人生经历的描写呢？这种三阶段的顺序，即生长、成熟、衰老，是我们大家所熟悉的人生过程，是生物世界的现象，我们自己也在亲身经历着这一过程。按照我们人类自己生命过程的规律来解读历史事件以及严肃的事实不是什么新发明，而是有着如同我们人类思想一样古老的传统。生命过程是一个很容易理解的比喻，文学艺术绝对不会放弃任何机会利用这个比喻和象征的。事实上，文学艺术从一开始就一直在使用这个比喻和象征。世界上的各种变化过程都是用这种语言来描述的。这种语言不是说不可以使用，只是在使用之前，至少应该先确认一下这种描述本身所使用的语言与它所描述的事实是否相吻合。但是实际情况是，这种使用进化论严厉语言的艺术理论常常迫使历史事实进行一定的扭曲变形来适应自己事先做好了的描述模式。最近一段时间确实出现了一些有一定价值的重新认识巴洛克艺术的理论，这些理论所使用的语言也从德语扩大到英语范围。或者换个说法更准确，它们实际上并不是真的开始欣赏巴洛克艺术，只是对于自己从前关于巴洛克艺术的偏见进行某些温和的调整而已。这些新的理论同过去的激情四射的理论相比，看起来略微有些苍白，但是，新色彩仍然重复着过去旧理论的结构与形式。生长期、高峰期、衰落期仍然按照从前的顺序继续发生着，这种顺序是无法改变的。建筑艺术在这种新理论的眼睛里，仍然是一种有机体，在按照自己的命运和时钟，继续不断地演变着。布鲁乃列斯基带来了新的希望；伯拉蒙特是取得最高成就的那个人；他们之后必然出现了伯尔尼尼，然后便是衰败。

　　现在让我们来重新追踪一下这种按照生物进化演变规律来解释艺术理论的轨迹，看看

immature—unskilled, but charming. This is true, in a sense, but already it is not exactly true. It asks us to regard Branelleschi's architecture as a less adept solution of Bramante's problem. It presents him as struggling with imperfect instruments after an ideal which later was fulfilled. We are bound to see his architecture in this light if our thoughts are on the *sequence*. In relation to the sequence, the description may be just. But this precisely was the fallacy of evolution. The values of art do not lie in the sequence but in the individual terms. To Brunelleschi there was no Bramante; his architecture was not Bramante's unachieved, but his own fulfilled. His purpose led to the purpose of Bramante: they were not on that account the same. There is in the architecture of the early Renaissance a typical intention, a desire to please, quite different from Bramante's monumental intention—his desire to ennoble. The immaturity of a child is spent in 'endless imitation' of the maturer world, expressed with unskilled thoughts and undeveloped powers. But the 'immaturity' of the Renaissance was rich with the accumulated skill of the mediaeval crafts: it was in some directions—in decorative sculpture, for example—almost too accomplished. And it was not spent in feebly imitating the mature, for the obvious reason that the 'mature' did not yet exist. True, the antique existed; but the Brunelleschian architecture was far from merely imitating the classic architecture of Rome. It had a scale of forms, a canon of proportions and an ideal of decoration that were all its own. The conception of immaturity, therefore, while it is appropriate in one or two respects, is in others misleading; and the parallel is so forced that it were best relinquished.

The first condition of aesthetic understanding is to place ourselves at the point of vision appropriate to the work of art: to judge it in its own terms. But its own terms will probably not be identical with those of the sequence as a whole. If we insist on regarding the sequence, we are forced to compare Brunelleschi with Bramante, and this can only be done in so far as their styles are commensurable—in so far as they have purposes in common. We shall compare them with regard to their command of architectural space and logical coherence, and here, no doubt, Brunelleschi is tentative and immature. But that does not exhaust his individuality: these qualifies were not his total aim. The more stress, then, that we lay on the sequence the less justice shall we do to *quattrocento* architecture. The habit of regarding Brunelleschi simply as Bramante's precursor long allowed his genius to remain in shadow. Not so very long ago the assertion of his independent rights, his unrepeated merit, was received as a paradox. He came first in a long sequence, and 'without experience'; how could he, therefore, be supremely great?

它到底有什么神秘之处。布鲁乃列斯基那个时期是探索性的,也还不成熟,就是说,技术还不完美,但是不乏讨人喜欢之处。在一定意义上讲,这个说法或许是成立的,但是,它不完全正确。这个说法实际上是在引导我们把布鲁乃列斯基的建筑艺术看作是伯拉蒙特的艺术早期不成熟的表现,它让我们相信布鲁乃列斯基似乎在努力解决伯拉蒙特的问题。它让布鲁乃列斯基看起来如同一个使用不称手工具的小孩子,跌跌撞撞,笨拙地尝试着后来在伯拉蒙特手里实现的理想。如果我们把思考的重点放在演变的过程上,那么,我们似乎不可避免地产生这样的想法。如果单看演变的过程,那么这种描述也可以说有它的道理。但是,这恰恰是进化理论扩展到艺术领域里的谬误之所在。艺术的价值不在于风格之间的演变,而在于每一个特定艺术风格本身。对于布鲁乃列斯基来说,伯拉蒙特根本就不存在。布鲁乃列斯基的建筑艺术绝对不是伯拉蒙特建筑艺术的尝试阶段,而是他自己的全部追求,并且是实现了自己艺术追求的完美作品。布鲁乃列斯基的艺术追求启发了伯拉蒙特本人的追求,但是这决不等于说他们二人的追求是一致的。在文艺复兴时期初期的建筑艺术中有一种明显的努力方向,那就是让建筑艺术为人服务,让人们感到建筑艺术给他们带来的愉悦和欢乐,这一点与后来的伯拉蒙特时代的努力目标根本不同。伯拉蒙特时代的追求是一种纪念碑的效果,追求一种尊贵的气势。不成熟的孩子在模仿大人的活动过程中所表现出来的是不成熟的思想和不熟练的技艺,但是,文艺复兴时期早期的那些被称为不成熟的艺术却是充满了从中世纪继承下来的炉火纯青的建筑技艺,例如在装饰性雕刻方面,这个时期的技艺几乎是前无古人后无来者的。这个时期的艺术家也不是像孩子那样在盲目地模仿着成年人的活动,原因很简单,因为那些"成年人"根本还没有出现呢。是的,古罗马的艺术早已出现过,但是布鲁乃列斯基的建筑艺术绝不是简单地模仿古罗马的建筑艺术,他有自己的造型尺度、比例原则以及艺术理想,这些都是属于布鲁乃列斯基个人的追求。至于说这个时期的艺术所谓的不成熟,在某一两个特定的方面或许成立,但是从整体上来说则完全是在误导人们的视听。在建筑艺术中与生物进化理论之间所找出的这种看似平行关系,人工斧凿的痕迹过于明显,我们最好把它扬弃。

　　美学的判断中的第一个条件是把我们看问题的立场和观点同所面对的艺术作品放在一个恰当的相互关系上:我们必须根据艺术作品自身的特点与性格来进行判断。艺术作品自身的特点与性格与它在整个艺术演变过程中所扮演的角色有可能是不同的,如果我们坚持强调演变过程,就是说我们被迫必须把布鲁乃列斯基拿来与伯拉蒙特进行比较的话,它只能是在他们二人艺术风格和艺术追求一致的情况下才能够进行。这样,我们可以来比较一下他们二人在建筑艺术空间的处理上与整体效果的和谐上各自采取了什么样的处理手段,这些手段所取得的效果又如何。从这个角度来看,布鲁乃列斯基的建筑艺术的确表现出来一些尝试性的特征,有些不成熟的成分。但是,在空间处理上与整体和谐效果上略逊色于伯拉蒙特绝对不能说明布鲁乃列斯基建筑艺术的全部,根本不能说明艺术家本人的全部个性与特点。我们越是片面强调历史的演变与进化,我们就越是不能正确理解整个十五世纪的建筑艺术。把布鲁乃列斯基简单地看作是伯拉蒙特的前导,长久以来一直让他的艺术天才埋没在阴影里面。布鲁乃列斯基的艺术成就和独特的艺术特点也就是到了最近才变得似乎难以接受。作为一个长期艺术演变过程的第一个主要艺术家,在"没有任何经验"的情况下,他的艺术成就怎么可能是伟大的呢?

The evolutionary criticism which belittled the period of Brunelleschi—and from the same unconscious motive—was something more than just to the period of Bramante: the 'prime and climax' of our architecture's life. Noble as it was in the hands of its finest architects, the central style of the Renaissance had, none the less, its vice. It is too terrified lest it should offend. Bramante, Raphael, Peruzzi, speak as having authority; but the *style* speaks as the scribes. A style has the right to be judged at its highest inspiration, yet, to be fully understood, must be watched at its common task. At moments—but at moments how infrequent!—this architecture makes concrete, as no other style has done, the mind's ideal of perfect humanism. But the authentic spirit of Bramante comes to us in how few examples; an element of weakness—an element of philosophy too rare and too exclusive—withered his inspiration at its birth. Of all the three stages of the Renaissance sequence, this central period was the most intensely academic. It could be as vacant as the Empire style, and as imitative. The spirit of life which, in spontaneous gaiety, never fails to play upon the sunny architecture of the *quattrocento*; the life which in the *seicento* flamed out and gave itself in prodigal abundance to a thousand ventures; the life which had been smiling and later laughed aloud, flickers too often in these intervening years to a dim, elusive spark. Much that was then built by admired masters—by the younger Sangallo, for example—would justify the 'evolutionary' strictures, had it been built later. If a servile attendance on the antique is a mark of declining force, Bramante himself must stand convicted of decadence, for no imitation is more self-effacing than his domed chapel of S. Pietro in Montorio. Here is the beauty of an echo: life, here, is scarcely stirring. The Roman civilisation, in that favoured moment, was the most brilliant that the Renaissance achieved, the most rounded and complete. But its architecture, for the most part, had a taint of too much thought, too incomplete a vigour. We do not seek to argue it *inferior* to that which followed or preceded: strictly, it is not comparable with either, and all three have their beauty. But even if it be preferred above them, the illuminating fact remains: the weakness that was in it is the weakness of a 'declining,' a too segregated art; a weakness which, if it did not thus impertinently intrude into the summer of the Renaissance, our historians would have signalised as the chill of its approaching winter.

进化论影响下的艺术理论在对布鲁乃列斯基进行矮化的同时,也下意识地在抬高伯拉蒙特时期艺术成就,把它说成是我们建筑艺术生命过程中"最辉煌的成就和高峰"。在那些最杰出的建筑艺术家手里,这个时期的建筑艺术获得了尊贵的效果,但是在那些同时期其他艺术家手里,这个时期的建筑艺术也是有各种各样的缺点与不足。这个时期有很多建筑表现出来的是咄咄逼人的架势,因为它们内心充满了恐惧。伯拉蒙特、拉菲尔、裴鲁齐在自己的作品中表现出自己的权威性,但是这个时期的建筑风格则表现为一种互相抄袭的风气。一种风格应当允许用它所达到的最高成就来判断自己的地位,但是,如果做到全面理解这种风格的价值,我们必须要看它最普通的使用目的。在个别的某一个时刻,这种建筑的确把人类的完美理想主义变得具体化,它所表现出来的人文主义精神是其他建筑艺术所没有的,但是,这种理想的时刻之间的距离太过遥远,非常偶然地出现一次,下一次不知道要过多久才能出现。代表伯拉蒙特理想精神的艺术作品太少了,我们知道的也就是凤毛麟角的那几个。他的理想中包含了一种不食人间烟火的艺术理念,这种艺术理念所包含的范围极其狭窄。这也是伯拉蒙特艺术理想与生俱来所具有的弱点。在整个文艺复兴时期的三个不同阶段中,中间的这个阶段表现得最具有学究气,最不讲究实际。它几乎已经具备了后来的帝国风格所表现出来的那种毫无实质内容,为了形式而形式,几乎是在不断地重复模仿抄袭从前空洞的样式罢了。十五世纪时期的建筑艺术充满了灿烂的阳光和旺盛的生命力,艺术家在尽情地根据人们的需要来表现自己。这种生命的火焰到了十七世纪的时候已经燃烧殆尽,取而代之的是千变万化且别出心裁的各种建筑花样。原先的微笑这时变成了开怀的大笑,原先明亮耀眼闪烁着的光芒,已经在这个演变过程中变成昏暗晦涩、忽隐忽现、偶然才会出现一下的火花。那个时期里出自受人尊敬的大师之手的很多建筑艺术作品,如果出现得再晚一点,就可以更加有力地佐证"进化论的艺术理论",例如,桑伽罗家族中属于晚辈的那一位,他的作品就是把一些按照"进化论理论"应该被划分为是后期的艺术特征,提前在成熟高峰期出现了。 如果说,我们把那些对待古罗马时期建筑艺术无条件接受、崇拜的做法看作是一种艺术衰落的特征,那么伯拉蒙特本人实际上正是这一衰落的始作俑者,因为那个时期的建筑艺术作品中,在模仿古代建筑艺术方面,没有一个比他的坦比哀多小礼拜堂以及整个蒙特利奥圣彼埃特罗教堂(S. Pietro in Montorio)更加惟妙惟肖,没有一丝违反古典原则的创意。这里的建筑艺术完全是古典建筑艺术经过十几个世纪后的回声:其中的生命已经失去了活力,我们感觉不到建筑中的脉动。这个时刻实际上是古罗马文明延续发展过程中最光彩夺目的时刻,建筑艺术形式在这个时刻最为圆满、最为完整。但是,关于其中的内涵,这个时期的建筑则表现为一种经过过多的前思后想而缺乏一种生命力的冲动;过于冷静、理性而缺少热情与活力。我们在这里不是说它与在它之前或者在它以后的建筑风格相比,它的艺术成就低人一等,绝对没有这个意思。这三者其实是没有任何可比性的,每一个时期的艺术有它自己的优点。即使是我们认为这个时期的建筑艺术比它前面和后面的两个时期里的建筑艺术要高明一些,我们还是不得不承认一个无法回避的事实:这个时期的建筑艺术的确包含着"进化论艺术理论"所说的"衰落"的特征,这是这个时期建筑艺术自身所具有的天生弱点,艺术作品只是不同局部构件的组合,不再是有机的整体。如果说这种弱点不是在文艺复兴最炎热的夏天突然冒出来的不合时宜的东西,那么根据我们的历史学家的理论,这种现象宣告了这个时期的艺术寒冷的冬天正在来临。

But, for architecture at least, winter was not approaching—rather, a scorching and resplendent heat. If the evolutionary sequence describes too little accurately the 'climax' and the 'birth,' it is forced to utter travesty for the 'decline,' If decadence means anything at all, it stands for loss of power, loss of self-confidence, loss of grip. It is a failure of the imagination to conceive, of the energy to complete, profound experiments—a wasting away of inherited capital no longer put to interest. The baroque style is the antithesis of all these things. Whatever faults it may have, these are not they. Intellect in architecture has never been more active; the baroque architects rehandled their problem from its base. Where the Brunelleschian architecture and the Bramantesque were static, this was dynamic; where those attempted to distribute perfect balance, this sought for concentrated movement. The expectation of repose, which there had been satisfied at every point, was here deferred, suspended to a climax. Architecture was considered, for the first time, wholly psychologically. So daring a revolution must needs be complex in its issue. The change of principle is so complete, its logic so perfect, that, if we fail to shift the angle of our vision, then virtues which the baroque architects passionately studied, must appear as vices; the very strictness with which they adhered to their aesthetic must seem an obtuse negligence of taste. A dangerous aesthetic, possibly: that is a point which need not here be argued;—but a *decadent* architecture—an architecture that lacked spontaneous force, energy of conception, fertility of invention, or brilliance of achievement —that the baroque style on no fair estimate can be called.

The art of painting—except in so far as it was merely, yet superbly, decorative and in closer subservience to architecture—did, on the contrary, show at this moment a real decline. For the genius of Michaelangelo, which in architecture had merely indicated a line of fruitful advance, had in painting fulfilled, and even passed beyond, the favourable limit. Thus, while the baroque architects were exploring in a veritable fever of invention the possibilities of their inheritance, their contemporaries in painting were marking time, and losing themselves in an empty, facile repetition of past phrases. This is true decadence. So little is it true that the energy of a race rises and falls in ordered sequence that even in artistic activity the most divergent results were simultaneous; and while architecture sprang forward, painting lost its nerve as an individual art, and its sole light was reflected from the conflagrating splendour of baroque architecture.

Even for the Empire style the charge of decadence—though here more

然而，事实是艺术的冬天根本没有来临，至少建筑艺术的冬天还十分遥远，相反地，我们感受到的是这一时期的炽热和辉煌。假如说，这个"进化论艺术理论"在描述文艺复兴时期艺术过程的"出生阶段"与"高峰阶段"有点力不从心、不很准确的话，那么，它在描述"衰落阶段"也只好被迫草草应付一下。什么是衰败？如果这个概念能够说明一点东西的话，它应该指的是力量的丧失、对自己不再具有信心、失去了对周边事物的掌控。它代表着人们丧失了对具有深远意义的艺术创作继续进行执着的追求，它也代表了人们不再具备追求那些艺术活动所必需的能量。从前辈那里继承下来的艺术遗产也不再加以利用，统统地被遗弃。假如这就是衰败没落的定义的话，那么，巴洛克时期的艺术所代表的精神则完完全全与它们背道而驰。我们可以给巴洛克时期的艺术罗织各种罪名或者缺点，但是无论如何也不会是衰败，因为根本不符合事实。建筑艺术中的思想活动从来没有像这个时期这么活跃过；巴洛克时期的建筑师从根本上重新面对建筑艺术中的各种问题。布鲁乃列斯基时期的建筑艺术和伯拉蒙特时期的建筑艺术如果说是属于静止的形式创作，那么巴洛克时期的建筑艺术已经是在追求动态的建筑形式；如果说前面两个时期所追求的是建筑上面各个部分之间的比例关系获得一种完美的均衡，那么后面一个时期所追求的是动感焦点的戏剧效果。从前人们努力在每一个角落都能寻得到的庄重效果，到现在变成了有节制地延迟到最后的高潮才出现。建筑艺术不再是静止的视觉活动，而是完全彻底的心理活动，这是前无古人的创举。这种新创举是如此地大胆，其中包含的问题无疑是一场革命性的改变。艺术创作的原则如此彻底地得到改变，它的逻辑关系也如此地完美无比，假如我们不能改变我们对这个时期艺术所持的观点和看问题的角度，那么，在看待巴洛克时期艺术的时候，巴洛克建筑艺术家们呕心沥血所追求的艺术特征与性质在我们眼里就会变成一堆堆充满罪恶的东西；那些艺术家眼中严格遵循自己艺术和美学规律的做法，在我们眼里就会变成大脑迟钝的举动，是没有审美感觉的表现。说巴洛克时期的艺术是具有危险性的美学追求，或许可以这样说，关于这一点，我们在这里不打算浪费笔墨；说巴洛克时期的建筑艺术是衰败的建筑艺术，也就是说它缺乏自然潇洒的力量，缺乏构思的能力，缺乏发明创新的本能，缺乏杰出的艺术成就，我们可以这样讲，这样的说法绝对不是用客观公正的态度来衡量巴洛克建筑艺术的。

当时的绘画艺术与建筑艺术相反，的确是出现了艺术创作中真正衰败的迹象。即使是这样，那些技术高超且完全隶属于建筑的装饰画和壁画等作品并不在此列。米开朗基罗的艺术天分在建筑艺术上是开创了后来众多富有成果的艺术探索之先河，但在绘画艺术上所取得的成就让后人无法逾越，他本人的绘画创作达到了登峰造极的地步。因此，当巴洛克时期的建筑师们在采用各种切实可行的手段，尝试着使用从前辈手中承传下来的各种传统，探索开创新领域的时候，同时代的画家则是在原地踏步，轻巧地重复着前人的成就，绘画作品因此变得空洞无物。这是真正的衰败、堕落。那种认为艺术发展过程是按照一起一落的规律在进行的想法几乎是站不住脚的，实际情况是各种艺术活动以及它们参差不齐的多元性结果都是同时发生的。建筑艺术向前冲刺的时候，绘画艺术不再以独立的艺术作品存在了，它满足于在一旁反射一点点巴洛克建筑艺术那熊熊燃烧的火焰。

甚至帝国风格的建筑艺术我们也不能说它是一种衰败、堕落的艺术，因为这种指控根本

plausible—is not convincing. Here, indeed, is displayed a preoccupation with a literary ideal that is never without menace to an art of form. Yet the forms of the style were congruous to a live tradition; they were beautiful; they were consistently applied. The judgment of decadence is here an *ex post facto* judgment. The Empire style did, in fact and as a point of history, mark the dissolution of Renaissance architecture. It had no future; it linked itself to no results. But this might well be accounted for on purely social grounds. A change of patronage in the arts, a profound change in the preoccupation of society, a collapse of old organisations, were necessarily, in France, the sequel of the Revolution and the Napoleonic wars. France, not Italy, was at this moment the holder of the torch of architecture. If the torch fell and was extinguished, we need not argue that it was burnt out.

Decadence is a biological metaphor. Within the field of biology it holds true as a fact, and is subject to law; beyond that field it holds true only by analogy. We can judge an organism by one constant standard—its power to survive: a power that varies in a known progression, a power of supreme importance. But even here—where the sequence of immaturity, prime and decay is a fact governed by predictable law—the power to survive is no test of aesthetic quality: the fragile unfolding of a leaf in spring, its red corruption in autumn, are not less beautiful than its strength in summer. And when we have to deal, not with a true and living organism but with a series of works of art, the tests of evolution are even more misleading. For here we ourselves define the unit which we estimate. We have to be sure that our sequence is really a sequence and not an accidental group. We have to be sure that there is a permanent thread of quality by which the sequence may at every point be judged, and that this quality is at each point the true centre of the art's intention. The mere power of an architectural tradition to survive—could we estimate it—might be a permanent quality but hardly a relevant one; for the successive moments of an art are selfjustified and self-complete. To estimate one by reference to another is a dangerous method of criticism. The archaic stage of an artistic tradition is not mere immaturity of technique. It implies a peculiar aesthetic aim and conception, and a peculiar relation between the conception and the technique. In the archaic stage, technique is as a rule adequate to the conception, and no more: it has no life of its own; it is no end in itself. And the period of so-called decadence, so far from showing a decline of technique—as the organism shows a decline of capacity—is often marked by a superabundance of technical resources, which stifle the conception. The atrophy is one of ideas. Our judgment, then, will have shifted its ground: it will have estimated, one period by its technique, and another by its conception. And, beyond this, it often falsifies

无法令人信服。乍看起来,这个指控好像有几分道理,因为帝国风格的建筑艺术充满了条条框框的约束,一切都追求严格的设计规则,一切都在追求着理想化的东西,这种教条式的东西从来都是对真正艺术生命的威胁。但是,这种风格的建筑形式却是延续着活着的传统;它的建筑形式不容否认地非常优美;它的应用也是十分严谨。我们说这个时期的艺术代表了一种衰败和堕落,这也是事后多年回头来看所得出的一个结论。帝国风格的建筑艺术标志着历史上的一个特殊的时刻,它代表了文艺复兴时期建筑艺术的终结。帝国风格没有产生出它直接的后续结果,可以说它没有自己的未来,但是这一现象完全是由于社会变革所导致的结果。随着法国大革命的兴起与拿破仑战争的不断进行,法国社会所关心的艺术问题重心有了根本的改变,艺术赞助人已经不再是从前的那些人了,旧有的社会与艺术组织已经瘫痪、瓦解。所有这一切已经不由人们的意志而改变了。而这个时期指引建筑艺术航向的火炬是在法国,不再是在意大利。如果火炬的火焰突然倒掉而导致火苗突然熄灭,我们不能得出结论说火炬自身已经燃烧成为灰烬。

  衰败本来是用一种生物过程拿来所作的比喻。在生物学领域里,这种说法是有其根据的,是一种事实,符合自然规律。超出生物学领域以后,它就只是一种类比而已。我们在对一个生物有机体做出判断的时候,遵循着一个普遍的规律法则,这就是生物的生存能力。这种生存能力在同一种过程中表现着不同程度的差异,代表着生存能力的不同。这种生存能力对于一个生物来说,它的重要性不言而喻是至高无上的。即便如此,这种由自然法则主导的从不成熟、到成熟、再到衰败的演变过程,也不能说明每一个阶段的美学品质:春天里含苞欲放的嫩芽,秋季中落英缤纷的残红,在美学上,一点儿也不比盛夏里茂盛的绿阴逊色。在讨论有生命的有机体时姑且如此,当我们所面对的是一系列艺术作品的时候,进化论理论就更加误导人们的思想。因为在面对艺术作品的时候,那一系列排列着的作品都是我们根据自己的判断而挑选、排列起来的,我们首先需要确认的是,这种排列的确反映着这些作品之间存在有前后顺序,确定它们并不是很偶然地组合在一起而已。我们也必须肯定的确存在着一种恒定的主线贯穿于这些全部的作品,而这条主线所代表的精神的确又是该作品在创作的时候成为每一件艺术品全部精力所集中的核心。对于一种建筑传统能够得以延续多年的那种能力,如果我们能够准确地加以评估的话,那么可以说这种延续能力的确具有一种永恒的品质在其中;但是,即便如此,这种品质与生物的繁衍过程几乎没有什么关系,因为艺术活动中每一个阶段都是独立于其他的,都是有其自身的道理和原因的,与其他阶段的艺术活动没有什么直接关系。在评估一个时期的艺术,把它与其他时期的艺术放在一起,按照其他时期的艺术标准来衡量,这种做法是很危险的。一种艺术传统中的远古阶段,它的艺术创作所达到的效果绝对不单单是因为在技巧上的不成熟。它表达的是一种特别的美学意识和追求,表达了一种在构思与技巧之间的特定关系。在古代,技巧只要能够满足构思的需求就足够了,古人不会去追求构思之外的技巧,这可以作为一条原则。技巧离开了艺术构思是不可能单独存在的,技巧不会以自身为目的,它一定是服务于其他目的的。所谓的艺术步入衰败的时期根本不像有机体衰败那样,开始丧失各种能力;相反,艺术的衰败时期通常都是它的技巧特别发达的时期,大量的技巧同时涌现和堆砌,反而窒息了艺术创作中的构思与灵感。对于这种现象的各种解释,把它说成是退化、衰败的,只是其中的一个说法而已。因此,我们的判断实际上是根据不同的理由所得出自己的结论的,有的时候是根据技巧方面的

both by relating each of them to the aesthetic purposes of the 'climax' that came between. In recent years it is true the independent value of archaic art has received a sudden recognition. To that extent the biological fallacy—at any rate in painting and sculpture—has been checked. But then a corresponding injustice is usually done to the later phases. For the critic's determination to take a comprehensive view, to use inclusive formulas, and to trace an evolutionary sequence beyond its proper limits, still causes him to read the whole series of his facts as related to a single ideal. Such an attitude had compensation when the tradition of architecture was alive, and taste was limited to a due appreciation of contemporary things; for then appreciation was *so far* perfect, and the past was merely ignored. Taste was specialised at every moment, and developed *pari passu* with creative art. No gift of imaginative flexibility was required. But for a modern criticism, which claims to judge with an impartial eye the whole sequence of architectural history, or even of one single 'style,' that gift, before all others, is demanded. The different aesthetic purposes possible to architecture are not necessarily equally worthy; but before their worth can be estimated it is necessary at least that they should be rightly distinguished and defined. A historical definition of architecture which traces the outward development of form from form will not of itself supply the needed definitions of aesthetic purpose. It will fail to strike the right divisions; it will be too unsubtle, too summary, too continuous. It will be intellectually simple but aesthetically unjust.

Criticism, based on historic evolution can no more afford a short cut to the problem of taste than criticism that is based on romantic formulas or on mechanical formulas or on ethical formulas. It is but another case of false simplification: another example of the impatience of the intellect in the presence of a living function that disowns the intellect's authority.

理由,有的时候则是根据艺术构思。而由于两者完美的结合所获得的所谓"艺术高峰"在与二者联系到一起的时候,也就是当我们在把二者与美学价值判断同时进行考虑的时候,我们常常误解这两个方面。最近一段时间,这方面有所改善。古代艺术的独特价值获得了相当程度的认识。正因为有了这些认识,这种根据生物进化规律而衍生出来的艺术理论,其荒谬性也就得到适当的暴露,至少在绘画和雕塑艺术上已经不再具有从前的影响力。但是,艺术理论家们对于一个艺术传统演变过程中属于后期的那些阶段,仍然存在着很大的误解。这是因为艺术理论家们总是在关注大范围内的演变规律,总是在使用自己的某种公式,结果在描述演变过程中每一个阶段的时候,总是根据主观的想法把过程中各个阶段之间的界限模糊掉了,使得理论家们相信自己所面对的对象仍然是根据同一个理念而发展起来的艺术演变过程。当一个建筑艺术的传统还在活生生延续过程中的时候,它其中的审美价值判断还会欣赏当代的很多艺术手法,那么这种看问题的态度是有好处的。因为这样的欣赏还是在关注目前艺术自身,这样的欣赏是完美的,它绝不关心过去的东西。审美倾向完全是由每一个具体的艺术对象所决定的,我们的审美习惯是公平对待艺术创作中的每一个阶段,并不需要什么特别的想象力才能理解某一时代的艺术作品。但是,现代的艺术理论虽然号称是不带任何偏见地对待建筑艺术历史的全过程,不带偏见地对待任何一种"建筑风格",然而在实际论述中,如果不具备这些理论家所说的特殊想象力,我们根本不知道他们在说什么。建筑艺术中需要面对的各种美学追求并不是具有同等价值的,但是在确认它们各自的价值之前,我们首先需要对它们加以区别,并给出明确的定义。建筑艺术历史描述了外在形式的演变过程,但是这种历史过程并不能够说明每一个时刻的建筑艺术形式所具有的美学目的。这种历史的方法将会错误地划分其中的各个阶段;它会失去很多其中非常含蓄微妙的内容,让划分变得十分僵硬,过于概括而失去很多细节,过于强调历史的延续性而忽略了各个阶段的个性。这样的做法,在理性分析上,它的确是方便人们的理解,但在艺术美学上,它却是不公平的。

根据历史进化演变思想而产生的艺术理论,它不像浪漫主义艺术理论,或者技术决定论艺术理论,以及用道德标准来进行艺术判断等艺术理论那样,可以在审美问题上走各种各样的捷径。但是,进化论艺术理论的错误也在于它把复杂的艺术规律进行过于简单化的概括。在活生生的艺术传统面前,当理论家们发现这些艺术传统的发展规律与艺术理论没有什么关系的时候,便开始急躁起来,变得没有了耐心,便开始不顾艺术规律地寻找自己的捷径。进化论艺术理论便是这种失去耐心的又一例证。

SEVEN
The Academic Tradition

I. 'There are in reality,' says architecture's principal historian, 'two styles of Architectural Art—one practised universally before the sixteenth century, and another invented since.' To the former belong 'the true Styles of Architecture,' to the latter 'the Copying or Imitative Styles.'[1]

Renaissance architecture is imitative. It is more imitative than any style of building that preceded it. It went further afield for its models and gave them greater honour. True, it is changeful, various, eager for experiment—this we have already seen: it presses forward. But also, and not less, it glances perpetually back. It has its own problems, but it is concerned, not less, with Greece and Rome. In the Renaissance for the first time the question asked is no longer merely, 'Is this form beautiful or suited?' but, 'Is it *correct*?' For the first time architecture canonised its past.

The outstanding mark of Renaissance architecture is a backward vision, a preoccupation with the antique. So much must be conceded even by those who have studied the variety and realised the vigour which the Renaissance style displays, who see most clearly how inevitable was this imitative impulse and how deep the inventive genius that accompanied it.

But, while this main fact is undeniable, the deductions which criticism has drawn from it are opposite enough. On the one hand it is said, Renaissance architecture, being imitative, has lost touch with life. It is a dead, an artificial, an 'academic' style. It lacks the originality, and it lacks the fitness of a style which springs unconsciously to suit a present need, as the mediaeval style sprang to suit monastic or civic institutions, or as the classic styles themselves, fitly and with originality, suited the ancient state. 'There is not perhaps a single building of any architectural pretension erected in Europe since the Reformation... which is not more or less a copy, either in form or detail, from some building either of a different clime or a different age from those in which it was erected. There is no building, in fact, the design of which is not borrowed from some country or people with whom

---

[1] Fergusson, *History of Modern Architecture.*

第七章
讲究学术理论的传统

一位重量级建筑历史学家这样说:"现实的建筑艺术中存在着两种主要的风格,一种是十六世纪之前的那些建筑风格,它们是在建造活动中从普遍实践中发展出来的,并且广泛地被人们使用着的建筑风格;另外一种就是十六世纪以后人们自己发明的各种新风格。"前者属于"真正的建筑艺术风格",而后者属于"模仿加上照抄照搬"。[1]

按照这个思路来看,文艺复兴时期的建筑风格是属于模仿一类的。它比它之前出现的任何一类建筑风格都更加具有模仿的特征。这个时期的建筑是经过翻山越岭、长途跋涉,从古代遗址中寻找灵感的,并且给予这些被模仿的对象极大的荣耀。这个时期的建筑形式多样、内容丰富,也不断地在尝试着新手法,它们在不断地向前探索。是的,这些都没有错。但是,归根到底,这个时期的建筑艺术从来都是不断地向后张望,从过去寻找灵感。这个时期的建筑艺术有自己需要面对的问题,但是,艺术家们最为关心的还是古希腊和古罗马的建筑艺术。在艺术的发展历史上,文艺复兴时期是第一个时期,它的艺术家们不单单是需要回答"这个形式合适吗?它美观吗?"这样的问题,他们还需要回答"这个形式是*正宗的*吗?"这样的问题。建筑艺术第一次把自己的过去变成一种后人必须遵守的法则。

文艺复兴时期的建筑艺术具有一个非常明显的特点,那就是它的视线范围始终注视着身后的过去,对古希腊古罗马时期的痴迷。即便是那些对于文艺复兴时期建筑艺术有过深入研究,并了解其中多样性与巨大活力的那些专家也不得不承认这一特征。这些专家十分清楚这种模仿古代风格的做法有其必然的理由;他们也十分明白,在这种从古代吸取创作灵感的做法背后,还有着深刻的创新天才。

虽然这个主要的事实是不可否认的,但是理论家们从这一事实推导出来的结论则出入甚大,几乎可以说是南辕北辙。一派说,文艺复兴时期的建筑艺术因为是以模仿为特征的,它们已经脱离了现实生活,因此是没有生命的,是人为造出来的,是一种"纯学术性"的建筑风格。这种风格缺乏原创性。它不像从具体时间和具体场合中根据需要自然发展出来的建筑风格那样,与环境与文化有着完美的结合;它甚至不能像古希腊、古罗马时期的建筑那样,既得体又具有原创性地来满足古代人的需求。"自从宗教改革发生以后,欧洲的建筑可以说没有一个不是在故意作出一副姿态……都是在抄袭以往的建筑风格,有的是抄袭建筑形式,有的是在抄袭建筑细节。而被抄袭的对象大多来自不同的自然环境,或者来自不同的时代。这期间出现的建筑物,或者其他设计作品,没有一件不是从其他的文化或者民族借鉴过来的,而这种借鉴的途径完全是从书本上得到的,完全不了解与原作密切相关的血与灵的感

---

[1] 佛格森(Fergusson),《现代建筑历史》(*History of Modern Architecture*)。

our only associations are those derived from education alone, wholly irrespective of either blood or feeling.'[1] That is to say, Renaissance architecture, like our modern 'revivals, ' lacks the merit that belongs to the natural products of a time and place. *It is too classical.*

On the other hand there is a school of critics who arrive at a diametrically contrary result. They do not complain that the Renaissance substitutes the ideal of 'correctness' for that of fitness and beauty, but that it is *insufficiently* 'correct.' They do not criticise the return to the antique: they applaud it; but they say that in the early Renaissance the classic manner was imperfectly mastered, and that in the later Renaissance it was deliberately misused. They approve Bramante and Palladio and the academic school; but for the rest—and above all for the baroque— they have one constant ground of censure: Renaissance architecture perverts the forms, and violates the 'rules' of classical design. It is *not classical enough.*

Among the prejudices which now affect our vision of architecture this point of 'imitation' must certainly be reckoned. Whether for praise or blame, we see, and we cannot help seeing, the Renaissance style as in some sense a transcript of classic style. The question is, in what sense? How are we to view this 'imitation, which for some critics is too servile, and for others too indifferent?

The answer is not easy, for at first sight the classic influence in Renaissance architecture takes wholly different forms. The classicism of Brunelleschi is in spirit a devout obedience to the antique; in result, it produced a style of rare originality. The 'seeker for buried treasure,' as the Romans called him, seeing him day after day bent eagerly among their ruins, returned to Florence to institute an architecture all grace and lightness and charm; slight in the projection of its mouldings slight in the body of its shafts, and wreathed with slender ornament: a style not rigid or of too strict a rule, seldom massive, and then more after the Etruscan manner than the Roman, and for the most part not massive at all, but lightly pencilled upon space. Yet to adopt the ancient style had been Brunelleschi's purpose, and to have restored it remained his boast. Later, at the height of its self-conscious power, and when, more than at any period, artists of original genius were concentrated in the capital, the Renaissance is satisfied, in architecture, with a merely reproductive effort. The little church of San Pietro in Montorio, already cited—save in a few details, a pagan temple merely—is a work of Bramante at his prime. His project even for St. Peter's is conceived in terms of ancient buildings: it is to raise the Pantheon upon the arches of Constantine's basilica. On the other hand it is in the great reaction when the neo-pagan culture is universally abused, and the academic 'rules' forgotten,

---

[1] Fergusson, *History of Modern Architecture.*

受。"[1]也就是说,文艺复兴时期的建筑如同我们当代的各种"复兴样式"一样,缺少一种与具体时间、地点相关联的自然真实感。*这种风格太过于古典。*

很有点戏剧性地,另一派的理论家则得出恰恰相反的结论。这一派对于文艺复兴时期的建筑艺术采用"正统"的建筑形式因而忽略了某些建筑形式的适当性与建筑形式的视觉美感是可以接受的;这一派的问题是认为这个时期的建筑形式还不够正统,不够正确。对于向古希腊、古罗马看齐的做法,这一派的理论家没有意见,反而是为它大声喝彩和叫好。但是,理论家们对于文艺复兴时期早期的建筑艺术有意见,认为它们还没能够掌握古希腊、古罗马建筑艺术的精华,而到了文艺复兴时期的后期,则是故意地滥用古典建筑语言。他们能够接受的做法只有伯拉蒙特和帕拉第奥的建筑,以及讲究学术研究的建筑形式。除此以外,尤其是巴洛克时期的建筑艺术,这一派理论家们只有一句话给他们:文艺复兴时期的建筑艺术糟蹋了古典建筑形式,违反了古典设计的规则。*这种风格还达不到真正的古典。*

目前影响我们正确认识建筑艺术的各种偏见中,这种"模仿"的观点必须要认真对待,并加以澄清。无论是褒义还是贬义,我们无法避免一种观念,那就是,文艺复兴时期的建筑从某种意义上讲就是在抄袭古典主义的建筑艺术。我们的问题是,这里所说的从某种意义上讲,指的是哪一种意义呢?这种以"模仿"为特征的建筑艺术,有人觉得它毫无创意,有人觉得模仿与否无关紧要,我们该怎样看待这件事呢?

答案没有那么简单。乍看起来文艺复兴时期建筑艺术中的古典主义影响表现为多种多样的形式。布鲁乃列斯基所遵从的古典主义是虔诚热情地遵守古代的建筑语言,其结果便是创造出具有十分强烈个性和独创性的建筑风格。罗马人说他是"寻找深埋地下宝藏的人",因为他们看见他日复一日地在罗马的废墟间低头寻觅,渴望发现古代作品的精华,然后把这些发现带回到佛罗伦萨去,让佛罗伦萨的建筑因此变得优雅、轻巧和妩媚。建筑上面的线脚出挑很小,柱子的柱身也非常纤细,柱子上的装饰花环也十分轻巧:整个建筑艺术风格灵活轻盈,没有特别强烈的规则法度感,几乎说不上是有什么厚重的体量,布鲁乃列斯基的建筑风格应该算作是更接近于伊特拉斯坎文明的,而不是古罗马时期的。从总的效果来看,布鲁乃列斯基的建筑体量不大,建筑造型元素也都是在空间采用清晰的线条轻轻地勾画。虽然如此,对于布鲁乃列斯基本人来说,追求古希腊、古罗马的建筑艺术是自己的目标,而且,再现古代建筑的辉煌是他本人一贯的追求。到了后来,当人们的自我意识达到最高峰的时期,那些具有非凡创造力的艺术家比以往任何时候都集中到了首都,文艺复兴时期的建筑艺术开始变得仅仅满足于简单地复制古代的建筑作品。比如在蒙特利奥的圣皮埃特罗教堂里的小礼拜堂坦比哀多(Tempietto),除了几处个别的细节之外,整个建筑群基本上属于非基督教教徒的一个去处,可以说是异教徒的东西。这个建筑作品是伯拉蒙特最高艺术水准的代表作。他甚至在为圣彼得大教堂构思设计的时候,也从复制古代优秀的建筑作品入手:他的思路就是在君士坦丁巴西利卡(Constantine's basilica)上面再加一个万神庙(Pantheon)。但是同时,伯拉蒙特的做法又是在当时异教徒文化盛行、甚至有泛滥趋势的时候,表现出一

---

[1] 佛格森,《现代建筑历史》(*History of Modern Architecture*)。

that the image of imperial Rome comes, in Christian architecture, most amazingly to a second life. The gates and aqueducts of the emperors, with their proud and classic inscriptions, rise again in the baroque city; the noble planning, the immense vistas, the insolent monuments, the scenic instinct, the grandeur and the scale are all the same. And this architecture, which might have satisfied the dream of Nero, is the work of Sixtus V., the Pope who so hated paganism that he could not look with patience on the sculptures of the Vatican, and in the Belvedere would frown on Venus and Apollo as he passed; who destroyed the ancient ruins which Pius II had protected, and valued what he spared only that he might plant upon it the victorious symbol of the cross. And at last, when these extremes of passion and revulsions of style had run their course, and architecture in the eighteenth century had brought classic example and modern needs to a natural consistency, the past once more recalls it to obedience, the Greek style supervenes, and the Renaissance dies after all upon a note of imitative fashion.

Sometimes it is the spirit, sometimes the letter of ancient architecture that the Italian style recalls. Now it indulges its thirst for novelty, and again at intervals does penance in Vitruvian sackcloth. The essence of the classic control is disguised beneath the variety of the forms which manifest it. In what did it consist?

II. The return to classic style in building forms part of the general movement of Renaissance Humanism—a phase of culture that touched life at every point and presents everywhere the same strange contradiction, spontaneous in its origin, profound in its consequence, yet in its expression often superficial and pedantic. Pedantry and humanism have in history gone hand in hand; yet humanism in its ideal is pedantry's antithesis.

Humanism is the effort of men to think, to feel, and to act for themselves, and to abide by the logic of results. This attitude of spirit is common to all the varied energies of Renaissance life. Brunelleschi, Macchiavelli, Michaelangelo, Cesare Borgia, Galileo are here essentially at one. In each case a new method is suddenly apprehended, tested, and carried firmly to its conclusion. Authority, habit, orthodoxy are disregarded or defied. The argument is pragmatical, realistic, human. The question, 'Has this new thing a value?' is decided directly by the individual in the court of his experience; and there is no appeal. That is good which is seen to satisfy the human test, and to have brought an enlargement of human power.

Power, in fact—a heightening of the consciousness of power as well as a widening of its scope—was the Renaissance ideal: and Greece and Rome, almost of necessity, became its image and its symbol. The Roman Empire had set the summit

种反潮流的精神,他讲究法则,反对滥用。在以伯拉蒙特为代表的建筑师手中,古代建筑"学术上"的法式和原则,亦即古代罗马帝国时期建筑艺术中的创作原则,在基督教的建筑上获得了第二次生命。古罗马皇帝兴建的那些凯旋门、输水道等大型建筑的细节,连同那上面那些经典、骄傲的题字,又出现在巴洛克时期的城市里:气势宏伟的城市规划设计,深远的视觉对景,宽阔的街道,神气十足的纪念碑和大型建筑物,开阔的景观,尺度巨大、气势宏伟的城市建筑设计与古罗马时期没有什么两样。这些看上去很可能是让古罗马帝国皇帝尼禄(Nero)感到满意的建筑实际上是为教皇西克斯图斯五世而建造的。这位教皇十分痛恨异教徒的任何文化,他甚至都不想看一眼收藏在梵蒂冈的那些雕塑。在梵蒂冈的另一个去处,贝尔维蒂尔(Belvedere)宫,当他路过宫殿里收藏的雕刻作品时,他会对着阿波罗和维纳斯的雕像皱眉头。西克斯图斯五世把庇护二世保护下来的古迹拆除殆尽,只保留那些他可以在上面置放十字架的建筑物。当这些极端狂热的宗教热情和仇恨终于退场的时候,时间到了十八世纪,这时的建筑艺术已经把古典主义建筑语言和当时的社会需求结合到一起,二者可以自然地并行不悖,历史再次让建筑艺术有所收敛,古希腊的建筑风格意外地出现了,文艺复兴时期作为一种模仿艺术终于寿终正寝。

意大利建筑力求再现古代建筑艺术,有的时候是在追求神似,有的时候又是讲究古代留传下来的文字资料。它一会儿是在不断地追求新意,一会儿却又在苦读钻研维特鲁威的陈旧说辞。古典主义精髓在各种各样的形式下被模糊了,虽然所有这些形式都在力争表现古典主义精神,但是,真正的古典主义精神到底包括哪些东西呢?

二、建筑艺术回归到古典形式,这是文艺复兴时期人文主义运动的一个组成部分,是文化发展过程中的一个特殊阶段,它影响到人们生活的许多方面。在很多情况下,这种人文主义给人的印象是与周围现状有点格格不入的,它们的起源也都是很偶然随意的,但是它们的影响却深远隽永。人文主义艺术的外在表现形式则常常是很肤浅的,也带有强烈的学究气息。历史上的人文主义和学究气从来都是形影不离的;但是,理想中的人文主义则恰恰是酸腐学究的对立面。

人文主义就是普通人从自身出发:他的一切思想、感受与活动都是为了自己的需求;他们的一切活动都在遵循着事物变化的规律。文艺复兴时期发生的各种变革都体现了这一精神与人生观。建筑家布鲁乃列斯基、政治理论家马基亚维利(Macchiavelli)、艺术家米开朗基罗、政治家齐萨利·波尔吉亚(Cesare Borgia)、科学家伽利略(Galileo)等人,从这个角度来看,他们都是一样的,都是人文主义者。这些人在各自的领域里,突然之间悟到了属于自己的一种新的方法,他们便对自己的方法加以试验,然后义无反顾地把新方法投入到实践中去,直到取得最终结果。过去的一切权威、习俗、正统,统统被忽略,甚至遭到蔑视。他们坚持的观点就是讲究实际,从实践出发,用真实的眼光看待真实的问题,坚持一切为人服务。如果遇到诸如"这种新东西有什么价值吗?"之类的问题,他们不会去进行纯思辨式地去长篇大论,而是根据个人的经验来作出最后的裁决,而裁决的结果是不允许再上诉的。满足人们需求的便被看作是好的东西,这种做法也极大地提高了人们对自身力量的认识。

力量,或者权威,就是对这种自身能力进行自我反省认识的结果,把自身能力加以提升并且扩大。力量,或者权威正是文艺复兴时期所追求的理想:正因为如此,古希腊和古罗马自然地就成为了这一时期的具体形象和象征。罗马帝国代表了取得巨大权威的顶峰,

of achieved power: the Holy Roman Empire had preserved its memory. The names of Greeks and Romans survived as names of conquest; even Virgil and Ovid were magicians, necromancers, kings. In their words, could the due sorcery be found, power still lay hidden. But most of all, because most visible, the stones which the Romans had built endured into the mediaeval world, dwarfing it by their scale and overshadowing it with their dignity. These were tokens of power which all could understand, and their effect upon the awakening mind of the Renaissance may be judged in the sonnets of Du Bellay. Humanism, therefore, inevitably fastened the imagination of architects upon the buildings of Rome.

The Renaissance style, we have already seen, is all architecture of taste, seeking no logic, consistency, or justification beyond that of giving pleasure. In this, clearly, it follows the natural bent of humanism, in its stress on liberty of will. And the baroque manner with its psychological method, its high-handed treatment of mechanical fact and traditional forms, is typically humanistic. But this claim of freedom involved architecture in a dilemma. For every art, and architecture more than any, requires a principle of permanence. It needs a theme to vary, a resisting substance to work upon, a form to alter or preserve, a base upon which, when inspiration flags, it may retire. So long as architectural art was closely linked to utility and to construction, these of themselves provided the permanent element it required. Greek architecture had on the whole observed the logic of the lintel, Gothic the logic of the vault. The restrictions which these constructive principles imposed, the forms which they helped to suggest, were sufficient for design. But when architecture, in the Renaissance, based itself on an experimental science of taste, and refused all extraneous sanctions, it felt for the first time the embarrassment of liberty. Baroque art, as soon as the creative energy deserts it, has nothing to fall back upon. It then becomes (as its failures prove) an unmeaning and aimless force, '*bombinans in vacuo.*'

Architecture, therefore, having denied the absolute authority of use and construction to determine its design, was led to create a new authority in design itself. And since Humanism, with its worship of power, had exalted Rome to an ideal, it was naturally in Roman design that this authority was sought. Roman buildings had to provide not merely an inspiration, but a rule.

Thus the mere aesthetic necessities of the case were sufficient to lead the tentative classicism of Brunelleschi towards the stricter manner of Bramante, and to recall the libertinism of the seventeenth century back to the academic yoke of Palladio.

SEVEN

后来的神圣罗马帝国（the Holy Roman Empire）继续保持了对古罗马帝国的记忆。那些从古希腊、古罗马留传下来的名字都被当作英雄的征服者；甚至传颂英雄故事的诗人维吉尔（Virgil）和奥维德（Ovid）也成为具有魔法的术士，掌握着向亡灵问卜的巫术，是至高无上的王者。他们留下的文字仍然具有魔力，力量和权威也隐藏在其中。但是，真正让当时人们感到那种巨大力量的并且是看得见、摸得着的东西，应该是罗马人建造的那些巨大的石头建筑。它们遗留了下来并且经历了整个漫长的中世纪，在它们巨大的尺度面前，中世纪的东西成了侏儒，同它们气宇轩昂的豪情相比，中世纪的东西根本抬不起头来。这些古罗马的巨大建筑才是真正权威的象征，这一点很容易被人们理解和接受。文艺复兴时期的人们在面对这种巨大建筑物时的觉醒，我们可以通过法国十六世纪的诗人杜贝雷（Du Bellay）的十四行诗来了解一些。因此可以说，人文主义思想抓住当时建筑师想象力的正是罗马城里遗留下来的古罗马建筑物。

我们在前面的讨论中已经看到，文艺复兴时期的建筑风格是一种根据人们审美爱好而兴起的形式，它并不拘泥于逻辑推理，也不坚持必须前后一致，或者什么理由，它只是为了满足人们的喜好。这样一来，这一时期的建筑风格就是在跟着人的自然习性在发展，它强调的就是人们的意愿、人们的意志，人们想怎样，它就怎样。这种情形发展到了巴洛克时期，艺术形式中充满了心理因素，人们对于工程技术的高度掌握以及对传统建筑形式的熟练应对，让人们的自由意志得到最充分地发挥。但是，这种对于自由的掌握带来了建筑艺术中的一个两难问题。每一种艺术形式，尤其是建筑艺术，都需要坚持一种永恒的艺术创作原则。它需要一种主题，在这个主题下追求各种形式的变化；它必须要有所依靠、有所依归；它也需要一种基本形态来加以变化或者保留，一种类似于根据地的地方作为自己的根基，当灵感得到启发而自由畅想的时候，仍然不会失去自己的根本。只要是建筑艺术不脱离它的实用性和建造技术的约束，那么建筑艺术就不会缺少这种根本。从宏观上来看，古希腊建筑艺术一直在遵循着梁柱结构的逻辑原理，哥特建筑一直在遵循着拱券技术的原理。这些建造技术原则强加在建筑艺术上面的这些约束，以及这些技术所带给我们的建筑形式，足以构成建筑设计的根据。但是，到了文艺复兴时期，主导建筑艺术发展的动力是人们的审美爱好，人们不再被动接受外部力量的约束，在建筑历史上第一次遇到了自由所带来的问题和窘境。当巴洛克艺术失去了创造力之后，一下子就变得根本没有立足之地。它就变成一种没有任何意义、没有任何目标的东西，像一只无头的苍蝇，在空中到处乱飞乱撞（bombinans in vacuo）。

因此说，当建筑艺术拒绝接受实用性和建造技术的绝对约束时，建筑艺术则必然是在形式设计中寻找自己的新原则。因为人文主义观念崇拜权威，它便把古罗马拔高成一种追求的理想，也就顺理成章地从古罗马人的设计中寻找自己需要的权威。罗马人的建筑不仅仅成为自己创作的思想源泉，而且也是衡量的法则。

单纯的美学追求因此变成了对权威的崇拜，布鲁乃列斯基富有试验特征的古典主义建筑被伯拉蒙特严格的手法所取代，十七世纪的放荡自由思想又再次回归到帕拉第奥严谨学术性的约束之下。

But other causes, still more powerful, were at work. Three influences, in combination, turned Renaissance architecture to an academic art. They were the revival of scholarship, the invention of printing, the discovery of Vitruvius. Scholarship set up the ideal of an exact and textual subservience to the antique; Vitruvius provided the code: printing disseminated it. It is difficult to do justice to the force which this implied. The effective influence of literature depends on its prestige and its accessibility. The sparse and jealously guarded manuscripts of earlier days gave literature an almost magical prestige, but afforded no accessibility; the cheap diffusion of the printing press has made it accessible, but stripped it of its prestige. The interval between these two periods was literature's unprecedented and unrepeated opportunity. In this interval Vitruvius came to light, and by this opportunity he, more perhaps than any other writer, has been the gainer. His treatise was discovered in the earlier part of the fifteenth century, at St. Gall; the first presses in Italy were established in 1464 ; and within a few years (the first edition is undated) the text of Vitruvius was printed in Rome. Twelve separate editions of it were published within a century; seven translations into Italian, and others into French and German. Alberti founded his great work upon it, and its influence reached England by 1563 in the brief essay of John Shute. Through the pages of Serlio, Vitruvius subjugated France, till then abandoned to the trifling classicism of Francois Ⅰ. ; through those of Palladio he became supreme in England. 'Nature, O Emperor, ' wrote the Augustan critic, 'has denied me a full stature: my visage is lined with age: sickness has impaired my constitution.... Yet, though deprived of these native gifts, I trust to gain some praise through the precepts I shall deliver. I have not sought to heap up wealth through my art.... I have acquired but little reputation. Yet I still hope by this work to become known to posterity.' Never was a hope more abundantly fulfilled. Upon this obsequious, short, and unprospering architect the whole glory of antiquity was destined to be concentrated. Europe, for three hundred years, bowed to him as to a god.

The treatise which has so profoundly altered the visible world was indeed exactly designed to fit the temper of the Renaissance. It is less a theory of architecture than an encyclopaedia of knowledge, general and particular, in easy combination. 'On the Origin of All Things According to the Philosophers' is the title of one chapter: the next is named 'Of Bricks.' The influence of older Greek treatises is everywhere apparent, particularly in the subtle observations upon optics, and a chapter on acoustics. Aesthetic distinctions are drawn in the manner of the Sophists, and Greek words are constantly employed. On the other hand, the author's first-hand experience is no less obvious, especially in his detailed directions for military architecture. The comprehensive scope of the book answers exactly to

SEVEN

但是还有其他的因素也在发挥着影响,而且也同样具有十分强烈的影响力。其中有三种主要的影响,它们的影响力合并到一处,就使得文艺复兴时期的建筑艺术变成了一种充满学术气氛的艺术。这三种主要影响是:对学问和理论的追求,印刷术的发明和应用,以及维特鲁威著作的发现。理论学问让追求古代艺术的做法得到正名,成为一种名正言顺的理想;维特鲁威的著作提供了一个典范;而印刷术则是散播种子的工具。这三种因素汇集到一起所形成的巨大力量,其影响难于估计。理论著作的实质有效影响力取决于著作的地位和普及程度。早期的手稿让人难得一见,保存手稿的人也生怕别人看见或者被损坏,不肯轻易示人,其结果是让古代文献手稿具有一种难以企及的高贵地位,但是,根本没有任何普及可言;到后来随着印刷术的推广,复制起来很便宜,让这些典籍变得十分普及,也因此让它们失去了原先的崇高地位。介于这两种极端情况之间的那个时期,是古代典籍空前绝后地最为具有影响力的时代。维特鲁威的著作就是在这个时期被发现的。同历史上任何理论家相比,维特鲁威大概是最大的受益者。他的著作是在十五世纪初期被人发现的,发现地点是在圣高尔(St. Gall)修道院。这本书第一次印刷发行是在1464年,在意大利;不出几年的工夫,维特鲁威的著作便在罗马畅销起来。在一百年内一共出现了十二个不同的版本,意大利文的版本就有七种之多,其余的版本是法文版和德文版。阿尔伯蒂正是在这本书的基础上写出他自己的那部不朽的著作的;维特鲁威是通过一个名叫约翰·舒特(John Shute)的人所写的介绍,才在1563年首次为英国人所熟悉。塞利奥(Serlio)的著作把维特鲁威带到了法国,一下子把法国给征服了。在那之前,法国出现的都是弗朗索瓦一世时期的那些花里胡哨的古典建筑;因为帕拉第奥的著作,维特鲁威在英国成为至高无上的绝对权威。"啊,伟大的罗马皇帝,造化没有给予我高大的身材,岁月也在我的脸上留下无数道痕迹,"这位奥古斯都时期的理论家这样写道,"疾病也让我的身体行动不便……虽然有这些不是很幸运的事情发生,但是我坚信我还是可以通过发表以下的论文来获得一些赞誉。到目前为止,我并没有通过我的才艺来获得什么财富……我只是通过我的技艺获得了一点点微不足道的声誉。然而,我仍然期望这部著作能够让我们的子孙后代知道历史上曾经有过我这么一个人。"历史上恐怕还没有任何许下的愿望得到像这个愿望一样的实现程度。这位阿谀奉承皇帝的建筑师,身材不高、生意又不好,结果是他集中代表了古典时期的全部荣耀。整整三百年,全欧洲把他当神明一样供奉。

这部深深改变了物质世界面貌的论著实际上完完全全是为文艺复兴时代的脉搏和精神而写成的。相对来说它关于建筑理论的论述很少,绝大部分是百科全书式的汇编;里面的内容既有概述,也有具体细致的描写,各类内容只是简单地汇集到一起。比如,某一章的标题是"论述哲学家们关于万物的起源",接下去的一章却是"关于砖瓦",章节和章节之前缺乏连贯性。古老的希腊论述方式在书中比比皆是,希腊文化的影响非常明显,尤其是关于视觉方面的微调来取得造型的最佳效果,关于声学的一般原理等,更是这样。关于美学特征的论述根本就是古希腊诡辩学者的论证方式,整本书中到处都是希腊的词汇。但是同时,这本书里也包括了许多内容,明显说明作者具有亲身建筑经验,尤其是在军事设施方面的建筑经验。这本包罗万象的著作满足了当时的各种好奇心,同时它里面的论述既可以照书本上的方法操作,也留有足够的任意发挥空间,十六、十七世纪欧洲人的思想都投入到这

the undiscriminating curiosity, at once practical and speculative, by which in the sixteenth and seventeenth century the mind of Europe was devoured. In and out of a vast store of useful, practical advice upon construction and engineering are woven a complacent moral philosophy, some geometry and astronomy, and a good deal of mythical history. We read of the Sun's Course through the Twelve Signs, and of Ctesiphon's Contrivance for Removing Great Weights. The account of the origin of the Doric Order is quoted by John Shute. It is a simple one: 'And immediately after a wittie man named Dorus (the sonne of Hellen and Optix the Nymphe) invented and made the first piller drawen to perfection, and called it Dorica.' And the history of the Corinthian Order—a charming fable—satisfied even some of the polished critics of the eighteenth century.[1]

All this was eagerly received, but most eagerly of all were welcomed the famous 'Rules.' 'The capitals must be such that the length and breadth of the abacus are equal to the diameter of the lower part of the column and one eighteenth more; the whole height (including the volute) must be half a diameter. The face of the volutes must recede by one thirty-ninth fraction of the width of the abacus, behind its extreme protection.' And so forth, through all the infinite detail of classic architecture. On those recondite prescriptions the humanist architects fastened; these they quoted, illustrated, venerated, praised; and these they felt themselves at total liberty to disregard.

Ⅲ. For it is too often forgotten by those who assail the influence of Vitruvius, how little in the curiously dual nature of the Renaissance architect the zeal of the scholar was allowed to subjugate the promptings of the artist. True, the zeal of scholarship was there, and it was a new force in architecture; but, fortunately for architecture, the conscience of scholarship was lacking. Pedantry, in that astonishing time, was an ideal; it was an inspiration; it was not a method. Vitruvius helped the architect to master the conventions of an art, of which the possibilities were apprehended but not explored. He wrapt it in the pomp and dignity of learning.

---

[1] I quote this story—like the last—in Shute's English: 'After that, in the citie of Corinthe was buried a certaine maiden, after whose burial her nourishe (who lamented much her death) knowing her delightes to have bene in pretye cuppes and suche like conceyts in her life time, with many other proper things appertayninge onely to the pleasure of the eye, toke them, and brake them, and put them in a littell preatie baskette, and did sette the basket on her grave, and covered the basket with a square pavinge stone. That done, with weeping tears she sayde, Let pleasure go wyth pleasure; and so the nourishe departed. It chanced that the baster was set upon a certain roote of an herbe called Acanthos, in frenche Branckursine, or bearefote with us. Now in the spring time of the yere, when every roote spreadeth fourth his leaves, in the encreasing they did ronne up by the sides of the basket, until they could ryse no higher for the stone, that covered the basket; and so grew to the fashion that Vitruvius calleth Voluta.' Calimachus of Corinth, passing by, borrowed the idea for the Corinthian Order.

本书里面去了。在这个装满了建造技术和工程经验的杂货店里,我们时常会碰到很多自以为是的道德哲学、几何学、天文学,以及数不清的历史故事和传说。其中有关于从十二星座看太阳运行规律的章节,也有关于古代波斯国泰斯芬(Ctesiphon)时期发明的巧妙的起重设计。约翰·舒特在自己介绍维特鲁威的著作时,也谈到了多立克柱式(Doric Order)的来源。它被描述得很简单:"一个名叫多洛斯人(Dorus)[他是希腊神话传说中的希腊祖先赫楞(Hellen)和掌管视觉的仙女(Optix the Nymphe)所生的儿子],他很有智慧,发明了第一根最完美的柱子,人们马上就用他的名字来给柱子命名,多立克柱式。"至于科林斯柱式(Corinthian Order)的来历,舒特简介中采用的那种童话故事般的叙述,甚至让十八世纪里最有学问的理论家也为之倾倒。[1]

所有的这些传说、真实的历史、各种理论,在这个时期均受到极大的欢迎,但是最为人们喜爱的莫过于那些著名的"建筑法则"。"柱头上面的方形顶石,它的长度和宽度必须等于柱子底部的直径再加上十八分之一的直径,而它的高度(包括漩涡装饰在内)必须是半个柱径。漩涡装饰的表面一定要从最外边缘缩进一些,缩进的尺寸应该是顶石宽度的三十九分之一。"诸如此类的法则,用极细微的尺寸,不厌其烦地详尽描述古典建筑的各种细节。这个时期的人文主义建筑师紧紧抓住这些法规的细节规定,他们到处引用这些法则,用图画来说明注释它们、崇敬它们、赞美他们。这些建筑师同时也坚信一点,只要他们自己愿意,所有这些法则条文都是可以忽略不计的。

三、那些极力贬低维特鲁威巨大影响的人常常忘记一个事实,那就是,在文艺复兴时期的建筑师所具有的既追求学术,又重视艺术创作的二重性当中,学术的研究对于艺术热情的限制实际上是微乎其微的。这个时期的建筑师热衷于学术研究,对于这一点,没有人会有疑问,这也是建筑艺术发展中的一股全新的力量;但是,这种对于学术的热衷绝对不是有意识地为了追求学术而追求学术,这可以看作是建筑艺术的万幸。在这个令人惊奇的时代,讲究学问是一种理想,是一种激励自己进步的动力,但它还没有成为一种做事情的方法。维特鲁威的著作帮助当时的建筑师了解并掌握了建筑艺术的传统技艺,这些技艺虽然已经为人了解,但是各种潜在的可能性并没有都被投入到实践中检验。维特鲁威通过一种学术的包装,把这些传统技艺装扮成华丽荣耀的学问,赋予它们崇高的尊严。但是在意大利,维特鲁威的

---

[1] 我在这里还是引用舒特的原话(老式的英文,拼写也和现代英文略有不同):"打那儿以后,在科林斯(Corinthe)城某一个地方的地下,埋着一位少女,她的保姆在她死后,极度地悲伤。她知道,这位少女在生前,每当看到美丽的杯子和花饰,每当看到很多让人眼睛感到愉悦的东西,就会开心不已。想到这些,保姆便把这些美丽的东西收集起来,放进一个花篮里,并且把这个盛着少女喜爱的那些宝贝东西的花篮,放在少女的坟前。保姆又在花篮的上面用一块正方形的石板盖起来。做完这些,保姆含着眼泪说,让快乐伴随着快乐吧,说完便离开了。保姆没有想到的是,这个花篮刚好放在一种名叫Acanthos的草根上,这种草在法国叫Branckursine,我们俗称它为狗熊裤子的一种草。现在刚好又是在春天,这些草根开始发芽、生长,很快,这种草便围绕着花篮长起来了,一直长到草叶碰到盖在花篮上的石板。也就因此有了维特鲁威所记载的Voluta。"生长在科林斯的卡利马索斯(Calimachus)刚好经过这里,便借用这个故事,给这种建筑柱式起名为科林斯柱式(the Corinthian Order)。

But in Italy when he was found at variance with the artist's wishes, his laws were reverently ignored. Even the austere Palladio, when it came to building, permits himself much latitude, and the motive of his written work is far less to propagate the canon of Vitruvius than to make known his own original achievements, which he reckons 'among the noblest and most beautiful buildings erected since the time of the ancients.' Vignola's outlook is no less practical. 'I have used this often, and it is a great success,' he writes against a classic cornice: '*riesce molto grata.*' And Serlio, the most ardent Vitruvian of all, admits the charm of novelty.

These were the masters of the academic school. The other camp—the architects of the style which culminated in Borromini—used the classical forms when and how they pleased, as mere raw material for a decorative scheme. They were consumed by a passion for originality that at times became a vice. Whatever their faults—and with the main charges against the baroque we have already dealt—no one could accuse them of imitativeness.

Academic art has its danger. Sometimes it implies a refusal to *rethink* the problem at issue. Sometimes, by a kind of avarice of style, it attempts to make the imagination of the past do service for imagination in the present. But this was not the case in Italy. The difference in the conditions which ancient and modern architecture had to meet, no less than the craving for originality that, after Michaelangelo, became so prominent in the art, were guarantees that the academic formula would not produce sterility. To the energy of Italian architecture, distracted as it was by insistent individualities, made restless with the rapid change of life, split by local traditions and infected always by the disturbing influence of painting, the academic code gave not a barren uniformity but a point of leverage, and a general unity of aim. If some needless pilasters and arid palaces were at times the consequence, the price was not too high to pay.

Outside Italy the value of the academic tradition was different but not less great. Here its function was not to restrain a too impatient and pictorial energy, but to set a standard and convey a method. The Renaissance was an accomplished fact: Europe had turned its back on mediaevalism, and looked to Italy for guidance. Italian architecture was the fashion: this was inevitable. But the 'Italian' styles which sprang up in France and England, while they sacrificed the unaffected merits of the old national architecture, were a mere travesty of the foreign. The spirit of fashion, as is commonly the case, seized on the detail and faild to grasp the principle. Ignorant builders, with German pattern books in hand, were little likely to furnish space, proportion and dignity. But capitals and friezes were the authentic

学问呈现出来的面貌完全是根据不同的艺术家所对它进行的解释而有所不同；在这些不同艺术家的各种解说中，维特鲁威的法则几乎在虔诚的崇拜中都被忽略了。即便是严肃认真的帕拉第奥，自己进行建筑创作的时候，也是允许自己根据需要和自己的艺术判断，来进行各种取舍，绝对不会拘泥于维特鲁威说过什么。帕拉第奥写作自己的著作的目的也是更多地在宣扬自己原创的建筑作品，而宣扬维特鲁威的准则则是次要的。他自认为自己的作品"是属于自从古希腊、古罗马以来最高贵、最美观的建筑物。"维尼奥拉（Vignola）的观点也同样是很讲究实用的。"我经常采用这种手法，每次都非常成功"，他在一座古典建筑的门楣上刻下了一行字"感恩不尽（riesce molto grata）"。这个时期意大利建筑师中最热情崇拜维特鲁威的人应当是塞利奥，他也承认自己的创新实际上带给他很多乐趣。

上面这些人还都是属于认真做学问的一派建筑大师。另外一派讲究艺术形式的建筑师则是把古典建筑的形式作为装饰性的材料，随心所欲地根据自己的美学原则加以利用。这一派的代表人物是巴罗米尼（Borromini）。这一派艺术家满腔热忱地在追求与众不同，追求新奇，这在当时被认为是一种罪恶。所有这些人的艺术创作，我们可以有各种各样的指控，可以说它们的各种错误，但是，无论如何，我们很难指控巴洛克时期的建筑艺术是在抄袭、模仿。

讲究学术的艺术存在着一种危险。它常常暗示着我们不去关注眼前需要解决的问题本身。有时候过于贪心地迷信某一种风格，把过去的某一种艺术形式风格不假思索地用来解决现在的问题，给今天的建筑戴上过去的面具。但是在文艺复兴时期的意大利，情况根本不是这样的。古代的具体情况与现代生活中的具体情况之间的不同，这一事实对于艺术创作的影响绝对不亚于艺术家追求原创性对艺术创作的影响。追求原创、追求新意自从米开朗基罗出现以来已经成为艺术创作的主要趋势。而古今具体情况的不同，可以说它保证了讲究学术的艺术公式也不至于产生出千篇一律的乏味作品。意大利的建筑艺术，在普遍的情绪高昂的同时，也受到每一位艺术家追求个性的影响，加上社会不断变化中的生活，各个城市区域的文化差异，再就是绘画艺术对于建筑艺术的影响，这一切避免了讲究学术的艺术风气把建筑艺术变成一个死气沉沉的复制工厂。学术在这时只是被看成是一种手段，一种杠杆，是一种大致上的全体追求目标。如果说在讲究学术的风气下产生出某些没有实际功能作用的柱子，或者没有什么表情的宫殿建筑，那么我们可以说，这其实算不得什么，这个代价并不算高。

在意大利之外的地区，讲究学术理论的传统与意大利有所不同，但是一点也不逊色。在意大利以外的地区，这种学术理论的作用，不在于约束艺术家的肆无忌惮地搞出一些自以为是的花样，而在于确立一些设计准则，教给当地建筑师一些设计方法。文艺复兴运动是一个有目共睹的事实：整个欧洲已经完全抛弃了中世纪流传下来的文化，当时全欧洲的人把目光转向意大利，希望从那里找到自己的指导方向。意大利的建筑艺术在这时是最时尚的设计，这已经是既成事实，向意大利看齐是不可避免的趋势。但是这种"意大利式"的建筑到了法国和英国则被看作是可笑的模仿外国的舶来品，在那些强烈的当地民族传统面前不免受到一些委屈。当时对意大利建筑的崇尚完全是追求时髦。在追求时尚的力量推动下，真正被抓住的就是一些细节和片断，而基本原理则是视而不见的。那些无知的工匠们，手里拿着德国建筑样式的书籍，根本不会注意到建筑空间的整体关系、建筑的比例关系和建筑的气质。

mode of Rome. Thus, with an ardent prodigality, little pilasters of all shapes and sizes were lavished, wherever they could find a footing, upon Jacobean mansions and the chateaux of Touraine. But the printed pages of Serlio and Palladio, when they came, were a pledge of orthodoxy. The academic influence rescued the architecture of England and France. It provided a canon of forms by which even the uninspired architect could secure at least a measure of distinction; and genius, where it existed, could be trusted to use this scholastic learning as a means and not an end. Wren, Vanbrugh, and Adam in England, and the whole eighteenth-century architecture of France, are evidence of the fact.

  The value of Vitruvius was relative to a time and place. After three hundred years of exaggerated glory and honest usefulness he became a byword for stupidity. Pope satirised him; archaeologists discovered that the Roman buildings corresponded but imperfectly to his laws; the Greek movement dethroned the authority of Rome itself; science turned its back on Greece and Rome together; and Romanticism, with its myth of 'untaught genius,' cast scorn on all codes, rules, and canons whatsoever, and as such.

  In this revulsion was born the current prejudice that Renaissance architecture is 'imitative, academic, unalive.' A measure of truth, slight but sufficient to give the Prejudice life, underlies the judgment. Fundamentally it is a confusion. An art is academic, in this harmful sense, when its old achievements crush down the energies that press towards the new. But the academic canons of the Renaissance did not represent the past achievements of the Renaissance, but of antiquity. To the Renaissance they were the symbol of an unsatisfied endeavour: the source, consequently, not of inertia, but of perpetual fruitfulness. The pedantry was superficial. Beneath this jargon of the 'Orders—to us so dead, to them so full of inspiration— the Italian architects were solving a vast and necessary problem. They were leading back European style into the main road of European civilisation—the Roman road which stretched forward and back to the horizon, sometimes overlaid, but not for long to be avoided. They were adapting, enlarging, revivifying the forms of the antique to serve the uses of the modern world. The change was deeply natural. Europe no longer recognised itself in the hopes and habits of its immediate past; it did recognise itself, on the contrary, in that remoter and more civilised society in which it had its origin. The mediaeval styles had run their course and outlived their usefulness To have resisted the logic of events, to have clung to the vestiges of local Gothic—vital and 'rational, as in their time they had been, picturesque and romantic as they are in their survival—this in truth would have been an artificial act

但是,柱头和柱廊上方的横梁部分可以把罗马的样式模仿得惟妙惟肖。因此,借助于工匠们的天才技艺和满腔热忱,各式各样的柱子都可以做到精雕细刻、花样繁多,无论是英国雅各宾式(Jacobean)的豪宅,还是法国土伦(Touraine)地区的城堡,只要能放下几个柱子的地方,就都会出现一些华丽造型的柱子。但是,当塞利奥和帕拉第奥的著作出现以后,正宗的建筑样式便被确立了。这种讲究学术传统的建筑艺术拯救了法国和英国的建筑,避免了恶劣时髦的东西泛滥成灾。塞利奥和帕拉第奥的理论确立了建筑形式的创作原则,这样即使是没有什么天分的建筑师也有据所依,使得他们的建筑设计不至于太走样。而到了天才艺术家的手里,这些理论和原则自然就会被当作是一种一般的原理,而不会变成碰不得的金科玉律。比如在英国建筑师伦恩、范堡卢(Vanbrugh)、亚当(Adam)的手里,以及整个十八世纪法国建筑师的手里,意大利的建筑理论只是一种指南、一种参考,不会被当成菜谱那样照本宣科。这一点有目共睹。

维特鲁威著作的价值因时间和地点而有所不同,不能一概而论。经过三百年的盲目崇拜与夸大其词的赞美,以及经过真正的实用检验之后,维特鲁威的名字成了愚笨的代名词。英国讽刺作家蒲柏就讽刺挖苦过他;考古学家也发现古罗马时期的建筑并不完全符合他在书中所陈述的法则;到古希腊建筑复兴运动开始的时候,古罗马的权威地位被推翻了;而到了崇尚科学技术的时期,古希腊、古罗马就统统被抛弃掉;浪漫主义带着自己天生的才华,嘲弄、蔑视一切法则、规定、经典,一切的一切都统统被废除。

正是在这种文化骤变中出现了一种偏见,认为文艺复兴时期的建筑艺术是"模仿性质的、讲究学术的、没有生命力的"。一定程度上的事实的确像这种偏见所说的那样,虽然只有很少的事实支持这样的说法,但是它们足以让这些偏见得以流传散布开来,成为一种主流判断。从根本上讲,这完全是出于误解,出于做出判断的人本身的思路不清晰。当一种艺术以往取得的成就,到后来成为阻碍新生力量进行探索的包袱,那么我们可以说,这种艺术已经演变成注重学术性质的艺术,而且是注重学术所产生的负面意义。但是,文艺复兴时期的追求学术的标准,不是宣扬它自己的过去成就,不是故步自封;而是宣扬古代的成就,是把古代的艺术实践当作是一个未完成的梦想来继续,是人类永恒成就的一种象征,不是过去的惯性,不是前进的包袱。文艺复兴时期的追求学术理论实际上都是很表面的,并不是深入、认真钻研的那样在做学问。这些柱式在我们今天看来是一些死的东西,但是在当时却是充满启发性的,是非常灵活的。在各种所谓的"柱式"理论名称之下,这个时期的意大利建筑师实际上正在解决一个巨大的历史问题:他们正在把欧洲的建筑艺术风格拉回到欧洲文明的正途上来,而正途就是罗马人的道路,这是一条通向未来的道路,它的起点来自遥远的地平线,这条大路在起点到未来的路途中,的确有过某些地方曾经和其他道路交织、重叠在一起,但是,过了不多久,这些岔路又分开了,这条大路又呈现出自己清晰的路线。文艺复兴时期的意大利建筑师们采用过去的建筑形式,把它们的内容扩充,让古代的形式重新获得生命来服务于现代的生活需要。这种变革是十分自然的。这时的欧洲,在刚刚过去的这段历史中,亦即中世纪的历史中,看不到自己的希望,不愿意再继续那种生活方式。相反,他们从遥远的古代文明中看到了自己的未来,看到了自己的希望,而那个古代文明原本就是自己的起点。中世纪产生的各种艺术风格到今天已经成为一种被动的重复,气数已尽,它们的外在

of style. It would have led, in a few generations, to a state of architecture as unalive, as falsely academic, as were the shams of archaeology three hundred years later.

That Renaissance architecture was built up around an academic tradition—that it was, in a measure, imitative—will not, if we understand aright the historical and aesthetic conditions of the case, appear to be a fault. The academic tradition will, on the contrary, be realised as a positive force that was natural, necessary, and alive. The Renaissance architects deviated from the canon whenever their instinctive taste prompted them to do so; they returned to the canon whenever they felt that their creative experiment had overreached its profitable bounds. And it should be realised that a convention of form in architecture has a value *even when it is neglected*. It is present in the spectator's mind, sharpening his perception of what is new in the design; it gives relief and accent to the new intention, just as the common form of a poetical metre enables the poet to give full value to his modulations. So, in Renaissance architecture, a thickening of the diameter of a column, a sudden increase in the projection of a cornice, each subtlest change of ratio and proportion, was sure of its effect. A new aesthetic purpose when it is ready for expression first shows itself and gathers force in a thousand such deviations, all tending in a sole direction. We may mark them, for instance, in the early years of the baroque, and realise how large a factor in their effect lies in the academic canon which they contradicted.

And if the inherited conventions of architecture assist the articulation of new style, they serve also to keep keen the edge of criticism. In Florence the advent of a new moulding could be the subject of epigrams and sonnets; the architect who ventured it risked a persecution.[1] The academic tradition ensured that the standard of taste was jealously guarded and critically maintained.

IV. An academic *tradition*, allied, as it was in the Renaissance, to a living sense of art, is fruitful; but the academic *theory* is at all times barren.

The view that, because certain forms were used in the past they must therefore

---

[1] Cf. the excitement which, according to Milizia, was roused by Baccio d'Agnolo's treatment of the windows of the Bartolini Place. The wrath of the Florentines might, in this case, have been appeased by a closer acquaintance with the Porta de' Borsafi at Verona, where Baccio has a classic precedent.

形式早已失去了内容的支持。如果违反事物的自然规律,抱残守缺地抓住几个属于地方性质的哥特建筑艺术做法不放,这必然成为一种不自然的艺术风格,完全是人工捏造出来的东西。不错,哥特建筑在它成熟的阶段有过旺盛的生命力,也是充满"理性"的艺术风格,到了后来追求图画般构图效果的艺术追求和浪漫主义艺术的风行,也让哥特建筑风格得以适当的保留。但是可以肯定的一点是,不出几代人,这种风格就会显露出自己的虚假学术性,暴露出它根本没有任何生命力。现在的这些东西和三百年后考古学家企图用虚假的东西骗人,本质上没有什么两样。

从某种意义上来说,似乎文艺复兴时期的建筑艺术的确是有一点讲究学术的味道。换句话说,就是它的确有点模仿前人的嫌疑。但是,如果我们正确地理解了这种现象的历史原因和审美原因,这种模仿前人的说法就会露出破绽;而且相反,我们还会从中看出文艺复兴时期出现的所谓追求学术的做法是有益处的,是正面的行为,是自然而然的行为,是必需的,也是有生命的。文艺复兴时期的建筑师,只要是自己的艺术敏感性告诉他们,自己必须与古典风格有所区别的时候,他们会毫不犹豫地扬弃古典主义的东西;如果他们发现自己的艺术追求让自己与古典艺术原则有了距离,而同时又认识到古典艺术在这时可以对自己的创作有所帮助,他们又会毫不犹豫地回归古典艺术原则。有一点应该得到充分的认识,建筑艺术中的传统形式即使是*被忽略、被遗忘*,仍然具有十分重要的价值。传统的形式是存在于观察者和欣赏者的脑子里的,因为这些传统形式的存在才使得他们清醒地认识到新元素的出现,新元素才因此被凸现和强调出来。这一点好比是诗歌的一般格式那样,诗人完全可以用传统的格式来填写出自己的新诗篇。所以说,在文艺复兴时期的建筑艺术中,柱子的直径略微加粗一点,线脚的出挑尺寸突然变大一点,每一部分的尺寸和比例出现一些微小的变化,这些艺术变化都在发挥着自己的作用。当一种新的美学观念需要表现自己的时候,它会在正式出现之前积蓄能量,积蓄的过程会有一些偏差,但是这些数以千计的尝试与能量的积累,都是朝着一个方向努力的。例如,我们从巴洛克时期早期看到这些能量的积蓄,也能够从中看出学术在这个巴洛克形成过程中到底发挥过多大的作用。而学术理论本来就是同巴洛克艺术风格是互相矛盾着的。

如果建筑艺术中传承下来的过去的传统,在新风格出现的过程中有所助益的话,那么这些传统也让建筑艺术理论保持着锐利的锋芒。在佛罗伦萨,一个新的建筑线脚出现的时候,艺术理论家会作诗赋词来描述它、赞美它,而创造出这个新线脚的建筑师同时也要准备好被攻击。[1]学术理论的传统就是要确保审美趣味标准不受任何侵犯,就是要批判地保持这样的传统。

四、一个具有学术精神的*传统*,再加上一种活生生的艺术与之配合,如文艺复兴时期那样,那么,这种传统是可以结出许多果实来的;但是单纯追求所谓的学术*理论*,这在任何时候都是贫瘠的,不会有什么结果的。

仅仅因为某些建筑形式曾经在过去出现过,那么在将来这种建筑形式再次出现的时候

---

[1] 根据米利兹亚(Milizia)的描写,意大利艺术家巴齐奥·当格尼奥拉(Baccio d'Agnolo)设计的巴托里尼宫(the Bartolini Palace)的窗户处理手法在当时引起了极大的争议,闹得沸沸扬扬。佛罗伦萨民众的不满后来因为得知,在维罗纳(Verona)的波萨利城门(Porta de' Borsari)上早在一千五百年前就出现过类似的做法,这才得到逐渐的平息。

be used without alteration in the future, is clearly inconsistent with any development in architecture. But that idea is, in effect, what the academic theory implies. And our modern cult of 'purity,' and 'correctness' in style reposes on the same presumption. 'By a "mistake,"' wrote Serlio, 'I mean to do contrary to the precepts of Vitruvius.' This happens now to sound absurd enough. But it is not more absurd than the taste which insists, in modern building, upon 'pure' Louis XVI. or 'pure' Queen Anne. Certainly every deviation from achieved beauty must justify itself to the eye, and seem the result of deliberate thought and not of mere ignorance or vain 'originality.' But deviations, sanctioned by thought and satisfying the eye, are the sign of a living art; and the cult of 'correctness,' is only to be supported on the assumption that architecture is now, and for ever, a dead contrivance to which our taste and habit must at all costs conform. Consequently, the judgment that Renaissance architecture is 'not classical enough is as ill-grounded as the judgment that it is 'too classical.'

This meticulous observance of 'pure styles' is a mark of a failing energy in imagination; it is a mark, also, of an inadequacy in thought: of a failure to define the nature of style in general. We cling in architecture to the pedantries of humanism, because we do not grasp the bearing upon architecture of the humanist ideal.

Criticism is in its nature intellectual. It seeks to define its subject matter in purely intellectual terms. But taste—the subject matter of criticism—is not purely intellectual. The effort of criticism to 'understand' architecture has done no more than add its own assertions to the confused assertions of mere taste. It has not rendered taste intelligible.

Of this tendency to over-intellectualize architecture we have already traced some typical examples. We have seen architecture reduced to purely mechanical terms, and to purely historical terms; we have seen it associated with poetical ideas, with ideas of conduct and of biology. But, of all forms of criticism, the academic theory which confines architectural beauty to the code of the Five Orders—or to any other code—is the most complete example of this excessive intellectual zeal. It is the most self-conscious attempt that has been made to realise beauty as a form of intellectual order.

Indeed, it is often stated that the beauty of classic architecture resides in Order. And Order, upon analysis, is found to consist in correspondence, iteration, and the presence of fixed ratios between the parts. Ratio, identity, and correspondence form part of the necessary web and fabric of our thought. Reason is compelled to seek them. When it finds them we feel conscious of understanding and control. Order is a desire of the mind; and Order is found in classic architecture. What more natural, then, than to say that architectural beauty—the beauty of classic architecture, at

SEVEN

也不能加以改变,必须照原样重新复制才行,这种观点显然不符合任何建筑艺术的发展规律。但是这种思想恰恰是追求学术理论所暗示的做法。而我们现代人在强调某种建筑形式的"纯粹"和"正宗"的时候,事实上也是在鼓吹这一观点。塞利奥在书中就曾写道:"我所说的'错误'指的就是不符合维特鲁威著作中所陈述的做法。"这个说法在今天听起来,我们都会觉得有些不可思议,会觉得它过于狭隘。但是,在当前的流行审美观念面前,我们常听到某人说要建造一个"纯粹"的路易十六时期风格的建筑,或者"纯粹"的大不列颠安妮女王(Queen Anne)时期风格的建筑的时候,其实是基于同一种思想,其荒谬性毫不逊色。如果在过去受欢迎的建筑风格基础之上加以任何改变,都必须首先要能够为我们自己的眼睛所接受,要能够经得起我们思想的检验,而不是出于无知,或者自以为是的"别出新裁"。那些经得起思想检验、经得起眼睛审视的艺术,说明仍然是一种具有活力的艺术,而强调"正宗"的那些人无非就是认为建筑艺术在现在、将来都不过是一堆没有生命力的东西,是一堆在任何情况下我们的审美爱好和审美习惯都必须无条件服从的装饰。因此,我们可以得出结论,那种认为文艺复兴时期建筑艺术还"不够经典"的观点,与那些认为同时期建筑艺术"过于古典"的观点,都是因为自己的立场有偏差而得出的错误结论。

这种全神贯注地盯住"纯粹建筑风格"细节的现象正说明想象力的衰落;也说明思想方面的缺欠。这两点加到一起,正说明持这种观点的人在整体上缺乏对艺术风格的一般定义。我们在建筑艺术中坚持重复人文主义的那些迂腐教条,其理由就是因为我们没有抓住建筑艺术中的人文主义理想。

艺术批评理论从本质上讲是一种属于理性的思辨活动,它需要从理性和思辨的角度来确定自己讨论的主题和范围。但是同时,人们的审美观念并不是单纯理性的行为。当艺术批评理论在面对自己的主题,亦即人们的审美观念这个问题的时候,在对于建筑艺术的"理解"方面,理论能够做的不过是把自己的一些判断结论摆放在审美爱好所得出的判断结论旁边而已;审美爱好中出现的各种困惑,艺术理论也没有办法让它们变得更容易理解一点。

在试图把建筑艺术进行超出常规的系统理论化的各种尝试中,我们已经看到了一些有代表性的例子。我们看到了有人把建筑艺术简化为单纯的建造技术演变过程,有人试图把它简化为历史现象,有人把它与诗情画意结合起来,有人则试图用行为道德准则或者生命进化规律来解释。在所有这些试图理论化的努力中,以追求学术为诉求的艺术理论把建筑的美归结为五种柱式的细致规定,是这类理论狂热的最极端的代表。这是认为通过理性分析便可以实现创造美的最自以为是的一种做法。

我们常常听见一种说法:古典建筑的美在于它采用了某种古典柱式。经过分析,我们知道,所谓的古典柱式是由各部分之间的固定比例关系所组成的,各个部分之间存在着彼此呼应、重复、对比关系。而比例关系、各部分的具体特征,以及部分与部分之间的彼此呼应构成了我们思想中的某种组织关系。理性被迫要对这些关系进行分析、比较,找出其中的规律。当我们从中看到了某些关系,我们就会有意识地感知它们的存在,就以为我们理解、掌握了它们。古典柱式便是一种我们的主观愿望;古典柱式于是便在古典建筑身上被发现了。还有什么说法能够比说古典主义建筑的美存在于古典柱式更加顺理成章的呢?柏拉图不是说

any rote—consists in Order? What higher or more perfect beauty, Plato asked, can exist, than mathematical beauty? And the academic criticism, with its canon of mathematical ratios, enforces the demand.

The intellectual bias of our criticism must be profound which allows this theory to be asserted. For this agreeable fancy—so flattering to the intellect, and so exalted—dissolves at the first brush of experience. It should at once be apparent that Order in design is totally distinct from Beauty. Many of the ugliest patterns and most joyless buildings—buildings from which no being can ever have derived delight—possess Order in a high degree; they exhibit fixed and evident ratios of design. Instances of this among the hideous flats, warehouses and other commercial buildings of our streets require no citation. Here is Order, and no beauty, but, on the contrary, ugliness.

Eighteenth-century critics, perceiving this difficulty, were fond of saying that beauty consisted in 'a judicious mixture of Order and Variety'; and this definition, for want of a better, has been a thousand times repeated. The emendation assists us little, for on the nature of the 'judicious' no light is thrown, save that it lies in a mean between the too much on the one hand, and the too little on the other. And, by a still more fatal oversight, it is not observed that almost every possible gradation of order and variety is found among things admittedly beautiful, and no less among things admittedly ugly. A certain minimum of order is implied in all *design*, good or bad; but, given this, it is clear that what satisfies the eye is not Order, nor a ratio between Order and Variety, but beautiful Order and beautiful Variety, and these in almost *any* combination.

Order, it is allowed, brings intelligibility; it assists our thought. But the act of quickly and clearly perceiving ugliness does not become more pleasant because it is quick, nor the ugliness beautiful because it is evident; and order combined with ugliness serves but to render that ugliness more obvious and to stamp it gloomily upon the mind.

So, too, with proportion. The attempt has constantly been made to discover exact mathematical sequences in beautiful buildings as though their presence were likely either to cause beauty or explain it. The intervals of a vulgar tune are not less mathematical than those of noble music, and the proportions of the human body, which artists like Leonardo (following Vitruvius) sought to describe within a circle and a square, are not most beautiful when they can be exactly related to those figures. It was realised that 'proportion' is a form of beauty: it was realised that 'proportion' is a mode of mathematics. But it was not realised that the word has a different bearing in the two cases. Criticism is not called upon to invent an aesthetic for disembodied minds, but to explain the preferences which we (whose minds

过的吗，还有什么样的美比数学之美更高尚更完美的呢？于是，这种追求学术的艺术理论便用各种数学比例关系来强化自己的发现。

艺术理论中的主观倾向性一定是非常深远的，所以像上面提到的这样的理论才会出现。这种对理论普遍适用性的想象与憧憬在理论接触到具体实践的一刹那便会立即瓦解。理论对于智慧具有非常强烈的诱惑，也受到智慧的崇拜。但是我们会立刻注意到，古典柱式与建筑之美是根本不相干的两件事情。很多外观布局难看、毫无趣味的建筑，任何人都无法从这些建筑上获得任何愉悦的感受，但是这些建筑却有着很规范的古典柱式，这些柱式准确地呈现出设计的各种比例关系。说起这样的建筑，我们就会立刻联想起我们城市大街两旁的那些令人恐怖的住宅楼、仓库、商店等，我想就不需要在这里具体列举出它们的名字了吧。在这些建筑上，我们看到了古典柱式，但是我们看不到任何美，相反，我们看到了丑陋。

十八世纪的艺术理论家注意到这种问题，感受到其中的难点，因此改口说：建筑之美在于"谨慎巧妙地使用古典柱式的秩序和各种变化手段"；由于缺乏更好的定义，这个说法被成千上万的人重复过。这个改变了的说法对我们也没有什么具体帮助，因为没有人告诉我们怎样才能做到"谨慎巧妙"，它大概是指在从过多到过少之间的某一个平均值吧。另外，这个新说法中还有一个更致命的问题大家都没有注意到，那就是，在所有美的建筑中，在所有丑陋的建筑中，各种程度不同的古典柱式秩序与变化手段，在所有的建筑中到处都是。只要是设计就必然具有一定的秩序，无关*设计*的好坏；这样说来，显然让我们的眼睛感到满足的绝对不是古典柱式的秩序，也不在于古典柱式变化的程度，可以说与它们之间的任何组合方式都没有关系。

说古典柱式可以让人们容易理解建筑，这一点是没有疑问的。它的确能够帮助我们厘清思路。它帮助我们及时准确地识别丑陋的建筑，却并不能因为它的迅速而让我们感到快乐，也不会因为它让我们识别出丑陋就让我们从中获得美感。古典柱式秩序加上整个建筑的丑陋只不过是让丑陋更加容易被识别而已，它让我们的思想情绪更加灰暗。

至于比例关系也一样。人们一直以来不曾间断地在美观的建筑上面寻找一些数学关系，希望借此体现美之所在，至少可以解释美感的来源。一段庸俗的小曲中所包含的数学比例关系绝对不比神圣的音乐缺少什么，而人体的比例关系，即便是能够找到一个人满足达·芬奇所画的维特鲁威人比例关系，即呈现正方形和圆形叠加在一起的比例关系，这个人也不是最美的。人们在过去已经认识到，和谐的"比例关系"是一种美的形式，人们也认识到，和谐的"比例关系"是一种数学形式。但是，人们没有认识到自己所说的"比例关系"一词，在两个不同学术领域中所代表的意义是不一样的。艺术理论不是用来发明美的东西以便满足主观思想的期待，而是用来解释我们主观思想在喜欢和欣赏美观的东西到底是发生了什么。我们的审美观念具有部分的物质性，不是全部。数学中所说的"比例关系"完全是抽象的，而美学体验中的"比例关系"则具有身体上的情感内涵。美学这里也是在说法则

are not disembodied) do actually possess. Our aesthetic taste is partly physical; and, while mathematical 'proportion, belongs to the abstract intellect, aesthetic 'proportion' is a preference in bodily sensation. Here, too, are laws and ratios, but of a different geometry. And there can be no sure criticism of architecture till we have learnt the geometry of taste.

Mass, Space, Line, and Coherence constitute, in architecture, the four great provinces of that geometry. When it has satisfied science with 'firmness,' and common use with its commodity, architecture, becoming art, achieves, through these four means, the last 'condition of well-building'—its 'delight.' By the direct agency of Mass and Space, Line and Coherence upon our physical consciousness, architecture communicates its value as an art. These are the irreducible elements of its aesthetic methods. The problem of taste is to study the methods of their appeal and the modes of our response; and to study them with an attention undiverted by the Romantic, Ethical, Mechanical, Biological or Academic Fallacies of the impatiently concluding mind.

和比例,但是它是一种不同的几何学。我们在搞清楚审美爱好这门几何学之前,建筑艺术是不可能有可靠的艺术理论。

　　体量、空间、线条以及和谐构成了我们审美爱好中物质基础的四个主要方面。当建筑满足了科学要求的"坚固"与日常生活的"实用"之后,建筑通过这四个方面的组合与互动,构成了"优秀建筑三要素"中的最后一个要素——美观,建筑便上升为艺术。但是,体量、空间、线条以及和谐,这四个方面在直接与我们的感觉系统互动的时候,建筑直接表现出来的就是它的艺术方面。这四个方面在讨论美学感受问题的时候是不能再加以进一步简化的。审美观念的问题就是要研究这些元素给我们的影响力,以及我们感知系统在接收到这些影响之后所作出的反应,而且在研究这些问题的时候,是全神贯注地在关注着这些问题本身,排除诸如浪漫主义、道德说教、建造技术、生命进化、追求学术等等谬误,不再受这些谬误对于人类急功近利思想的影响。

EIGHT
Humanist Values

I. Architecture, simply and immediately perceived, is a combination, revealed through light and shade, of spaces, of masses, and of lines. These few elements make the core of architectural experience: an experience which the literary fancy, the historical imagination, the casuistry of conscience and the calculations of science, cannot constitute or determine, though they may encircle and enrich. How great a chaos must ensue when our judgments of architecture are based upon these secondary and encircling interests the previous chapters have suggested, and the present state of architecture might confirm. It remains to be seen how far these central elements—these spaces, masses and lines—can provide a ground for our criticism that is adequate or secure.

The spaces, masses and lines of architecture, as perceived, are appearances. We may infer from them further facts about a building which are not perceived; facts about construction, facts about history or society. But the art of architecture is concerned with their immediate aspect; it is concerned with them as appearances.

And these appearances are related to human functions. Through these spaces we can conceive ourselves to move; these masses are capable, like ourselves, of pressure and resistance; these lines, should we follow or describe them, might be our path and our gesture.

Conceive for a moment a 'top-heavy' building or an 'illproportioned' space. No doubt the degree to which these qualities will be found offensive will vary with the spectator's sensibility to architecture; but sooner or later, if the top-heaviness or the disproportion is sufficiently pronounced, every spectator will judge that the building or the space is ugly, and experience a certain discomfort from their presence. So much will be conceded.

Now what is the cause of this discomfort? It is often suggested that the top-heavy building and the cramped space are ugly because they suggest the idea of instability, the idea of collapse, the idea of restriction, and so forth. But these *ideas* are not in themselves disagreeable. We read the definition of such words in a dictionary with equanimity, yet the definition, if it is a true one, will have conveyed the idea of restriction or collapse. Poetry will convey the ideas with vividness. Yet we experience from it no shadow of discomfort. On the contrary, Hamlet's

第八章
人文主义的价值

在我们面对建筑艺术的时候，最简单、最直接感受到的东西，就是空间、体量和线条的某种组合，通过光和影呈现在我们面前。就是这几样东西构成了建筑艺术体验的全部核心内容。这种建筑艺术体验是无法借助于文学幻想、历史想象、良知道德的判断或者科学的计算等方法来加以决定的，也不能借助那些手段来构成。这些手段只能在问题的周边打转，丰富对问题的理解，但是不能切中要害。我们对建筑艺术所作的判断是根据周边这些各种次要的方法来进行的，我们的建筑艺术理论能够产生出多么混乱的局面，前面几章已经作了一定的解说，而我们眼前的建筑艺术现状可以证明这些解说。还没有解决的问题是，被围绕在当中核心位置的那些元素，即空间、体量、线条，能够提供什么样的基础，让我们建立起一个充分的，并且经得起考验的建筑艺术理论。

建筑艺术中的空间、体量、线条，都是显而易见的，是我们能够直接感受到的元素。我们可以由此引申出一些我们不能直接感受到的内容，例如有关建筑施工建造方面的内容，有关历史文化和社会的内容。但是，建筑中有关艺术的内容却又都是与这几种元素直接相关的，与它们直接呈现出来的效果有关。

这些呈现在我们面前能够为我们感受得到的内容都与我们人类自己的一举一动有关。那些建筑的空间让我们感受到在其中活动的情形，建筑的体量让我们联想到我们自己对重量的承受能力以及对压力的抵抗能力，如果我们把建筑中的那些线条加以描绘的话，那正是我们的行走路线或者身体的动作姿态。

我们假想面前有一个"头重脚轻"的建筑物，或者一个"比例关系很差劲的空间"。我们对这种恶劣建筑的反应到底会有多强烈，这取决于观察者对于建筑艺术的敏感程度；但是，只要这头重脚轻与差劲的比例关系达到相当的程度，那么，人人都会觉得这个建筑很难看，这样的空间会很不舒服。这一点不需要多说，大家都不会有异议。

但是，是什么引起我们这种内心不舒服的感觉呢？最常听说的解释是，头重脚轻的建筑物，或者压抑的空间之所以让人觉得难看，是因为它们让我们感到失去稳定性，会想到坍塌，会联想到受到制约而行动不便，等等。但是，这些联想本身并不会让我们感觉到有什么不适。我们在字典里查找这些词汇的时候，心情也是很平静的。如果字典里的定义是准确的话，它们也早已经把那些不便或者坍塌的意思传达给我们了。诗歌中这类的描述更加生动，但是我们不会因为诗歌的生动描述而感到任何的不安与恐惧。同样因为准确的表达，莎士比亚话剧中哈姆雷特的"我现在却被关起来、囚禁起来，限制在一个狭小的地方"（作者

'cabined, cribbed, confined' delights us, for the very reason that the idea is vividly conveyed. Nor does Samson painfully trouble *our* peace, when

'Those two massie Pillars
With horrible convulsion to and fro
He tugged, he shook, till down they came and drew
The whole roof after them with burst of thunder
Upon the heads of all who sate beneath. '

Clearly, then, our discomfort in the presence of such architecture cannot spring merely from the idea of restriction or instability.

But neither does it derive from an actual weakness or restriction in our immediate experience. It is disagreeable to have our movements thwarted, to lose strength or to collapse; but a room fifty feet square and seven feet high does not restrict our actual movements, and the sight of a granite building raised (apparently) on a glass shop-front does not cause us to collapse.

There is instability—or the appearance of it; but it is in the building. There is discomfort, but it is in ourselves. What then has occurred? The conclusion seems evident. The concrete spectacle has done what the mere idea could not: it has stirred our physical memory. It has awakened in us, not indeed an actual state of instability or of being overloaded, but that condition of spirit which in the past has belonged to our actual experiences of weakness, of thwarted effort or incipient collapse. We have looked at the building and identified ourselves with its apparent state. *We have transcribed ourselves into terms of architecture.*

But the 'states' in architecture with which we thus identify ourselves need not be actual. The actual pressures of a spire are downward; yet no one speaks of a 'sinking' spire. A spire, when well designed, appears—as common language testifies—to soar. We identify ourselves, not with its actual downward pressure, but its apparent upward impulse. So, too, by the same excellent—because unconscious—testimony of speech, arches 'spring', vistas stretch, domes 'swell,' Greek temples are 'calm,' and baroque facades 'restless.' The whole of architecture is, in fact, unconsciously invested by us with human movement and human moods. Here, then, is a principle complementary to the one just stated. *We transcribe architecture into terms of ourselves.*

This is the humanism of architecture. The tendency to project the image of our functions into concrete forms is the basis, for architecture, of creative design. The tendency to recognise, in concrete forms, the image of those functions is the true

误把麦克白的台词写成哈姆雷特的台词)却让我们能够陶醉其中,没有一丝的恐惧与厌恶。而关于力士参孙(Samson)的诗歌描写也没有引起*我们读者*的任何不安:

"那两根巨大的柱子
在巨大的震动中前后摇摆
他在用力拖住一根,他试图扶持着另一根,直到两根柱子倾倒,连带着
整个屋顶跟着一起坍塌在地,爆发出的声响
如同雷电,在屋顶下面进餐人的头上轰鸣。"

从上面这些例子看到,我们在面对这种头重脚轻的建筑和比例关系恶劣的空间所产生的不愉快感受,绝对不是因为我们由此联想到建筑的不稳定或者我们自己行动的不自由。

这种不愉快的感受也肯定不是来自建筑物本身真实的不坚固或者真实具有的不方便,限制了我们的在其中的行动。实际情况是,这些建筑物并没有阻碍我们在其中的行走,建筑本身也没有失稳或者坍塌;一个五十英尺(约15.2米)见方、高七英尺(约2.1米)的空间,根本不会限制我们在其中的行动;一个花岗岩的建筑物最下面地面层都是商店的玻璃窗也不会引起建筑物的坍塌。

我们讨论的建筑,看上去是有一定的视觉上不稳定特征,但是,那是建筑物自身的视觉不稳定性,而那种不舒服的感觉却是真实地发生在我们自己身上。到底怎么引起这种不舒服的感觉呢?结论似乎很显然:我们眼前看到的东西对我们产生了一种作用,而这种作用是其他文字、思想所不能产生的:它们引发我们对自己亲身经历的记忆,眼前的景象唤醒了我们记忆深处的亲身经验。不是因为眼前建筑物的不稳定状态,也不是它承受的重力过大,而是我们思想深处过去的不安经验被唤醒,这些不安的经验有的是因为不够坚固,有的是不便于我们的行动,有的是坍塌的前兆。我们在面对建筑物的时候,把它的外在状态与我们自己的亲身经历联系起来。*我们已经把我们自己转换成建筑语言,投射到建筑身上了。*

但是,当我们把自己与建筑艺术中的某种"态势"对应起来的时候,建筑艺术中的这种外在状态不一定是真实的物理状态。一个尖塔的受力因为重力的缘故显然是向下的,但是从来没有人会用"向下沉没的尖塔"来描述它。一个尖塔如果设计得恰当,看上去总是给人向上腾飞的印象,人们最常见的描述语言可以证明这一点。我们把我们自己的感受同建筑的外观印象联系到一起,而不是它们的真实物理特性。有些俗话和习惯用语,因为在说的时候并没有刻意去想,所以,更能说明人们的真实感受。比如说拱券,我们用"弹跳、出挑",视觉范围,我们用"延伸",穹顶,我们用"膨胀",古希腊的神庙,我们用"静谧",巴洛克建筑的正立面,我们用"无休止的动感"。实际上,建筑艺术的全部内容在无意识的情况下已经被我们用我们自己的身体的动作与情绪统统描述过了。在这里,我们可以引申出一条与上面结论互补的另外一条基本原则:*我们把建筑艺术转换成我们自己的语言,用描述我们自己身体动作的语言来描述建筑。*

这就是建筑艺术中的人文主义。一方面,把我们自己的动作和情感投射到建筑上面,用具体真实的建筑形式加以表现,作为创造性设计的真实基础。另一方面,在真实的建筑形式中,我们看到了我们自己的动作与情感,用我们自己的动作与情感来理解建筑的形式,作为

basis, in its turn, of critical appreciation.[1]

II. To this statement several objections may be expected. This 'rising' of towers and 'springing' of arches, it will be said—these different movements which animate architecture—are mere metaphors of speech. No valid inference can be drawn from them. Again, the enjoyment of fine building is a simple and immediate experience, while this dual 'transcription,' by which we interpret the beauty of architecture, is a complicated process. And not only—it will again be objected—is the theory too complicated; it is also too physical. The body, it will be said, plays no part—or a small and infrequent part—in our conscious enjoyment of architecture, which commonly yields us rather an intellectual and spiritual satisfaction than a conscions physical delight. And it will be further said that such a theory is too 'farfetched'; we cannot readily imagine that the great architects of the past were guided by so sophisticated a principle of design. And, if some such process has indeed a place in architecture, it may be doubted finally how far it can account for all the varied pleasures we obtain. It will be convenient to consider these objections at the outset.

The springing of arches, the swelling of domes, and the soaring of spires are 'mere metaphors of speech.' Certainly they are metaphors. But a metaphor, when it is so obvious as to be universally employed and immediately understood, presupposes a true and reliable experience to which it can refer. Such metaphors are wholly different from literary conceits. A merely literary metaphor lays stress on its own ingenuity or felicity. When we read

'Awake, for Morning in the bowl of Night

---

[1] The theory of aesthetic here implied, is, needless to say, not new. It was first developed by Lipps twenty years ago, and since then has been constantly discussed and frequently misunderstood.

In what follows I owe a debt to many suggestive points in Mr. Berenson's studies of Italian painting, where this view of aesthetics found its most fruitful concrete application. With this exception the present chapter has been derived wholly from the author's own immediate experience in the study and practice of architecture, and is intended to satisty rather an architectural than a philosophical curiosity. Time-honoured as Lipps's theory now is, and valid as it appears to me to be, its influence upon purely architectural criticism has been negligible. In English architectural writing it is totally ignored; even Mr. Blomfield, the most philosophical of our critics, gives it but a frigid welcome. (*The Mistress Art*, p. 118.) Yet its architectural importance, both for theory and practice, is immense; and it is for lack of its recognition that the Fallacies of Criticism still flourish so abundantly. For some theory criticism must have, and in the absence of the true, it makes shift with the palpably false.

I have avoided, as far as clearness seems to permit, all purely psychological discussion. Those interested in this aspect of the matter will find in the recent writings of Vernon Lee the most extensive survey of the question which has appeared in English, together with all necessary references to the foreign literature of the subject.

建筑艺术批评理论的真实基础。[1]

二、对于我们刚刚得出的结论，一定会有人有不同的看法。他们会说，尖塔的"高耸"和拱券的"弹跳、出挑"等说法，不过是一些修辞学中比喻罢了。我们不能从这些比喻中引申出什么有实质意义的结论。再者说，对于优秀建筑作品的欣赏和从中获得愉悦的感受其实是很简单的一件事，不过就是直接的感受，而上面结论中所说的"双向"转换过程实在是太麻烦了，没有必要。这个理论不仅太过复杂，而且太过于具体，一切都在用具体的身体来加以说明。他们会说，人们的身体在我们欣赏建筑艺术中根本没有发挥任何作用，退一步讲，即便是有一定的作用，也是很小、微不足道的作用，在欣赏过程中真正发挥作用的是人们智慧上和精神上的满足，肯定不是身体上的。他们会继续说，这个理论扯得太远了，我们无法想象过去的伟大建筑师会用这么复杂的理论来指导他们的设计。即使是有某些过程的确如同这个理论这样说的，在建筑艺术创作中具有一定的影响，我们仍然怀疑这个影响对于我们从建筑上感受到的结果有多大关系。为了论述的方便，现在让我们从这个反对意见的最外围开始来一点一点分析它。

拱券的弹跳、出挑，穹顶的膨胀，尖塔的高耸，都不过是"修辞学中的比喻"手法。的确没有错，它们确实都是比喻的说法。但是，这种比喻不是一般的比喻，而是一种所有的人都能认同的说法，而且是所有的人能够立刻理解并且毫不怀疑的说法。这样一种比喻就是不言而喻地把一种真实的经验作为自己的前提。这样的比喻与文学作品中的比喻完全不同。文学作品中比喻依赖于比喻的贴切与否。例如当我们读到下面这两行古波斯诗人海亚姆的诗句的时候：

"醒来吧，因为黑夜笼罩的天空下的黎明

---

[1] 这里所包含的美学观点根本不是什么新发明，这一点无须赘述。它是由利普斯（Lipps）在二十年前首先提出，从那以后，就从来没有停止过对它的讨论。多数情况下，这个观点总是被误解。

下面的讨论内容，有很多是受到贝伦森先生关于意大利绘画艺术的研究成果的启发才得以形成的。贝伦森先生的研究应该是这一观点的具体应用并取得卓越成就的研究。除了这一项之外，这一章里的其他内容则都是作者本人根据自己的亲身经历与建筑实践所得到的结论，拿出来与大家分享的目的主要是回答一些建筑艺术的疑问，并不打算进一步的美学哲学讨论。利普斯的理论，到今天已经经历过了时间的检验，现在看起来仍然具有说服力，但是它对建筑艺术理论的影响却是微不足道的。至于英国本土的建筑艺术理论家则根本就无视它的存在，甚至连最有宏观思维的理论家布龙费尔德先生（Mr. Blomfield）在提到这个理论的时候，也是一副冷冰冰的面孔。（《主流艺术》，第118页。）但是这个理论，无论是从理论方面讲，还是从实践方面讲，都具有重大的意义。正是因为没有认识到这一理论的意义，建筑艺术理论中的这些谬论到今天仍然大行其道。因为没有真正理解它，有些理论甚至扭曲本意来附和一些错误的观点。

在不影响论述清晰的情况下，我已经尽可能地回避了所有有关纯粹心理学方面的讨论。如果读者对这方面的问题有兴趣，可以参考最近由维尔农·李（Vernon Lee）所写的有关书籍。她的著作应该是英文出版物里面关于这个话题最为全面的资料，同时里面也可以看到引用的很多相关的外文参考书籍。

Has flung the Stone which puts the stars to flight,'

we are first arrested by the obvious disparity between the thing and its description; we then perceive the point of likeness. But when we speak of a tower as 'standing' or 'leaning' or 'rising,' or say of a curve that it is 'cramped' or 'flowing,' the words are the simplest and most direct description we can give of our impression. We do not argue to the point of likeness, but, on the contrary, we are first conscious of the fitness of the phrase and only subsequently perceive the element of metaphor. But art addresses us through immediate impressions rather than through the process of reflection, and this universal metaphor of the body, a language profoundly felt and universally understood, is its largest opportunity. A metaphor is, by definition, the transcription of one thing into terms of another, and this in fact is what the theory under discussion claims. It claims that architectural art is the transcription of the body's states into forms of building.

The next point is more likely to cause difficulty. The process of our theory is complex; the process of our felt enjoyment is the simplest thing we know. Yet here, too, it should be obvious that a process simple in consciousness need not be simple in analysis. It is not suggested that we *think* of ourselves as columns, or of columns as ourselves. No doubt when keen aesthetic sensibility is combined with introspective habit, the processes of transcription will tend to enter the field of consciousness. But there is no reason why even the acutest sensibility to a resultant pleasure should be conscious of the processes that go to make it. Yet some cause and some process there must be. The processes of which we are least conscious are precisely the most deep-seated and universal and continuous, as, for example, the process of breathing. And this habit of projecting the image of our own functions upon the outside world, of reading the outside world in our own terms, is certainly ancient, common, and profound. It is, in fact, the *natural* way of perceiving and interpreting what we see. It is the way of the child in whom perpetual pretence and 'endless imitation' are a spontaneous method of envisaging the world. It is the way of the savage, who believes in 'animism,' and conceives every object to be invested with powers like his own.[1] It is the way of the primitive peoples,

---

[1] Thus it has of late been more fully realised that children and primitive races are often capable of very remarkable achievement in expressive art, while the scientific perception of the world for the most part undermines the gift. If the child or the savage is incapable of appreciating great architecture, it is not because they lack the aesthetic sense (for a child the general forms, for instance, of a piece of furniture are often charged with significance and impressiveness), but because the scope and continuity of their attention is too limited to organise these perceptions into any aesthetic whole, still more to give them concrete realisation. None the less, it is on this half-conscious or subconscious, yet not quite undiscoverable world in which, more than ourselves, they live, that architecture, like all the arts, depends for its effect.

已经把石头抛出,凭借这块石头把所有的星星都赶跑了。"

我们第一个直觉就是,诗歌中直接说出的语言和它们描写的内容之间,存在一种明显的不同;随后我们可以感受到二者之间的相似关系。但是,当我们说一个高塔"站立"在那里,或者"倾斜依靠着",或者"上升",这些描写是最真实、最直接把我们的感受表达出来;再比如说,我们说一个曲线"很局促",或者说"在流动",这些描写也是把我们的感受和印象在用简单、直接的语言叙述出来。我们根本不需要考虑比喻的贴切问题,因为我们的思想意识中清楚地知道,这些词语是十分准确的,只是事后才意识到,原来我们这是在使用比喻的手段。艺术在同我们交流的时候都是通过最直接的感知和印象,并不是通过理性的思考,这种普遍为人接受的肢体语言和用人体作为比喻,是人人都曾经感受到的,是人人都可以很轻易地理解的,因此它最容易为人们所接受。根据定义,比喻就是把一种事物转换成另外一种事物来进行说明,而这正是我们的理论所采用的方法,它的中心思想就是认为,建筑艺术是在把人体的各种状态转换成建筑的各种形式。

下面的问题回答起来会有一些困难。有人说,我们的理论具有一个很复杂的过程,而我们从建筑艺术中获得享受却是一个非常简单直接的过程。然而在这里我们必须了解一点,简单直接的感受并不能说明我们对它的分析也必须是简单直接的。我们不是说要把自己*设*想成柱子,或者把柱子设想成自己。毫无疑问,当敏锐的美学感受力与自我反省习惯结合到一起的时候,上面说的那种转换过程很可能进入到我们的清醒的意识当中,就是说,这种转换过程有可能是在有意识的情况下发生的。但是,即便是我们具有最敏锐的感受力,面对的结果是实实在在的快乐愉悦,只是这仍然不能让我们有任何理由说服我们自己,我们一定非要清楚地了解这个感觉快乐的过程到底是如何发生的不可。虽然不需要去了解,可是这种结果一定是有它的某些原因和过程的。这个过程中那些让我们最无法清醒准确地把握的地方正是潜藏最深刻、最普遍、最常出现的内容,如同我们的呼吸过程。把我们自身的动作和姿态投射到外面的物质世界上去,通过我们自身的语言来解读外部世界,这样的一种思考习惯是自古有之的,是十分普遍的现象,也是意义深远的一个现象。实际上,这是再*自然不过*的来感知和解释我们所看到的现象的方式。这正是孩童用自己的眼睛看待周围世界的时候所表现出来的方式;他们自己不是很懂,却总是在根据自己的本能装出很懂的样子,或者不停地模仿。这也是原始部落的人根据自己所能理解的方式来理解周围世界的方法,他们相信任何物体都具有"灵性(animism)",在各个方面就和我们人类自己一样。他们对待身边的物体如同对待自己人那样。[1]原始人的舞蹈动作正是通过舞蹈和身体动作来表达

---

[1] 到最近我们已经逐渐明确地认识到,儿童和早期的原始人类在艺术表现上其实是具有非凡能力的,只是因为科学的思维模式对于世界的认识在很大程度上限制了这种能力的开发。假如说,儿童或者原始人类不能够充分欣赏建筑艺术的话,这绝对不是因为他们缺少对美的感受(对于一个儿童来说,一个家具的基本形状就可以让这个孩子发挥出无限的想象力,这说明一个简单的家具对孩子来说就有很大的意义,也能够给孩子留下极深刻的印象。仅举一例),而是因为他们的注意力和联想范围过于狭窄,以至于不能把自己的感受放进一个更大的美学框架里来叙述,更关键的是他们并没有清醒地认识到自己的这些感受。但是,无论怎样讲,儿童和原始人类生活在这种半有意识,甚至无意识的状态下,所产生的对于这个还没有完全被自己掌握的世界的认识,才是建筑艺术,乃至所有艺术,所追求的效果,而不是我们今天自己已经了解的那些认识。

who in the elaborate business of the dance give a bodily rendering to their beliefs and desires long before thought has accurately expressed them. It is the way of a superbly gifted race like the Greeks, whose mythology is one vast monument to just this instinct. It is the way of the poetic mind at all times and places, which humanises the external world, not in a series of artificial conceits, but simply so perceiving it. To perceive and interpret the world scientifically, as it actually is, is a later, a less 'natural,' a more sophisticated process, and one from which we still relapse even when we say the sun is rising. The scientific perception of the world is forced upon us; the humanist perception of it is ours by right. The scientific method is intellectually and practically useful, but the naïve, the anthropomorphic way which humanises the world and interprets it by analogy with our own bodies and our own wills, is still the aesthetic way; it is the basis of poetry, and it is the foundation of architecture.

A similar confusion between what is conscious in architectural pleasure, and what is merely implied, seems to underlie the objection that our theory lays too great a stress on physical states. Our pleasure in architecture, it is true, is primarily one of the mind and the spirit. Yet the link between physical states and states of the mind and the emotions needs no emphasis. Our theory does not say that physical states enter largely into the spectator's consciousness; it says that they, or the suggestion of them, are a necessary precondition of his pleasure. Their absence from consciousness is indeed a point of real importance. Large modifications in our physical condition, when they occur, alter our mental and emotional tone; but, also, they absorb our consciousness. A person, for example, who is taking part in an exciting game, will reel exhilaration and may enjoy it; but the overtones of gaiety, the full intellectual and emotional interest of the state, are drowned in the physical experience. The mind is not free to attend to them. It is precisely because the *conscious* physical element in architectural pleasure is so slight, our imitative self-adjustment to architectural form so subtle, that we are enabled to attend wholly to the intellectual and emotional value which belongs to the physical state. If we look at some spirited eighteenth-century design, all life and flicker and full of vigorous and dancing curves, the physical echo of movement which they awaken is enough to recall the appropriate mental and emotional penumbra; it is not sufficient to overwhelm it. No one has suggested that the experiences of art are as violent or exciting as the experiences of physical activity; but it is claimed for them that they are subtler, more profound, more lasting, and, as it were, possessed of greater resonance. And this difference the theory we are considering assists us to understand.

Any explanation of the workings of the aesthetic instinct, however accurate,

自己关心的事物，表达自己的思想情感，而人类找到准确表达这些情感的方式是很久很久以后的事情。这种表达方式也是天才的古希腊民族通过自己的神话故事所表达出来自己的直觉。古希腊神话是人类文化的一座高耸的丰碑。任何时代、任何民族的诗人也都会采用这种表达方式：它把外部的世界人性化，而人性化的过程不是借助于诗人自己自以为是的主观想象，而是简简单单地接受着世界感染。我们今天按照科学的观念来了解世界、解释这个世界都是很久很久以后的事情，是十分复杂深奥的过程，一点儿也不自然；即便是那些完全接受了科学认识世界的人，也还会旧病复发地说出诸如太阳正在升起这样的话。科学感知世界的方法不是我们人类固有的，是迫于压力才勉强接受的，而人文主义的观念则是与生俱来的。科学的认识方法是理性分析的结果，可以解决不少实际问题；而采用幼稚拟人化手段看待我们的世界则是把世界人性化，把周围的物质世界同我们的人类的身体形态、我们的思想活动进行类比，这种幼稚的拟人化手段仍然是我们的审美途径，它构成了诗歌艺术的基础，它是建筑艺术的起点。

从建筑艺术中获得愉快的感受，有些是我们可以清醒地感受到它是如何发生的；有些则是并没有真实地感受到任何东西，只是让我们间接地产生出那些联想。这两者之间的混淆让一些人不认同我这里的理论，认为它过于强调人体状态对于我们审美体验的影响。对于建筑艺术的愉悦体验主要还是通过我们的大脑和精神感受，这是没有疑问的。然而在人体状态、思想状态和情绪之间的紧密联系是不需要特别强调的。我们的理论并不是说我们的人体状态会直接进入到审美观察者的清醒意识中来；它只是在说，这些人体的状态是我们体验美感愉悦时的前提条件。我们在体验美感的时候并没有意识到这些前提条件的存在，这一事实具有非常重要的意义。当我们的身体出现重大的变化时，我们的精神状态和情绪也会跟着发生剧烈变化；但是同时，身体的状态也会针对我们主观意识作出反应。举例来说，假如一个人正在从事一项激烈刺激的比赛，他会感到十分兴奋，也因此十分地投入；但是，他的兴奋、他的全神贯注所包含的智力活动，以及他的全部感情都因为身体的剧烈运动而被掩盖住了。他这时的注意力已经顾及不到这些方面。正是因为在建筑艺术的愉悦感受中，有意识的身体活动很有限，我们自我调整身体的姿态来呼应建筑形式的过程又非常微妙，因此，我们便可以把充分的精力投入到理性分析与情感的体验，这些活动在一定程度上也还仍然属于身体活动。如果我们注意观察十八世纪里活泼的设计，里面包含的所有的生命力、闪烁的火花，以及充满活力的舞蹈曲线等，它们所表现的运动感觉都会在我们身体上产生回响，而这种回响刚好足以让我们回想起头脑中与情绪中的边缘区域；它还不足以对这个边缘区域造成巨大冲击。没有人说过艺术的体验要和体育运动一样剧烈，也不可能那样紧张兴奋；我们的理论曾经断言，艺术的体验会更含蓄、更意味深长、更持久、更具有大量的共鸣。我们的理论正试图帮助我们理解这里的不同。

对于美学直觉到底是如何产生的这个问题所进行的任何解释，无论它是多么准确，都

must inevitably have a modern ring. It must seem incongruous when applied to the artists of the past, for the need and the language of such explanations are essentially of our own day. It would not therefore—to pass to the next objection—be a serious obstacle to our theory if the conception of architecture, as an art of design based on the human body and its states, had been wholly alien to the architects of the past. But this is not altogether the case. The Renaissance architects were, in fact, frequently curious to found their design upon the human body, or, rather, to understand how the human body entered into the current traditions of design. Among their sketches may be found some where the proportions of the male form are woven into those of an architectural drawing and made to correspond with its divisions. An elaborate, though uninspired, rendering of the Tuscan, Ionic, and Corinthian Orders into human forms was published by John Shute in the earliest printed work on architecture in England. And in this connection the ancient, though seldom felicitous, habit of actually substituting caryatides and giants for the column itself is not without significance. It was realised that the human body in some way entered into the question of design. But habits of thought were at that time too obiective to allow men any clear understanding of a question which is, after all, one of pure psychology. What they instinctively apprehended they had no means intellectually to state; and that correspondence of architecture to the body, which was true in abstract principle, they sometimes vainly sought to prove in concrete detail. Thus they looked in architecture for an actual reproduction of the proportion and symmetries of the body, with results that were necessarily sometimes trivial and childish. Vasari was nearer the truth when he said in praise of a building that it seemed 'not built, but born'—*non murato ma veramente nato*. Architecture, to communicate the vital values of the spirit, must appear organic like the body. And a greater critic than Vasari, Michaelangelo himself, touched on a truth more profound, it may be, than he realised, when he wrote of architecture: 'He that hath not mastered, or doth not master the human figure, and in especial its anatomy, may never comprehend it.

III. But, how far, it is natural to ask, can such an explanation be carried? Granting its truth, can we establish its sufficiency? Our pleasure in architectural form seems manifold. Can one such principle explain it? A full answer to this question is perhaps only to be earned in the long process of experiment and verification which the actual practice of architecture entails. How minutely Humanism can enter into the detail of architecture, how singularly it may govern its main design, could not, in any case, be demonstrated without a mass of instances and a free use of illustration. A study of these, drawn from Renaissance architecture, could form the matter of another volume. But the main divisions of the subject—

不可避免地具有我们现代的约束。如果试图把它应用到古代的艺术家身上则一定是不恰当的,因为这种解释的必要性和使用的语言,从根本上讲,完全是我们这个时代的产物。明白了这一点以后,我们再来看看下面一个针对我们的理论所提出的质疑。他们说,我们把建筑艺术理解为是把人类身体的各种状态来作为建筑艺术设计的根据,但是,古代的建筑师根本就没有听说过这种概念。要驳回这个质疑实际上应该不会有什么特别的困难。说过去的建筑师从来没有过这种概念实际上是完全不准确的说法。文艺复兴时期的建筑师就经常琢磨着要从人体比例关系中获得设计上的灵感,换句话说,这个时期的建筑师想要了解到当时为止所出现的建筑设计发展过程,人体的影响在其中到底发挥了多大的作用和影响。在那个时期流传下来的草图中,有些草图明显地可以看出,在建筑设计图旁边穿插着一些男性人体的速写,建筑设计上的一些主要特征对应着人体的各种比例关系。由约翰·舒特介绍到英国的建筑书籍中,也有大量的精美插图,显示出人体的形态与塔司干(Tuscan)、爱奥尼、科林斯等柱式的比例关系。插图很精美,但是缺乏灵气。按照这个思路往前追溯,古代的人干脆用女像柱或者巨人的人像柱直接替代柱子本身。这个事实也是非常能够说明问题的。这说明,人体已经正式进入到艺术设计的考虑范围之内了。但是,当时摆在匠人、艺术家面前的具体任务和当时人们的思考习惯,没有让他们深入地探讨这一问题;这个完全属于心理学范围的问题,没有让他们产生一个明确的认识。我们现在可以明白一点,古代的人实际上已经凭直觉清醒地认识到了这些问题,只是找不到恰当的语言来记录和表达。建筑艺术与人体的对应关系无疑是一种完全抽象的对应关系,但是,古代人总是在想着要用具体手段表现这一点。所以他们曾经尝试着在建筑造型上复制出人体的对称关系和各部分间的相互比例,当然,这样试验的结果必然是幼稚可笑的。文艺复兴时期的艺术家和理论家瓦萨里(Vasari)在赞美一座建筑的时候,用了"它不像是人们建造起来的,而是建筑物自己生长起来的(non murato ma veramente nato)"一句,这应该是最能接近这种认识的表达。建筑艺术在表达自己内在精神价值的时候必须表现为如同人体一样的有机形态。比瓦萨里更加伟大的理论家,米开朗基罗本人说过一句意味深长的话,直接触碰到真理的核心,只是他自己并没有清醒地认识到这一点。他说建筑师"如果还没有掌握人体绘画技巧,或者他不想掌握人体绘画技巧,特别是有关人体解剖构造的内容,那么他就永远不会理解建筑艺术。"

三、然而,我们自然要问,这样的一种解释能够走多远呢?就算它所说的内容全部正确,它能够足以解释所有的问题吗?我们在建筑艺术中获得的愉快感受不是单一的,是多个方面的,这样的一种解释能够回答全部的问题吗?这个问题的全面完整回答大概要靠建筑艺术的全部实践过程才能够概括。至于人文主义理论观点能够在怎样的细微层面上指导设计实践,怎样主导建筑设计的主要走向,无论怎样讲解,无论从哪一方面讲解,没有大量的事例和图片加以辅助说明,这两个问题都是无法说清楚的。如果把文艺复兴时期的建筑艺术沿着这一思路作一番整理和总结,完全可以再出版另外一本书。在这里,我们就这个话题的几个主要线索先简单地概括一下:我们要谈的是空间、体量、线条以及和谐一致,外加它们的一些主要应用情况。

space, mass, line and coherence, with their more obvious applications—may here be singled out.

The principle is perhaps most clearly to be recognised in *line*. Lines of one sort or another always form a large part of what is visually presented to us in architecture. Now in most cases, when we bring our attention to bear on one of these lines, its whole extent is not seen with absolute simultaneity; we 'follow' it with our eye. The mind passes successively over points in space, and that gives us movement. But when we have got movement we have got expression. For our own movements are the simplest, the most instinctive, and the most universal forms of expression that we know. Identified with ourselves, movement has meaning; and line, through movement, becomes a gesture, an expressive act. Thus, for example, the curves of a volute are recognised as bold or weak, tense or lax, powerful, flowing, and so forth. It is by such terms as these, in fact, that we praise or condemn them. But we must recognise them as having these qualities by unconscious analogy with our own movements, since it is only in our own bodies that we know the relation of the line—or movement—to the feeling it denotes.

Movement is most obviously communicated by curved lines; but it is conveyed also by lines which are straight. No doubt the straight lines which bound the rectangular forms of architecture, its doors and its windows, are chiefly realised, not as sensations in themselves, but as definitions of the shapes they enclose. Their chief use is to determine the position of a patch upon a given surface; and the aesthetic value of this will be considered in a moment. But any emphasis upon vertical lines immediately awakens in us a sense of upward direction, and lines which are spread—horizontal lines—convey suggestions of rest. Thus the architect has already, in the lines of a design, a considerable opportunity. He controls the path of the eye; the path we follow is our movement; movement determines our mood.

But line is not the sole means of affecting our sense of movement. Space, also, controls it. Spaces may be in two dimensions or in three. We may consider the simpler case first. A large part of architectural design consists in the arrangement of forms upon surfaces, that is to say, within spaces. The part which movement here plays will be clear from a common instance. A man who is arranging pictures on a wall will say that one is 'crowded' or 'lost' in the space it occupies, that it 'wants to come' up or down. That is to say, the position of forms upon a surface is realised in terms of our physical consciousness. If a certain patch 'wants to come' down, we ourselves, by our unconscious imitation of it, have the sense of a perpetually thwarted instinct of movement. The arrangement of the scheme is imperfectly *humanised*. It may be picturesque, it may be useful, it may be mechanically superior; but it is at variance with our ideal movement. And beauty of disposition

能够最清楚地识别出这个原则的地方是在建筑艺术的线条里。建筑艺术呈现在我们面前的视觉内容,很大一部分就是表现为各种各样的线条。在大多数情况下,当我们把我们的注意力放在这些线条上面的时候,整个线条的全貌不会一下子展现出来,眼睛会跟着线条移动。我们的理智跟随我们的眼睛走过空间里的一系列的点,这个过程带给我们一种运动的感觉。而当我们获得了一种动感的时候,我们便获得了一种表现手段。因为我们自己身体的运动方式是我们所熟悉的最简单、最为直接和本能的,是人人都可以理解的表现方式。因为这种运动是和我们自己的活动联系到一起的,因此,每一个动作都是有意义的。通过这些运动,这里的线条就会成为了一种身体的姿态,也就因此变成了一个表现的动作。因此,建筑细部中的漩涡就会产生或粗犷或纤细,或紧张或轻松,或孔武有力或行云流水的曲线。正是由于这些对于线条性格的描写和刻画,我们便对携带这些线条的建筑主体加以赞美或者贬抑。但是,我们必须认识到一点,这些线条之所以获得这些性格完全是由于我们在潜意识中把建筑上面的线条与我们自己身体的动作作过了一个类比,只有通过我们自己身体的这些动作才使得我们能够把这些线条与那些情感和性格特点联系到一起的。

曲线最能够清楚地表现出运动感觉,但是直线也可以用来表达出运动的感觉。构成矩形建筑形式的直线,构成门窗形状的直线等,虽然没有曲线那样动人,但是无疑可以很明确地规定出很多建筑元素的形状。它们的主要功能就是确定每一个元素在平整的表面上的位置,而这种功能的美学价值也是由这些元素形成的动感规律所决定的。但是,任何竖向垂直的直线都会唤起我们上升的感觉,而向两旁展开的水平线则带给我们一种休息的联想。仅仅凭借着线条,建筑师已经可以做多事情了。他控制着眼睛移动的路径;而眼睛移动的途径代表了我们的身体动作;而我们的身体动作又决定了我们的情绪。

线条不是唯一一种影响我们身体运动的手段。空间也决定了我们的身体运动状态。空间可以是二维平面的,也可以是三维立体的。先让我们从最简单的情况入手吧。建筑设计中有很大一块儿工作内容是在一个表面上安放一些形状,也可以说是在二维空间里工作。在这样的情形里,运动所能够发挥的作用可以从一个最简单的实例中清楚地说明。比如,一个人打算在墙面上挂上一些画,他或许会说墙面"太拥挤了";或者画挂上去以后就自己"消失了";墙面上的这些画,有的希望自己能够"再向上或者向下移动一点"。这说明墙面上的某些画,它们的位置已经在我们的主观意识中被转换成我们自己的身体语言了。如果某一幅画"需要向下移动一点",我们在潜意识中开始模仿这个向下的动作,我们自己便因此产生出遭受到不断挫败的那种动作的直觉感受。这时墙面的布置已经开始成为一种人文的东西,在一定的程度上已经开始具有了人类的性格。这时墙面的布局可以是按照绘画构图原则来完成的,可以是按照实用原则来完成的,也可以是受技术方面原因所左右的;但是,它最后都是根据不同的身体理想动作来作出最后判断的。建筑艺术构图的美感,正像线条所表

in architecture, like beauty of line, arises from our own physical experience of easy movement in space.

But not all movements are pleasant or unpleasant in themselves; the majority of them are indifferent. Nevertheless, a *series* of suggested movements, in themselves indifferent, may awaken in us an expectancy and consequent desire of some further movement; and if the spaces of architecture are so arranged as first to awaken and then falsify this expectation, we have ugliness. For example, if a design be obviously based on symmetry and accustoms us to a rhythm of equal movements—as in the case of a typical eighteenth-century house—and one of the windows were placed out of line and lower than the rest, we should feel discomfort. The offence would lie against our sense of a movement, which, when it reaches that point of a design, is compelled to drop out of step and to dip against its will. Yet the relation of the window to its immediately surrounding forms might not in itself be necessarily ugly.

A converse instance may here be given. Classic design—the style which in Italy culminated in Bramante—aims at authority, dignity, and peace. It does this by conveying at every point a sense of equipoise. The forms are so adjusted amid the surrounding contours as to *cancel all suggested movement*: they are placed, as it were, each at the centre of gravity within the space, and our consciousness is thus sustained at a point of rest. But the baroque architects rejected this arrangement. They employed space adjustments which, *taken in isolation*, would be inharmonious. In their church facades, as Wölfflin has pointed out, they quite deliberately congested their forms. The lower windows are jammed between the pilasters on either side; they are placed above the centre of gravity; they give the sense of lateral pressure and upward movement. This, taken alone, would leave us perpetually in suspense. But in the upper part of the design our expectancy is satisfied; the upward movement is allowed to disperse itself in greater areas of lateral space, and makes its escape in a final flourish of decorative sculpture; or it is laid to rest by an exaggerated emphasis upon the downward movement of the crowning pediment and on the horizontals of the cornice. Here, therefore, a movement, which in the midst of a Bramantesque design would be destructive and repugnant, is turned to account and made the basis of a more dramatic, but not less satisfying treatment, the motive of which is not peace, but energy.

IV. But besides spaces which have merely length and breadth—surfaces, that is to say, at which we look—architecture gives us spaces of three dimensions in which we stand. And here is the very centre of architectural art. The functions of the arts, at many points, overlap; architecture has much that it holds in common with sculpture, and more that it shares with music. But it has also its peculiar province

现的美感那样,是来自我们身体在空间移动的动作所带给我们的感受。

身体的动作本身,并不是一定会带给我们以快乐的感觉,或者不快乐的感觉。绝大部分的动作都是属于中性的,不会带有明显的倾向性。但是,尽管每一个动作自己没有倾向性,当我们把一系列的动作组合到一起的时候,这一系列的动作便带给我们一种倾向性,让我们对接下去的动作有所期待。如果建筑空间的布局开始时布置得让我们产生某种期待,到后来却又显示这种期待不过是空欢喜一场,那么我们说这样的建筑就是丑恶的建筑。比如说,一个建筑物的格局呈明显对称布置,而且行进节奏是均匀且有规律的——如同十八世纪典型豪宅的布局那样——当我们发现其中一个窗户没有和其他窗户在一条线上、比其他窗户低矮了一些、没有对齐,我们对它就会感到不舒服。这种不舒服的感觉应该是来自我们的对于身体运动的感知,当身体运动行进到与那个低矮窗户对应的位置时,我们不得不把脚步向下探伸,或者身不由己且不情愿地把脚步下沉一点。这种节奏上的变化让我们觉得不舒服。但是每一个具体的窗户本身,以及窗户与周边的关系或许什么问题都没有,一点儿也不难看。

在这里,我们再给出一个反例。古典主义建筑设计的目的是力图造就一种权威性,一种具有尊严、安详的建筑风格,这在意大利伯拉蒙特的建筑艺术中达到了顶峰。这种效果的取得是因建筑师力争在任何一点上都创造出一种四平八稳的对称建筑造型。这种形式与周边所有方向上的地形平稳地结合到一起,*任何能引起动感的元素都被清理掉*:每一个部分都布置在各自空间里重心的位置上,我们的感觉就处在一种静止状态,形成一种安逸休息的氛围。而巴洛克建筑艺术则全面抛弃了这种手段,每一部分如果从整体中分离出来,*单独地来看*,它都是不和谐的构图。沃尔夫林(Wölfflin)教授就曾经指出,这个时期教堂建筑的正立面上,建筑师们有意识地大量堆砌了很多形式,让正立面感觉很拥挤。靠近下面一点的窗户被故意用两根高大的壁柱从两边夹住;这个组合又被放在重心之上的位置;整个构图给人的感觉是两边的壁柱向中间挤压,中间的窗子被迫向上提升。就这一个细节,它已经给我们带来一种挥之不去的悬念。从整个建筑设计的上半部来看,我们的期待在那里得到了回应。这种从下面产生的向上升起的动感到了上半部的时候开始向周边横向分散,而这种分散的细节则用华丽的装饰加以渲染,形成装饰性雕塑;或者用一个具有向下动感的巨大而夸张的山花,山花又用横向的檐口线脚作为装饰和支撑,这样来平衡自下而上的升腾的感觉。在这里,动感成了不可或缺的富有戏剧性的效果,同时又是十分令人满意的建筑处理手段的基础;这里的主题不再是平和,而是能量。但是这种动感如果出现在伯拉蒙特式的设计中则成了破坏性的手段,与它的建筑艺术的整体效果是格格不入的。

四、只有长度和宽度的这类空间(亦即平面空间)为我们提供的是只能用眼睛看看的空间。除了这类空间之外,建筑艺术还有三维的空间,我们可以站在其中,这才是建筑艺术的核心所在。从任何角度来看,各种艺术之间总有相通的地方;建筑艺术有很多地方和雕塑艺术相似,建筑艺术与音乐艺术相同的地方更多。但是,建筑艺术也具有仅仅属于它自己的独特领地,从建筑艺术中获得的愉悦感受也具有它自己的特殊性。独霸建筑艺术领域的主要

and a pleasure which is typically its own. It has the monopoly of space. Architecture alone of the Arts can give space its full value. It can surround us with a void of three dimensions; and whatever delight may be derived from that is the gift of architecture alone. Painting can depict space; poetry, like Shelley's, can recall its image; music can give us its analogy; but architecture deals with space directly; it uses space as a material and sets us in the midst.

Criticism has singularly failed to recognise this supremacy in architecture of spatial values. The tradition of criticism is practical. The habits of our mind are fixed on matter. We talk of what occupies our tools and arrests our eyes. Matter is fashioned; space comes. Space is 'nothing'—a mere negation of the solid. And thus we come to overlook it.

But though we may overlook it, space affects us and can control our spirit; and a large part of the pleasure we obtain from architecture—pleasure which seems unaccountable, or for which we do not trouble to account—springs in reality from space. Even from a utilitarian point of view, space is logically our end. To enclose a space is the object of building; when we build we do but detach a convenient quantity of space, seclude it and protect it, and all architecture springs from that necessity. But aesthetically space is even more supreme. The architect models in space as a sculptor in clay. He designs his space as a work of art; that is, he attempts through its means to excite a certain mood in those who enter it.

What is his method? Once again his appeal is to Movement. Space, in fact, is liberty of movement. That is its value to us, and as such it enters our physical consciousness. We adapt ourselves instinctively to the spaces in which we stand, project ourselves into them, fill them ideally with our movements. Let us take the simplest of instances. When we enter the end of a nave and find ourselves in a long vista of columns, we begin, almost under compulsion, to walk forward: the character of the space demands it. Even if we stand still, the eye is drawn down the perspective, and we, in imagination, follow it. The space has suggested a movement. Once this suggestion has been set up, everything which accords with it will seem to assist us; everything which thwarts it will appear impertinent and ugly. We shall, moreover, require something to close and satisfy the movement—a window, for example, or an altar; and a blank wall, which would be inoffensive as the termination of a symmetrical space, becomes ugly at the end of an emphasised axis, simply because movement without motive and without climax contradicts our physical instincts: it is not humanised.

角色就是空间。在所有艺术中真正体现、见证空间的全部价值的场所就是在建筑艺术领域里。空间可以用它三维的虚空把我们罩住；而从这个虚空的环境中获得的任何快乐都是由建筑艺术独立完成的，它不需要借助于任何其他辅助手段。绘画可以描绘表现空间；诗人雪莱的诗歌让我们联想到空间的形象；音乐艺术可以带给我们它与空间的类比关系；只有建筑艺术的直接对象就是空间本身。建筑艺术把空间当作是自己的基本素材，同时把我们这些欣赏建筑艺术的人围在其中。

艺术理论家们一直以来就恰恰没有认识到空间在建筑艺术中的价值和绝对的主导地位。艺术理论的传统是很讲究实用的，我们的思维习惯总是集中在具体实物上的。我们关心什么样的工具可以做什么，关心什么东西会抓住我们的视线。实物是可以具体地给出形式的，而空间则是随之而来的。空间是"什么都没有"的那个部分（就是我们古时候的思想家老子所说的"无"），是实在的对立面。因此，它很容易被我们视而不见。

我们或许会忽略它，但是空间却影响着我们，左右着我们的精神和情绪。这时从建筑艺术中获得的快乐似乎是无缘无故的，我们也似乎不在意这时的快乐到底来自何处。这些快乐中的很大一部分实际上来自于空间对我们的作用。甚至从纯粹的使用角度来看，空间也应该成为建筑艺术的目的。用围合手段分离出一部分空间正是建筑的基本目的；我们建造一个建筑物，就是把一部分适当的空间从空旷中切割出来，归我们自己使用。我们把这块分割出来的空间加以割断并保护起来，建筑艺术就在这个分割过程中产生了。但是，从美学的角度来看，空间的意义更加重大。建筑师像雕塑家塑造自己的雕塑形象一样地塑造自己的空间。建筑师把空间当作是一个艺术品来塑造刻画。就是说，建筑师通过自己的手段来创造一种空间形象，这样，进入其中的人都会被空间的气氛所感染。

建筑师所使用的方法又是什么呢？他的追求在于人体的动感。空间实际上就是无拘无束的运动场所；这也是空间对于我们人类所具有的价值；它也是凭借着这种特点而进入到我们的记忆当中的。当我们站在一个空间里的时候，我们会本能地把自己看成是那个空间的一部分，想象着我们在其中的位置，以及在那个空间里所能进行的各种运动。让我们来看一个最简单的例子。当我们从一个尽端走进一座教堂狭长的中央大厅里的时候，我们会发现自己站在一个由两排柱子构成的深远的视觉走廊。一走进这个中央大厅，我们就会感觉到被迫必须向前走去：这个空间的特征强迫我们必须这样。就算是我们打定主意站着不动，我们的眼睛也会自动地向前方看去，而我们的脑子里已经开始想象，想象着我们自己的身体已经跟着眼睛向前走去了。这个空间在告诉我们它所暗示的移动方向。只要是一个空间里的移动趋势被有效地建立起来，那么任何物体进入到这个空间里之后，便开始受到这种趋势的驱动。任何物体只要与这个趋势具有一致的倾向性，那么这个物体就会加入进来帮助我们。若任何物体的倾向性与这个大趋势相违背，那么这个物体就会让人感觉到它是不恰当的，感觉它很丑。不仅如此，在这个趋势的尽端我们需要用一个重要一点的东西来把它收住，比如一个大窗户或者祭坛，都可以。一个光秃秃的墙面作为一个单纯对称空间的尽端是没有什么问题的，但是，因为在这个方向性强烈的空间里，当你走到尽端的时候，如果没有一个重要的物体来强调这里是这条路线的终点，就让整个行进过程失去了意义、失去了目的地；在本来需要出现高潮的地方，结果什么都没发生，这不符合人们的期待。换句话说，这个空间没有被人性化。

A symmetrical space, on the other hand, duly proportioned to the body—(for not *all* symmetrical spaces will be beautiful)—invites no movement in any one direction more than another. This gives us equipoise and control; our consciousness returns constantly to the centre, and again is drawn from the centre equally in all directions. But we possess in ourselves a physical memory of just the movement. For we make it every time we draw breath. Spaces of such a character, therefore, obtain an additional entry to our sense of beauty through this elementary sensation of expansion. Unconscious though the process of breathing habitually is, its vital value is so emphatic that any restriction of the normal function is accompanied by pain, and—beyond a certain point—by a peculiar horror; and the slightest assistance to it—as, for example, is noticed in high air—by delight. The need to expand, felt in all our bodily movements, and most crucially in breathing, is not only profound in every individual, but obviously of infinite antiquity in the race. It is not surprising, then, that it should have become the body's veritable symbol of well-being, and that spaces which satisfy it should appear beautiful, those which offend it ugly.

We cannot, however, lay down fixed proportions of space as architecturally right. Space value in architecture is affected first and foremost, no doubt, by actual dimensions; but it is affected by a hundred considerations besides. It is affected by lighting and the position of shadows: the source of light attracts the eye and sets up an independent suggested movement of its own. It is affected by colour: a dark floor and a light roof give a totally different space sensation to that created by a dark roof and a light floor. It is affected by our own expectancy: by the space we have immediately left. It is affected by the character of the predominating lines: an emphasis on verticals, as is well known, gives an illusion of greater height; an emphasis on horizontals gives a sense of greater breadth. It is affected by projections—both in elevation and in plan—which may cut the space and cause us to feel it, not as one, but several. Thus, in a symmetrical domed church it will depend on the relation of the depth of the transepts to their own width, and to that of the span of the dome, whether we experience it as one space or as five; and a boldly projecting cornice may set the upward limit of space-sensation instead of the actually enclosing roof.

Nothing, therefore, will serve the architect but the fullest power to *imagine* the space-value resulting from the complex conditions of each particular case; there are no liberties which he may not sometimes take, and no 'fixed ratios' which may not fail him. Architecture is not a machinery but an art; and those theories of architecture which provide ready-made tests for the creation or criticism of design

一个集中式对称、同时又具有符合人体规律的比例关系的空间,不会有任何一个方向能够形成强烈的方向性,因为各个方向都一样。不是说这样的空间就一定美观,但是它给我们一种平稳对称的感觉,一切都在我们的控制之中。我们的主观意识总是不停地回到这个空间的中心,然后再从中心不断地移动到四周所有的方向。但是,我们的身体本身就在记忆中有过类似这样的运动,因为我们每次呼吸的时候正是在重复这样的动作。因此,这样的空间在我们的感受中,又多了一层含义;在体会它的视觉美感的时候,多了一层由于呼吸所带给我们的那种最基本舒畅与快乐。这种习惯性的呼吸过程是在无意识的情况下进行,而它的关乎生命的重要性又是那样地无法替代;因此任何影响到这种机能的正常运作都会让人产生痛苦,当这种痛苦超过一定的程度,就变得特别恐怖。而任何对于正常呼吸有所帮助的事和物,哪怕只有一点点,也是大受欢迎的,比如人到了高空中,面对空旷的天空就神清气爽。我们身体的运动对于舒展空间的要求,尤其是呼吸对这一空间特征的要求,对我们每一个人来说意义都不容小觑,甚至很明显地,对古代人的意义也是深远的。因此舒展的空间很具体地象征了我们身体健康正常生长;满足我们身体需要的空间就是美好的空间,反之就是丑陋的。

但是,我们不能够从建筑艺术角度来规定一个固定的比例关系,说那就是正确的建筑艺术。在确定建筑艺术空间价值的判断过程中,占第一位的无疑是空间的具体尺寸大小;但是在第一位因素的旁边还有上百个其他因素也在发挥着自己的作用。空间的价值要受到光线照明、阴影的位置等因素的影响;光源的位置和光线强度能够很强烈地吸引我们的眼睛视线,光源的位置暗示了我们前进行走的方向,它不同于空间形态所示意的那个方向。空间的价值也受到色彩的影响:深颜色的地面加上浅淡颜色的天花板所形成的空间氛围与深暗颜色的天花板加上浅颜色的地板所形成的空间氛围,让人有着完全不同的感受。空间的价值也受到我们本身心理期待的影响,这种影响来自于进入这个新的空间之前我们所经过的那个空间对我们的作用,它让我们产生了一种期待。空间的价值也受到占主导地位的线条对我们的感知体验所产生的影响:强调竖向线条让空间有一种很高大的幻觉,强调横向线条让同一个空间具有一种很宽阔的幻觉。空间的价值也受到空间内凸出体块的影响,包括凸出于水平面的体块,也包括凸出于竖直墙面的体块。这些凸出物会把整个空间进行某种切割,而我们的感知体验是能够感受到这种切割的,而且在感觉上,一个空间或许会因此被切割成几个空间。因为这个缘故,在一个集中式教堂的顶部再添加一个穹顶,这样的空间效果完全取决于穹顶底部的直径和下面大厅通道的尺寸相互关系,因为穹顶与大厅通道的尺寸关系不同,这里的空间或许会形成一个单一的高耸空间,也可以形成五个不同的空间。而在屋檐高度出现的建筑造型线脚,如果出挑过大,就会让我们产生一种错觉,误以为这个空间的高度就在这些线脚的地方结束,而不是在它之上的真实屋顶。

根据上面的讨论,真正能够帮助建筑师的东西莫过于他自己的*想象力*,用想象力来畅想在每一个既定的具体情形下他所能取得的空间效果。没有任何东西可以阻止他随心所欲地发挥自己的想象力,也不可能有什么"固定的金科玉律"般的比例关系保证他不犯错。建筑艺术不是机械生产,而是艺术创作。凡是给艺术创作准备一些半成品或者公式,希望拿来便可以运用,这种货色的建筑艺术理论根本就是死路一条。不但是在艺术创作方面如此,在艺

are self-condemned. None the less, in the beauty of every building, space-value, addressing itself to our sense of movement, will play a principal part.

V. If voids are the necessary medium of movement, solids are the essential instrument of support; and a dependence upon physical firmness and security is not less fundamental to our nature than that instinctive need for expansion which gives value to architectural space. Any unlooked-for failure of *resistance* in tangible objects defeats the vital confidence of the body; and if this were not already obvious, the pervasive physical disquiet which the mildest tremor of earthquake is sufficient to excite, might show how deeply organised in our nature is our reliance upon the elementary stability of mass. Weight, pressure and resistance are part of our habitual body experience, and our unconscious mimetic instinct impels us to identify ourselves with apparent weight, pressure, and resistance exhibited in the forms we see. Every object, by the disposition of the bulk within its contours, carries with it suggestions of weight easily or awkwardly distributed, of pressures within itself and upon the ground, which have found—or failed to find—secure and powerful adjustment. This is true of any block of matter, and the art of sculpture is built upon this fact. But when such blocks are structurally combined, complex suggestions of physical function are involved—greater in number, larger and more obvious in scale. Architecture selects for emphasis those suggestions of pressure and resistance which most clearly answer to, and can most vividly awaken, our own remembrance of physical security and strength. In the unhumanised world of natural forms, this standard of our body is on all hands contradicted. Not only are we surrounded by objects often weak and uncompacted, but also by objects which, being strong, are yet not strong in our own way, and thus incapable of raising in ourselves an echo of their strength. Nature, like the science of the engineer, requires from objects such security and power as shall in fact be necessary to each; but art requires from them a security and power which shall resemble and confirm our own. Architecture, by the value of mass, gives to solid forms this human adequacy, and satisfies a vital instinct in ourselves. It exacts this adequacy in the detail of its decoration, in the separate elements that go to make its structure, in the structure itself, and in the total composition. The Salute at Venice—to take a single instance—possesses the value of mass in all these particulars. The sweeping movement suggested by the continuous horizontal curve of the Grand Canal is brought to rest by the static mass of the church that stands like its gate upon the sea. The lines of the dome create a sense of massive bulk at rest; of weight that loads, yet does not seem to crush, the church beneath; as the lantern, in its turn, loads yet does not crush the dome. The impression of mass immovably at rest is

术理论方面，这类的货色也是没有出路的。虽然这样说，我还是要作一个总结，结论便是：在每一个建筑作品中，它的空间效果因为反映了我们人体的动作，将会在建筑之美的创造过程中扮演最主要的角色。

五、如果虚空的空间是我们运动在其中的媒介，那么，空间外面的实体则是支撑整个建筑物的必需手段。对于我们人类来讲，在根本上，对于建筑物的坚固与安全性质的注重程度绝对不亚于对于呼吸空间的要求。呼吸的经验让室内空间有了价值；而建筑物本身中任何让人意想不到的结构缺陷则是对人身安全信心的致命打击。这样讲如果还不够直观明白的话，那么你就看一下另一个情形：当一个级别很小的地震发生的时候，它所引起的骚动和慌张则是明显可见的。这一点也说明我们人类的本性对于建筑主体的稳定性到底有多强烈的依赖。肩负的重量、承受的压力和抵抗这些重量和压力的努力，都是我们身体十分熟悉的习惯性经验，也因此在我们的潜意识中形成了一种直觉，让我们本能地对于某种形状产生出的对它固有的重量、压力和承载力作出一个判断。任何物体凭借着自身外表的组合形状和质感，向我们传达着它的与重量相关的信息：显示了其中的重力分布是自然的还是强扭不自然的；显示了内部构件之间的压力以及整体对于地面的压力，这些压力让我们感受到整个建筑体系是否是安全稳定的，是否已经得到强有力的平衡。这一点，在任何有体积感的物体造型中都具有相同的意义，雕塑艺术就是体现了这个结论。当这个有体积感的物体不是单一结构系统，而是复杂的组合，那么，这个组合复杂的形状就会传达出一些复杂的信息；信息的数量不但多，而且给人的感受也强烈。建筑艺术就是在这些数量众多的信息中，有选择地重点强调建筑整体造型所具有的压力与平衡之间的关系，通过这样的手段来使得我们确信，我们在自己记忆中对于安全与承重力的认知在整个建筑造型中得到满足，而且这种满足明显地表现在我们面前。自然界的形式是不以人们的意志为转移的，它的形式还没有人性化，所以到处可见与我们基本认知相矛盾的形状。我们不但总能看见虚弱、松散的东西，而且也总见到一些物体，虽然它们真实的物理性能是很强硬的，但是它们的强硬并不是我们能够感受得到的，和我们的期待还有距离，因此仍然无法在我们的认知上对它们的强度获得信心。自然界对物体的要求，如同工程师的计算那样，对物体自身的强度和稳定有所要求，但那都是针对物体本身的；艺术对物体的强度和稳定也有所要求，但是，艺术的这些要求是为了满足我们人对物体强度和稳定性的感受。建筑艺术通过对体量的表现，把人类的这种满足感带给了建筑实体，让我们的直觉感受得到了确认。人们从建筑的造型中获得那种心理上的满足感与建筑的结构体系、构件以及构造组合等科学计算中获得满足是根本不相同的两件事情。我们仅用一个例子来说明一下。威尼斯的安康圣母圣殿就具有我们所说的体量应该体现出来的所有特征。威尼斯大运河（Grand Canal）那舒展连续的曲线向我们暗示了一种动感和运动趋势，而这种动感正是用这个教堂那稳定、静止不动的体量来作为一个结束。教堂也作为象征大运河通往海洋的一个大门，矗立在那里。巨大穹顶的外轮廓线给人的感觉是分量很重的实体，岿然不动。虽然穹顶给人的感受是它拥有巨大的重量，但是，却一点也没有它会把下面的教堂建筑群压垮那样的感觉。同样地，穹顶上面的小亭子也给人以厚实的感觉，但是，也不会对穹顶造成任何压力上的伤害。所有的重量都呈现出一种静止稳定的状态，这种稳定的感觉又因为那十六个巨大漩涡状的支撑而得到进一步的加强。这十六个巨大漩涡是从圆形穹顶底盘到下面八边形建筑平面之间的过渡手段，因为有了这种过渡，穹顶的圆形

strengthened by the treatment of the sixteen great volutes. These, by disguising the abrupt division between the dome and church, give to the whole that unity of bulk which mass requires. Their ingenious pairing makes a perfect transition from the circular plan to the octagonal. Their heaped and rolling form is like that of a heavy substance that has slidden to its final and true adjustment. The great statues and pedestals which they support appear to arrest the outward movement of the volutes, and to pin them down upon the church. In silhouette the statues serve (like the obelisks of the lantern) to give a pyramidal contour to the composition, a line which more than any other gives mass its unity and strength. Save for a few faults of design in the lower bays, there is hardly an element in the church which does not proclaim the beauty of mass, and the power of mass to give essential simplicity and dignity even to the richest and most fantastic dreams of the baroque.

In architecture, then, the principal conditions of mass are these. In the first place the effect of the whole must predominate over that of the parts; the parts must enforce the general character of the whole and help us to realise its bulk; they must not detach themselves from the mass in such a way as to detract from its apparent unity. This, for example, is the ground of the Renaissance insistence upon crowning cornices and other devices for tying the elements of a building, and forcing it as a single impression on the eye.

Secondly, the disposition of the whole must conform to our sense of powerfully adjusted weight. Hence the careful study which the baroque architects gave to the effect of receding planes, and the influence of upward perspective upon mass. Hence also, obviously, the use of rusticated bases, battered plinths, pyramidal composition and the subordination of the Doric to the lighter Ionic and Corinthian Orders.

Finally, it is necessary that the several parts of a building should be kept in proper 'scale.' Scale, in any design, is that relation of ornament (or minor features) to the larger elements, which controls our impression of its size. In any building three things may be distinguished: the bigness which it actually has, the bigness which it appears to have, and the feeling of bigness which it gives. The two last have often been confused, but it is the feeling of bigness which alone has aesthetic value. It is no demerit in a building that, it should fail (as St. Peter's is said to fail) to 'look its size.' For big things are not, as such, more beautiful than small, and the smallest object—a mere gem for example—if it satisfies the three conditions just stated, may convey a feeling of dignity, mass, and largeness. On the other hand, a building which looks big may fail to convey a *feeling* of bigness. No one, for instance, looking at the new Museum at South Kensington, could fail to realise that its dimensions are

和下面的八边形才不至于显得生硬。这种过渡手段让整个体量有了和谐统一的感觉,体量和重量需要的就是这种和谐统一的整体感。巨大的漩涡状装饰成对地出现,精巧的造型恰到好处地把上面的球形体积与下面的八角形体积连接到一起。漩涡的轮廓线看上去好像是从上面流淌下来,向外伸出并旋转回来;它的重量让它固定在那样一种位置上,之后肯定不会再发生任何的移动。矗立于其上的高大雕像,以及雕像的基座刚好位于圆形漩涡装饰物的最上面的顶点,犹如一根一根的钎子一样,把每一个漩涡紧紧地插在下面的建筑上,被固定得再也动弹不得。从整个建筑的外轮廓来看,这些高大的雕像凭借着它们的剪影,加上顶上小亭子的剪影,形成了一个金字塔形的构图。这些雕像所形成的空间体量比其他任何元素的作用都更加明显,形成一个非常完整的统一整体,表现出稳定又有力量的构图。除了在下方有个别的细节看起来与表现整体的力量关系不大以外,其他所有的构件和所有的组成部分,无一例外的都在渲染整个构图的稳重感;这种厚重的体量所产生的力量,甚至让这个巴洛克建筑艺术最华丽喧闹的代表性作品充满了一种简单明快的造型和落落大方的庄严气氛。

因此,关于建筑艺术中的体量问题,我们可以概括为以下几点。第一,整体的效果一定要主导每一个局部;局部必须服从、协助整体的主要特征,并且让我们能够体会到整体作为一个完整的作品所拥有的体量;局部不可以从整体中分离开,造成建筑的整体视觉效果受到分散和降低。这也是为什么文艺复兴时期的建筑艺术一定坚持要在建筑物的最上端加上一个统一的檐口线脚,这样,建筑物的各个部分就会通过这个统一的线脚被整合到一起,对眼睛来说,就形成了一个整体的印象。

第二,建筑在整体上的组合关系也一定要协助创造整体上追求的厚重和稳定的感觉;因此才会产生出巴洛克建筑师花费精力认真研究墙面后退的视觉效果,利用这种效果帮助造成体量向上所形成的透视视觉感受;也因此才会出现粗大石块的基础、出现略微倾斜一点的柱础、才会出现呈金字塔形状的构图,才会出现让多立克柱式配合轻巧的爱奥尼柱式和科林斯柱式的情况。

最后一点,让建筑中的各个部分之间保持一种合适的"尺度"。在任何设计中,所谓的尺度指的就是它上面的装饰构件或者细节,相对于大一点的组成部分所保持的固定比例关系,这种尺度左右着我们对于建筑到底是大还是小的直觉感受。建筑的大小可分为三类:一种是建筑本身的真实尺寸的确很大,一种是建筑看起来很大,还有一种就是建筑给人的感觉很大。后面说的两种情况有点容易让人感到迷惑,它们的不同在于只有最后一种情况包含有美学因素在里面。如果一个建筑看上去与自己的真实尺寸有差距,这不一定就是它的一项缺点(据说圣彼得大教堂就没有成功地表现出自己的真实尺寸);因为大建筑不见得一定会比小建筑更美观些。即使是最小尺寸的物体,如果能够满足我们刚刚在上面说的这三个基本要点,那么它也可以给人以庄严、结实和高大的感觉;反之,一个真实尺寸巨大的建筑物看上去却不会给人以高大的感受。例如面对伦敦的南肯辛顿的博物馆(Museum at South Kensington),没有人会误解它的巨大尺寸,就是说这个建筑看上去和自己的真实尺寸还是相匹配的。但是,这个博物馆的整体效果没有控制住它的局部,局部的细节过多,它产生的尺

vast; it looks its size. But the whole does not predominate over the parts, the parts are many and the scale is small. Hence, while we perceive this building to be large, it conveys a feeling not of largeness, but of smallness multiplied.

Small scale, no less than large, may be employed to emphasise effects of mass, as, for example, when fine mouldings are used in combination with large, unbroken surfaces. In transcribing ourselves into such a building we instinctively take its detail as our unit of measurement, and this gives us an increased sense of the grandeur and simplicity of the unbroken mass. Broadly speaking the *quattrocento* architects employed this method, while the baroque architects sought to emphasise mass by the magnitude of the parts themselves. But in both cases the conditions of success were the same: the whole must predominate over the parts, the weight seem powerfully adjusted, the scale be consistently maintained.

Ⅵ. The humanist instinct looks in the world for physical conditions that are related to our own, for movements which are like those we enjoy, for resistances that resemble those that can support us, for a setting where we should be neither lost nor thwarted. It looks, therefore, for certain masses, lines, and spaces, tends to create them and recognise their fitness when created. And, by our instinctive imitation of what we see, their seeming fitness becomes our real delight.

But besides these favourable physical states, our instinct craves for order, since order is the pattern of the human mind. And the pattern of the mind, no less than the body's humour, may be reflected in the concrete world. Order in architecture means the presence of fixed relations in the position, the character and the magnitude of its parts. It enables us to interpret what we see with greater readiness; it renders form intelligible by making it coherent; it satisfies the desire of the mind; it humanises architecture.

Nevertheless order, or coherence, in architecture stands on a different plane to the values of mass, space, and line; for these, of themselves, give beauty, while order (as was shown in the last chapter) is compatible with ugliness. Yet it is clear that in all the architecture which descends from Greece and Rome, order plays a principal part. What then is its place and function?

Order—a presence of fixed ratios—will not give beauty, nor will a mixture of order and variety, but so much order, merely, and of such a kind, as is necessary for the effects which humanised mass and space and line are at any point intended to convey. Thus, in making the masses, spaces, and lines of architecture respond to our ideal movement and ideal stability, a measure of symmetry and balance are constantly entailed. Not perfect symmetry, necessarily. We in our bodies have a sense of fight and left, and instinctively require that architecture should conform to this duality. Without it we could not so smoothly read or interpret architecture

度很碎。因此,我们感到这个建筑的尺寸不小,但是它给人的气势却不够大,整个建筑是一堆琐碎的、细小的东西堆砌而成的。

小尺度的东西也如同大尺度的东西一样,都可以用来渲染建筑整体的分量,例如精细的线脚用来搭配大面积的墙面就是一例。在把我们自己转化投射到建筑物上面的时候,我们本能地把建筑上的某些细节作为我们的衡量单位,这个衡量单位使得我们感觉到那个完整的建筑整体相对地变得高大而简洁。宏观地讲,*十五世纪*的建筑师采用的就是这种办法,而巴洛克时期的建筑师则是采用把局部放大的手法来强调整体的体量。但是,这两种情形都很成功,原因就在于他们成功地让建筑整体控制住了局部,整体的重量感和厚重感获得有力的支撑与抵抗,建筑的整体尺度从头到尾得到了统一。

六、坚持人文主义理念的人,本能地在这个物质世界中寻找与我们人类生活有关的具体情形,寻找可以与我们人类喜爱的运动相类似的运动状态,寻找那些与我们人类承受压力时的状态相似的受力状态,让我们在这种环境里,既不会丧失自己,也不会违背我们自己的意愿。因此,这些人凭借本能所寻找的是某种体量、某种线条、某种空间;他们试图创造出这几样元素,并且需要确认自己创造的这些东西恰当合适。当确认这些创造都是恰到好处的结果的,便会在心里产生出由衷的快乐。

在这些具体物质性质的实物创造之外,我们人类的本能也需要创造出一种秩序,秩序是人类思想的一种规律模式。不仅仅是人体活动规律可以生动地反映在真实的世界里,人类思想的模式也是可以反映在真实的世界里的,而且后者的影响力绝对不比前者差。当秩序表现在建筑艺术领域里的时候,它代表了建筑组成部分之间的某种固定比例关系,代表了建筑组成部分的特征和分量。秩序可以让我们根据自己所看到的建筑局部,很有把握地对整个建筑进行一个认识判断;秩序通过建筑艺术整体的一致性来让每一个局部的形式很容易被人理解接受;秩序满足了人们智力活动的要求;秩序让建筑艺术变得人性化。

无论从哪一个角度来讲,秩序或者和谐一致,在建筑艺术中具有不同于体量、空间、线条的性质。秩序与这后面三者处在不同的层次上面。后面三者本身都会直接创造出美的形体,但是只有秩序一条却能够与丑陋相兼容(我们在上一章讨论了这一点)。然而有一点十分明显,那就是从古希腊、古罗马以来的所有建筑艺术中,秩序扮演了一个最主要的角色。那么,秩序到底具有怎样一种地位呢?它的作用又是什么?

秩序就是呈现在建筑外观上的某种固定比例关系。秩序本身是不会创造出美来的,各种不同秩序的混合搭配也不会产生出美来。但是经过人性化的建筑体量、空间和线条所试图表达的效果,离开秩序这个因素在其中发挥作用的时候,也是不可能创造出美的。只有借助于秩序,借助于某种特殊的组织关系,体量、空间、线条等基本元素才能创造出我们所需要的美。因此,为了让建筑物实体中的体量、空间、线条与我们自身的活动和思想的理想形式取得呼应,某种对称和平衡手段就经常被使用。但是,我们这里所说的对称并不是数学意义上的那种严格对称关系。我们指的是一种与我们身体相呼应的左、右平衡关系。因为我们自己身体的这种对称关系,我们便自然地、本能地要求我们的建筑艺术也具有这种成双成对

in our own terms. Dissymmetry in an object involves an emphasis or inclination to one side or the other in the movement it suggests, and this sometimes may be appropriate to the mood of the design. But, whenever architecture seeks to communicate the pleasure of equipoise and calm, or to impart a sense of forward, unimpeded movement, symmetrical composition and axial planning must result. Symmetry and Balance are forms of Order; but they are beautiful, not because they are orderly, but because they carry with them a movement and stability which are our natural delight. Then, since architecture is a monumental art, surrounding us with an influence never relaxed and not to be escaped, calm and unthwarted movement will here most often be desired. Thus Order, though it cannot ensure beauty, may follow in its wake.

Yet Coherence in architecture, distinct though it is from beauty, has a function of its own. Humanised mass, space, and line are the basis of beauty, but coherence is the basis of style. Mass, space, and line afford the material of individual aesthetic pleasures, of beauty isolated and detached. But architecture aims at more than isolated pleasures. It is above all else an art of synthesis. It controls and disciplines the beauty of painting, sculpture, and the minor arts; it austerely orders even the beauty which is its own. It seeks, through style, to give it clarity and scope, and that coherence which the beauty of Nature lacks. Nature, it is true, is for science an intelligible system. But the *groups* which the eye, at any one glance, discovers in Nature are not intelligible. They are understood, only by successive acts of attention and elimination; and, even then, we have to supplement what our vision gives us by the memory or imagination of things not actually seen. Thus, Order in Nature bears no relation to *our* act of vision. It is not humanised. It exists, but it continually eludes us. This Order, which in Nature is hidden and implicit, architecture makes patent to the eye. It supplies the perfect correspondence between the act of vision and the act of comprehension. Hence results the law of coherence in architecture; what is simultaneously seen must be simultaneously understood. The eye and the mind must travel together; thought and vision move at one pace and in step. Any breach in continuity, whether of mood or scale, breaks in upon this easy unison and throws us back from the humanised world to the chaotic. The values of mass, space, and line are as infinite as the moods of the spirit, but they are not to be simultaneously achieved, for they are mutually conflicting. Style, through coherence, subordinates beauty to the pattern of the mind, and so selects what it presents that all, at one sole act of thought, is found intelligible, and every part re-echoes, explains, and reinforces the beauty of the whole.

的特征。不成双成对地出现,我们在理解建筑艺术的时候就不那么容易用我们自身的词语来描写它们。当一个造型中出现非对称的构图的时候,也就出现了某一部分受到强调的机会;有一侧受到更多的青睐,那么就会对人们的行进方向产生出某种暗示,这一点在某些情况下是十分有助益的。但是,当建筑艺术想要表达平稳、宁静的轻松欢乐气氛,或者引导前来的人们继续前行,让行进过程不受到任何阻碍和干扰,对称的布局、围绕轴线进行的布局就成了一种必然的结果。对称和平衡就是秩序的表现形式;但是,这种布局所产生的结果之所以美观,根本原因并不是因为它们有了这种布局,而是这种布局让整个建筑构图获得了某种动感与稳定感,这两点使我们人类自身很容易从中获得快乐的状态。通过上面的讨论,我们可以得出这样的结论:由于建筑艺术具有很强的纪念性,它的存在对于我们的影响极大,而这种影响是环绕在我们周边的,是赶也赶不走的,它让我们无法摆脱它的影响;所以说,安静的性格加上流畅的空间应该是最受欢迎的布局。建筑艺术中的秩序尽管自己不能确保美观的设计出现,但是秩序却是紧随美观之后的重要特征。

  和谐一致不同于美观,建筑艺术中的和谐一致有着自己独特的功能。建筑美的基础是经过人性化的体量、空间和线条,而和谐一致是艺术风格的基础。体量、空间、线条构成了美学享受的基本材料,但是这些材料是各自独立于其他基本材料而单独存在的。建筑艺术绝对不只是把这些各自独立的美观材料简单地拼凑在一起的结果。建筑艺术的目的是通过艺术手段把所有这些材料有机地组合到一起。和谐一致控制并主导艺术作品中的美感元素;它体现在绘画、雕塑,以及所有的艺术创作活动。它甚至严格地控制着美本身的面貌。通过艺术风格的语言,和谐一致性控制着美学表达方式的清晰以及美学的适用范围,而这种艺术创作中的和谐一致性是大自然中所见不到的。对于科学来讲,自然是可以为人们所掌握的,这没有什么疑问。但是,从任何角度、任何时刻看,我们用肉眼所看到的自然都是*一团团*彼此没有什么关系的物体。这些自然景象需要经过一次次地筛选过滤、认真观察,才可能从中找到一定的联系;即便是这样,这些联系也不可能完全是依赖于我们肉眼所看见的东西,而是在相当大的程度上借助于我们过去的某些记忆,加上我们当时希望看到的某些内容。因此说,自然界所呈现出来的秩序与*我们*所期待看到的景象没有直接的关系。真正的自然景象还没有被人性化。自然的秩序当然是存在的,但是它仍然是一个谜,仍然没有被我们掌握。自然界的秩序是深藏不露的,但是建筑艺术却让它明显地呈现在我们面前。建筑艺术中的秩序把创作时的想象与对艺术作品的欣赏理解直接联系起来。建筑艺术中的和谐一致的定律就这样形成的;人们能够用眼睛看到的东西必须同时可以用智慧来理解。眼睛和大脑必须是同步前进的。人们的思想与视觉体验必须是同步的,任何一方都必须跟上另外一方的脚步。无论是整体气氛还是具体造型的尺度,如果这种连续性在某些地方被打断,那么这种同步性和一致性就会遭到破坏,这时的建筑也就不再具有人性化的特点,我们也就从人性化的世界被抛回到混乱之中去了。体量、空间、线条的价值与作用,在变化上与我们的精神、情绪所可能具有的状态一样无法估计,这些无法计算的状态也是不可能同时出现的,因为有些状态根本就是互相矛盾的。因此,艺术风格借助于一致性的手段,让美观的标准符合我们智慧的判断模式,借此来选择哪些应该出现,哪些不应该出现。这个选择过程完全是我们人类智慧的理性挑选活动。被挑选出来的全部内容都是容易为我们的智慧所理解的,每一个组成部分都在呼应、解释、强化整体的美感。

Of all the styles of building that yet have been created, the forms of Greece and Rome, with those of the Renaissance after them, were in this point the most exact and strict. They are by consequence the fittest instruments for giving clarity to sharp ideas, however varied, of function and of scale. Other instruments, doubtless, there will be in the future. For if the scope of classical design could be perpetually enlarged until the eighteenth century, it is not probable that its history is closed. But first we must discard a century of misplaced logic. Architecture must be perceived sensitively but simply, the 'theories' of the art have blunted sensitive perception without achieving intellectual force. Architecture that is spacious, massive and coherent, and whose rhythm corresponds to our delight, has flourished most, and most appropriately, at two periods, antiquity and the period of which antiquity became the base—two periods when thought itself was humanistic. The centre of that architecture was the human body; its method, to transcribe in stone the body's favourable states; and the moods of the spirit took visible shape along its borders, power and laughter, strength and terror and calm. To have chosen these nobly, and defined them clearly, are the two marks of classic style. Ancient architecture excels in perfect definition; Renaissance architecture in the width and courage of its choice.

在人类到目前为止已经创造出来的全部建筑艺术中，古希腊、古罗马的建筑艺术，加上以古希腊、古罗马为基础的文艺复兴时期的建筑艺术，它们最能准确、严格地体现我们刚刚说的这些结论。这些建筑艺术也能够成为最准确的工具，通过最严格、清晰的手段，来满足各种追求、用途和尺度的不同要求。我们不否认，将来一定还会有更多的这类工具出现。然而我们认为，如果古典主义建筑设计的应用范围直到十八世纪之前一直是处于不断扩大的趋势的话，那么，我们不认为它的历史就这样地结束了。但是，当务之急是抛弃最近一个世纪的错误逻辑。建筑艺术需要艺术敏感性去感受，但是，感受过程应该是非常简单的过程，不应该有任何神秘；过去的各种所谓的艺术"理论"只是让我们的感受能力变得迟钝，也根本没有加强我们的理性分析能力。在空间、体量、和谐统一、韵律等方面曾经让我们人类从中获得快乐的那些建筑艺术，它们的成就也就最为辉煌，这也是很自然的结果。取得这样辉煌成就的显然在两个时期：古典时期本身和以古典时期为基础的文艺复兴时期。这两个时期都是以人文主义理念作为自己思想的基本点的。这种建筑艺术的核心是人类的身体；这种建筑艺术的方法就是把人体的最佳状态用石头摹写到建筑艺术上面；这种建筑艺术所包含的精神和氛围都反映在建筑的视觉形式上，通过建筑形式反映了其中的严肃与欢乐、坚强与恐惧，反映了其中的平静。古典建筑艺术风格有两点要求：一个是用骄傲的心情看待这种风格，一个是明确地了解这种风格的内涵。古典时期的建筑艺术出色地完成了这种建筑艺术风格的定义，文艺复兴时期的建筑艺术拓展了这种古典风格的应用范围并且勇敢地向世人宣布了自己的选择。

NINE
Conclusion

Such are the four great elements of building from whose laws the finest masters of the Renaissance, however various their impulse and achievement, did not deviate. Theirs is an architecture which by Mass, Space and Line responds to human physical delight, and by Coherence answers to our thought. These means sufficed them. Given these, they could dispense at will with sculpture and with colour, with academic precedents and poetic fancies, with the strict logic of construction or of use. All those, also, they could employ: but by none of them were they bound. Architecture, based on Humanism, became an independent art.

Architecture a humanised pattern of the world, a scheme of forms on which our life reflects its clarified image: this is its true aesthetic, and here should be sought the laws—tentative, at first, but still appropriate—of that third 'condition of well-building,' its 'delight.' To combine these 'laws of delight' with the demands of 'firmness' and 'commodity' is a further problem: in fact the practical problem of the architect. To trace how this union has been achieved, and by what concessions, is the task of the historian. But these three studies are distinct. And the crucial, the central, function of architectural criticism is the first.

This principle of humanism explains our pleasure in Renaissance building. It gives us, also, some final links that we require. It forms the common tie between the different phases—at first sight so contradictory—of Renaissance style. It accounts for its strange attitude, at once obsequious and unruly, to the architecture of antiquity. It explains how Renaissance architecture is allied to the whole tendency of thought with which it was contemporarys—the humanist attitude to literature and life.

Man, as the savage first conceived him, man, as the mind of science still affirms, is not the centre of the world he lives in, but merely one of her myriad products, more conscious than the rest and more perplexed. A stranger on the indifferent earth, he adapts himself slowly and painfully to inhuman nature, and at moments, not without peril, compels inhuman nature to his need. A spectacle surrounds hims—sometimes splendid, often morose, uncouth and formidable. He may cower before it like the savage, study it impartially for what it is, like the man

第九章
结论

上面我们所谈的是建筑艺术中四种最重要的元素,与这四种元素相关的各种原则是文艺复兴时期里那些最伟大的建筑艺术家所坚持的,不管这些艺术家的动机和艺术成就有多么不同,他们都没有偏离这些原则。这些伟大艺术家的建筑艺术都是通过体量、空间、线条呼应着我们身体的动作,通过和谐一致满足了我们理智的需求。这几种手段完全满足了这些艺术家的要求。有了这些,艺术家们根本不再依赖于雕塑和色彩,不再依赖于亦步亦趋地模仿前人,不再幻想诗人的浪漫,不再盲目地局限于建造技术的束缚。艺术家们不排除使用所有这些手段,但是,他们不会因为其中任何一个手段而让自己变得缩手缩脚。坚持人文主义理念的建筑艺术是具有独立精神的艺术。

建筑艺术是把世界人性化之后所得到的结果,它的形式把我们人类的生活清晰地进行了形象化:这是人文主义建筑艺术美学的真谛。正是在这里,我们需要对优秀建筑中的第三个条件,亦即"使人愉悦"的法则进行一番探索,开始的时候一定具有很多不确定的因素,但是这种探索仍然是有道理的。至于这里产生出来的"愉悦法则"如何与"坚固法则"和"适用法则"完美地结合到一起是未来的问题。也可以说那是实践建筑师需要面对的问题。清楚地描述这三者之间通过什么方式结合到一起的,这是历史学家的任务。这三个方面的研究是各自独立的。而建筑艺术理论最关键、最核心的任务就是要解决"愉悦法则"的问题。

人文主义的这一基本原理说明了为什么文艺复兴时期的建筑艺术会让我们产生愉悦和快乐。这一原理也给我们提供很多线索来理解很多现象之间的内在联系。比如在整个文艺复兴时期里的各个阶段之间的联系,乍看起来这些不同的阶段毫无瓜葛,甚至彼此对立、矛盾,但是人文主义原则帮助我们看到了这些阶段之间的共同纽带。这个原则也解释了文艺复兴时期的艺术家对待古典时期建筑艺术的奇怪态度,既五体投地地敬佩,又肆无忌惮地篡改。这个原则也说明了为什么文艺复兴时期的建筑艺术与当时的主导思想在大趋势上是一致的,人文主义理念在当时贯穿于所有的理论和生活。

最初原始人时代认为自己是自己所生活的这个世界的中心,现代科学则到目前为止仍然还在证明着我们人类不过是整个宇宙中无数物体中的一个,根本不是它的中心。人类只是比其他物体和生物具有更高级的自我意识,思想更为复杂而已。作为地球上的一个生物,开始的时候,人类也是同其他生物一样,是这个世界上的陌生来客。人类逐步地让自己适应这个无情的世界,经过痛苦的努力,付出极大的代价之后,利用某一个偶然的机会,强迫这个无情的自然世界让步,来满足人类自己的需求。在我们人类的身边于是出现了众多的奇迹,个别的情形明亮耀眼,大多数的情形则是充满了困惑、笨拙与力不从心的痕迹。在自然界面

of science; it remains, in the end, as in the beginning, something alien and inhuman, often destructive of his hopes. But a third way is open. He may construct, within the world as it is, a pattern of the world as he would have it. This is the way of humanism, in philosophy, in life, and in the arts.

The architecture of humanism rose in Greece; and of the Greeks it has been said that they first made man 'at home in the world.' Their thought was anthropocentric: so also was their architecture. Protagoras, who first made humanity the centre of a metaphysic, and 'the measure of all things'; the poets who, in the labours of Heracles and Theseus and the strife of the gods with centaurs, celebrated the conquest by human reason of a corner in the darkened world; Socrates, who drew down speculation from the flattery of the stars to the service of the conscience; the dramatists, who found tragedy a savage rite and left it a mirror of life, not as it is but as our mind demands: these were the first humanists. Among these men, and to satisfy this same proclivity, was created an architecture whose several elements were drawn indeed from primitive necessities, but so ordered and so chosen that its constructive need and coarse utility were made to match the delight of the body and mock the image of the mind. Matters—the very antithesis of spirits—matter with its mere weight and mass and balance; space, the mere void we recognise as nothing, became, for them, the spirit's language. Within the world of concrete forms indifferent to man, they constructed a world as man desires it, responsive to his instinct and his stature.

But humanism has its practical aspect as well as its ideal; and the values which the Greek defined and founded, the Roman fixed impregnably upon the earth. Roman architecture, less fastidious than the Greek, and less restricted, preserved the principles of mass, space, line and coherence for rougher uses, wider and more general. It ensured their survival, their independence of the time and place whence they had sprung. In architecture as in thought it is to Rome, not Greece, that humanism owes its deep and racial hold upon the West.

The architecture which thus rose with humanism was with humanism eclipsed and with humanism restored. To pass from Roman architecture or that of the Renaissance to the fantastic and bewildered energy of Gothic, is to leave humanism for magic, the study of the congruous for the cult of the strange. It is to find that the logic of an inhuman science has displaced the logic of the human form. It is to discover resplendent beauty of detail in glass and bronze and ivory and gold; it is to lose architecture in sculpture. Here is structure, certainly—daring, intricate, ingenious; but seldom humanised structure. Here is poetry, curious craftsmanship, exquisite invention. But the supreme, the distinctive quality of architecture—that

前,今天的人类仍然会像原始野蛮人那样畏惧不安;也会像科学家那样认真地研究摸索自己所面对的问题;到最后,仍然还会有很多无法理解的问题,甚至会让自己绝望。但是,人类学会了第三种办法。在自己面对的这个世界里,人类可以根据自己对于这个世界的认识,建造出自己心目中的理想世界。这就是人文主义的世界观,从哲学角度、从生活角度、从艺术角度来看,都是如此。

人文主义建筑艺术起源于希腊。人们都说是希腊人最早让人类感觉到"这个世界如同自己的家"。希腊人的思想就是按照人类的标准来衡量整个宇宙的,人类是宇宙的中心。希腊人的建筑艺术也体现了这一精神。古希腊哲学家普罗泰格拉(Protagoras)就把人文主义当做自己抽象论述的核心问题,"是衡量一切事物的一把尺子"。在大量赞美大力神赫拉克勒斯(Heracles)和雅典国王忒休斯(Theseus)的英雄故事的诗篇中,在描写众神与半人马的怪兽交战的诗篇中,古代诗人们讴歌了人类的智慧征服了邪恶世界中黑暗的角落;苏格拉底把当时占星卜卦的空谈拉回到探讨与人们生活相关的良知问题;古希腊戏剧作家们用悲剧表现出野蛮的仪式,把它当成一面反映人生的镜子,不是简单地照搬,而是根据人类的思想需要来进行创作加工。这些人都是最早的人文主义思想的传播者。正是在这些人的影响下,为了追求同样的理想,古希腊的建筑艺术诞生了。这时的建筑艺术当然是受到当时原始条件的制约,有些手段的确是没有选择的选择,但是,就是在这样的制约条件下,利用这样粗糙的技艺,古希腊建筑艺术创造出了让我们的身体感到快乐、让我们的智慧感到满足的建筑形象。作为精神的对立面,物体仅仅具有重量、体积和平衡;而原本什么都没有、只是空虚的空间,对于这时的人文主义者来说,成为一种精神语言。就这样,在这个本来对人类没有什么意义的物质世界里,人类根据自己的直觉,用自己的身体作为参照,创造了一个自己的理想世界。

人文主义不但具有勇于追求理想的色彩,同时也具有非常实用的方面。古希腊人建立了价值观,古罗马人在全球范围内不可动摇地确立了这种价值观的地位。古罗马的建筑艺术不像古希腊那样严谨、挑剔,但是古罗马的建筑艺术在不拘小节的创作中,自始至终保持了注重体量、空间、线条和谐一致的艺术原则,并把这些原则推广到更广泛的领域,让这些原则成为更为普遍遵守的一般原则。罗马人让这些原则得以不断的延续,不再局限于它们的起源地点和特定的时代。建筑艺术中的人文主义如同思想界的人文主义一样,是依赖于罗马人才能够得以在西方世界取得根深蒂固的地位的,而并不是依赖于古希腊人。

因此说,伴随人文主义思想而兴起的建筑艺术,它所跟随的是一种曾经被屏蔽过的人文主义,是后来的人们重新建立起来的人文主义。抛弃了古罗马建筑艺术,或者说抛弃了文艺复兴时期的建筑艺术,转而以极大的狂热来推崇哥特风格的建筑,这种现象正是背离了人文主义思想的正途,却狂热地追求一种奇异幻想的不可思议的举动。我们从中可以看到,这种奇怪的转变正是由于非人性的所谓科学逻辑取代了以人为本的形式逻辑所造成的结果。我们也发现这是因为人们陶醉于用玻璃、青铜、象牙、黄金等贵重材料精雕细刻的那些华丽细节,建筑艺术在对雕刻艺术的赞美欢呼声中丧失了。这时的建筑物不乏大胆的构想,技艺精湛、手法奇妙,但是,很少有人性化的建筑物出现。这时的建筑物充满了诗意的、令人称奇的做工和新颖的处理手法,但是建筑艺术中的那种宏伟气魄和超越现实的品质却很少有人问

pure identity between the inner and the outer world—is unattempted. The lines of this amazed construction are at one moment congruous with our movement, at the next they contradict it with a cramped and angular confusion. Mass is too often lost in multiplicity. Space and coherence come, if at all, unsought and unregarded; and when they come it is most often because the ritual of the Church, preserving something of the pagan order it inherits, imposed a harmony upon the plan. Divorced from this ritual, Gothic, as its domestic building and its streets suffice to prove, admits its deep indifference to ordered form. It is entangled, like the mediaeval mind itself, in a web of idle thoughts of which man as he is has ceased to be the centre.

When, in the Renaissance, that centre was recovered, and humanism became once more a conscious principle of thought, Roman design in architecture came with it as of right. But there was now a difference in its intent. Humanism has two enemies—chaos and inhuman order. In antiquity humanism strove principally against the primitive confusion of the world: its emphasis was laid on order: it clung to discipline and rule. Hence Greek architecture is the strictest of all styles of building, and Rome, in whatever outposts of Spain or Britain her legions were remotely quartered, there set a tiny Forum, and preserved without concession the imperial order of its plan. But in the thought of the Renaissance humanism was pitted, not against chaos, but against the inhuman rigour of a dead scholastic scheme, whose fault was not lack of logic, but excess of logic with a lack of relevance to man. Thus the emphasis of Renaissance humanism, in all its forms, was less on order than on liberty. And, in architecture, while it rebelled against the mere constructive logic of the Gothic style, while it returned with passion to the aesthetic logic of antiquity, it makes that logic serve the keen variety of life. It is no longer content to rest for ever in the restraint of classic equipoise and calm. It has learned the speech of architecture from Greece and Rome, but the Renaissance itself will choose what things that speech will say. Every value, every avenue of promise, it will explore, enjoy, express. Hence the insatiate curiosity, the haste, the short duration of its styles; hence the conversion of classic forms to the gay uses of baroque and rococo invention; hence the pliancy and swift recoveries of taste, of which our first chapter took account. But not the less does the Renaissance employ the language of Humanism; and hence its unsevered ties with classic architecture, its reliance on the 'Orders,' its perpetual study of the past. Still, as in antiquity, it speaks by mass, space, line, coherence; as in antiquity, it still builds through these a congruous setting to our life. It makes them echo to the body's music—its force and movement and repose. And the mind that is responsive to that harmony, it leads

津,就是说,同时代表了内心的精神世界与外在的物质世界的那种建筑艺术在这时很少有人去尝试。这种建筑风格中对于线条的运用,有的时候与我们所强调的含义是非常一致的,但是,马上会在另外一个时刻使用一些拥挤或者角度混乱的线条与我们的理想相矛盾。这种风格的建筑整体体量常常会因为过多的堆砌而丧失整体感。这种建筑风格中的空间关系与整体的和谐一致性也是意想不到的结果,并不是有意识地主动追求而来的。而出现和谐一致的主要原因,也多半是因为教堂本身对于仪式的要求强加给建筑设计一些必须遵守的内容,也就造成了建筑设计的某种和谐,保留了某些教堂建筑上通过传统留传下来的属于异教徒的建筑特征。当哥特建筑艺术应用于教堂类以外的建筑上,例如在民用建筑上的时候,也就是它不再受到宗教仪式制约的时候,这种建筑艺术则对艺术形式的秩序毫无清醒的认知,艺术家和中世纪的人一样,思想混乱,在各种懒散、无用的思想交错中,清理不出自己需要的一个头绪。这些混乱的思想具有一个共同点,就是人类已经不再是问题的核心。

到了文艺复兴时期,人类作为一切活动的核心这一思想又重新被发现,人文主义再一次有意识地成为人们思想的基本原则。罗马人的建筑艺术随之成为一种法则,但是在内容上,这时的建筑艺术与古代的艺术则有所不同。人文主义实际上面对着自己的两个敌人,一个是单纯的混乱,另一个是非人性的秩序。在古代,人文主义所面对的主要敌人是原始时期对于物质世界的迷惑与懵懂,它的主要目的在于建立起理想的秩序,让一切变得有序可循;因此,古希腊的建筑艺术准则是所有建筑准则中最为严格的一种,古罗马帝国时期的军团无论是驻扎在西班牙还是不列颠等边远地区,在布局上都严格保持着帝国的布局规律,都在驻扎地建立一个小小的集市中心。但是到了文艺复兴时期,人文主义的思想具有了具体的使命,不再是以整顿混乱为主要目的,而是纠正从过去经院哲学流传下来的各种反对人性的谬误。这种反人性的谬论不是因为违反逻辑规律才成为谬误理论,而是因为过于强调所谓的逻辑而背离了人性,变得与人们的生活脱节。因此,文艺复兴时期的人文主义所重点强调的不是严格的秩序,而是人们的自由。这时的建筑艺术一方面反对像哥特建筑那样单纯片面地强调建造技术的做法,一方面以满腔的热忱追求古代的美学逻辑,文艺复兴时期的建筑艺术强调的是逻辑必须为生活服务。这时的人们不再满足于一劳永逸地接受古代建筑中的那种典雅和宁静,他们从古希腊和古罗马建筑艺术中学到了艺术修辞手法;而文艺复兴时期的艺术家则根据自己的想法来决定哪些修辞手法会被采用,他们会坚持说出自己想说的话。任何手段、任何可能性都会在这些艺术家的手中得到尝试,都会被拿来用于自己的享受,用于表达自己的思想。因此才会出现这个时期里的那些无尽的好奇心,各种艺术风格来得急,去得快,转眼之间就改变了刚刚还在流行的风格;因此才会出现把古典建筑语言进行变形,热情地创造出巴洛克和洛可可的艺术风格;因此才会出现人们的审美趣味在不停地变化这一现象,关于这一点我们在第一章里进行了讨论。但是,无论怎样讲,文艺复兴时期的建筑所使用的艺术语言只有人文主义的词汇;所以它与古典主义建筑艺术的纽带仍然十分牢固,它仍然依赖于各种建筑"柱式",它也一直在不断地研究古代艺术,从中汲取营养。和古代建筑艺术一样,文艺复兴时期的建筑艺术也是遵循建筑体量、空间、线条和谐一致等艺术规律;和古代建筑艺术一样,它也是通过这些具体的手段来满足我们生活的需要。文艺复兴时期的建筑艺术也是通过这些具体的手段来呼应我们身体语言所谱写的音乐,呼应着我们身体所表达的力量、动作和姿态。而我们的理智在面对这种和谐建筑形式的时候,也会陶醉于这种用

enchantingly among the measures of a dance in stone.

Virgil attends on Dante, and St. John, in the solitude of the Adriatic shrine he shares with Venus,[1] may ponder if ascetic energy is not best mated with a classical repose. The architecture of humanism has on its side the old world and the new; it has this repose and this energy. The spirit of perpetual change awoke in Europe, and architecture through four centuries gave to each change some shape of pagan beauty. A beauty of paganism, not its echo: Renaissance architecture is misconstrued wholly when we dismiss it as an imitative art. It served antiquity, not with the abject duty of a slave, nor always even with a scholar's patience, but masterfully, like a lover, with a like kindling of its proper powers. Brunelleschi, Bramante, Michaelangelo, Bernini had, as few can have it, their originality. But they followed on the past. The soil they built in was heavy with the crumbling of its ruins.

Yet every art that finds a penetrating pathway to the mind, and whose foundations are profoundly set, must needs have precedent and parallel, ancestors and heirs. For the penetrating paths are few; and, despite their baroque liberty of fancy, we can forget, as from the Palatine we watch the domes that overpeer the Forum, and see the front of San Lorenzo rise through the grey portico of Antoninus, how sheer an interval, with how vast a change of life, sunders two forms of art so congruous and familiar. Where classic power once stood, its shadow lingered: Mantegna, in the fifteenth century, painted men as Caesars and made splendid with antique frieze and column the legends of the Church. The architects of humanism built deep. Like the heroes of Mantegna, they performed their labour in a Roman panoply, and in the broken temples of Rome dreamed their own vision, like his saints.

---

[1] *San Giovanni in Venere*—the Baptist lodged with Venus—is a deserted church on the Abruzzi coast. The structure is Romanesque; the name more ancient still; but not until the Renaissance can its patrons have achieved their perfect reconciliation, which now the browsing goats do not disturb.

石头记录下来的舞蹈动作。

古罗马诗人维吉尔侍奉着但丁（Dante）行走在地狱的道路上，圣徒约翰（St. John）孤独地在亚德里亚海岸与维纳斯女神（Venus）共享一处祭奠的灵台[1]。圣徒约翰在自己灵台的角落里一定会困惑地想，禁欲主义的毅力与古典主义的恬静到底是不是真的彼此不相容呢？人文主义建筑艺术有古代建筑和现代建筑站在自己一方；人文主义建筑也同时具有这种恬静和毅力。一切都在不停地变化着，这一思想到了这时开始唤醒了欧洲，四个世纪的建筑艺术演变，在每一个时期都引进了一些原本属于世俗生活的美感的东西。文艺复兴时期的建筑艺术就是异教徒的艺术，它并不是在间接地反映着异教文化，而是它本身就是地地道道的异教生活的一部分。所以，当我们说这个时期的建筑艺术只是在模仿别人，我们实际上误解了它。文艺复兴时期的建筑艺术从古典主义时期的艺术中获得灵感，但绝对不是像奴隶那样的没有自由，同时也不像固执的学究那样，不厌其烦地追求着惟妙惟肖。这个时期的艺术家在学习古典建筑的时候，更像恋爱中的情人那样，通过唤起自身的热情，通过杰出的艺术手段，让自己的固有力量得到释放。布鲁乃列斯基、伯拉蒙特、米开朗基罗、伯尔尼尼，这些伟大的艺术家都发挥出历史上几乎无人可以比拟的创造性；但是，他们却又都在追随着古人的脚步。在这些伟大艺术家建造自己优秀建筑作品的土地上，仍然到处可以看见古代遗迹的残垣断壁。

如果一种艺术能够排除许多障碍，达到唤起我们严肃认真思考，同时这种艺术的基础又是非常牢固的话，那么它一定会有自己的先导和同类，有自己的长辈和后来的继承人。由于能够让我们为之思考的艺术引起我们注意的途径有限，虽然巴洛克建筑艺术有充满想象力的自由个性，但是我们常常会忘记，在两种如此相似的形式之间的距离是多么遥远，它们所代表的生活多么不同。就好比我们从罗马城里的帕拉丁山（Palatine）上观望城市中心市场（Forum）区域里的那些此起彼伏的穹顶，我们是透过安东尼神庙（Antoninus）灰色柱廊的空隙，才看到了圣劳伦佐（San Lorenzo）宫殿的正脸。在古典主义艺术曾经发挥过影响的地方，那里也一定会有它们留下的阴影。十五世纪的画家蒙塔尼亚（Mantegna）把自己时代的人物描绘成古罗马皇帝恺撒时期的人，在自己的绘画中采用了古代神庙建筑檐口部分常常采用的手法，他用古代浮雕系列组画和柱式的华丽方式，渲染着自己教堂的传奇故事。具有人文主义理想的建筑师在建造自己的作品时也具有同样深刻的意义。正如蒙塔尼亚绘画作品中所刻画的主角人物那样，他们身着古罗马时期的服装从事着自己的工作，借助刻画罗马城里的残破神庙，表达着自己幻想中的愿景，就像他作品中的圣徒那样。

---

[1] 在意大利阿布鲁佐海边（Abruzzi coast）曾经有一座被废弃的修道院，名字叫 San Giovanni in Venere，意思是"洗礼约翰与维纳斯女神在一起"。建筑物是罗马帝国开始衰败的那个时代的，修道院的名字比建筑物更要早一些。但是，直到文艺复兴时期，信徒们才完全接受这个名称，在那之后才开始整修这座修道院。放牧的羊群从此以后也就不会随意地走进来吃草了。

Epilogue, 1924

I wrote these chapters ten years ago in a hope of which I was too young to realise the full temerity, conceiving that what I had to say might interest those who practise architecture, and also those who deal in philosophy; yet desiring also to conciliate those who require that a book, whatever else it may attempt, should not cease to be a book.

To carry my subject a stage further would be to enter a field where I could no longer entertain any such ambition. If I ever now attempt this, it will be in a separate and more technical form.

Since, however, the book as it stands has been fortunate enough to evoke some discussion, I would like here to say one or two things by way of epilogue and in reply.

It has been remarked with truth that the destructive portion of the book overweighs the constructive. But if the conclusions at which I arrive are rightly appreciated this will, I think, be seen to be inevitable.

My contention is that 'theory'—the attempt to decide architectural right and wrong on purely intellectual grounds—is precisely one of the roots of our mischief. Theory, I suppose, was what made the chatter on the scaffolding of the Tower of Babel. It is the substitute for tradition, and it has thriven and multiplied in England since our tradition terminated with Nash. I set out to show how untenable were the 'first principles' to which the teaching and the criticism of architecture usually make appeal. And I sought to indicate how these fallacies arose, and why they are still believed; for in these matters it is not enough to argue against an opinion: the opinion will remain unless the roots of it are exposed.

Moreover, since every error draws part of its vitality from some measure of truth which is embedded in it, I sought in each case to lay bare that element of truth, and to show that these half-truths really derive, not from the theories they have been held to justify, but from that general principle of Humanism of which I speak in the eighth chapter.

To do this, however succinctly, was a long task. But the length of this portion is not of my choosing: it is determined by the number of the fallacies.

后记，1924年

这本书里的这些章节是我十年前写成的。当时因为年轻,还没有完全认识到自己的行为有多么鲁莽,自以为我打算落实到纸面上的这些话对于实践建筑师来说或许能让他们感到有些兴趣,或许对于从事艺术哲学研究的人来说也有一定的参考价值。与此同时,也幻想着这本书对于那些认同本书观点的人来说,不会因为时间的流逝而失去它作为一本书的存在价值。

如果就这个话题继续深入探讨下去,进而超出当初的设想,那么它就超越了我的能力,我肯定会有力不从心的感觉。假如现在我还有这种类似的计划,那一定会是另外一本书,而它的内容也一定会是更加注重技术方面的探讨。

我们所看到的这本书非常幸运,它发行之后曾经引起人们的关注,并且引起广泛的讨论。正因为如此,我想借这个机会以后记的形式作一两点说明,同时对于若干讨论作一点回应。

人们说,我在这本书中采用指责口吻进行的论述多过建设性论述,的确是这样的。但是,如果我在书中探讨之后所得出的结论能够获得大家的正确理解和欣赏,那么,我想,这些所谓的指责将被视为是不可避免的唯一论述途径。

我的基本论点就是:所谓的"理论"正是我们建筑艺术所面临困境的根本原因。我这里所说的"理论",指的是试图通过各种单纯理性分析来寻找出能够据此判断出建筑艺术对与错的体系。在我看来,建筑艺术理论就好比是建造巴比伦通天塔脚手架上工匠们的语言,上帝为了阻止工匠们彼此交换想法,便让人类说不同的语言,因此彼此无法直接沟通,建造巴比伦塔通天塔的工程也就只好作罢。理论替代了传统,而在英国,自从纳什(Nash)终结了我们的建筑传统之后,建筑理论就一直是层出不穷,不胜枚举。我在开始的时候便试图说明,各种所谓的"第一理论和原则"是根本站不住脚的,但是关于建筑艺术的各种说教和评论却总是围绕着这些原则打转。我努力指出这些谬误理论之所以出现的原因,指出为什么到现在还会有人相信它们。关于这些谬误,仅仅进行驳斥还是不够的,必须把它们的根源暴露出来,不然这些错误观念还会继续流传下去。

不仅如此,因为每一种谬误都是从某些片面的内容开始,利用其中固有的一点点正确的理由,才使得人们相信其中貌似正确的道理,最后让谬误的结论大行其道。为了揭露这一点,我针对每一个具体的理论,把其中所包含的正确理由摊开,让它赤裸裸地暴露在世人面前,让人们看到这局部真理的真正源头,不是那些理论所鼓吹的各种思想,而是人文主义思想的最基本的原则,关于这个原则我在本书的第八章里专门作了讲解。

然而为了达到这个目的,不管进行怎样的精简和概括,都肯定是要进行长篇大论的工作。但是,具体的长度并不是我个人的主观意愿能够决定的,它取决于谬误理论的数量到底

And if I have planted no full-blown theory on the ground thus cleared, this follows from the very nature of my contention. The most one can here do is to clear the ground: and then to indicate where the creative instinct lies, and in what it consists; and this I have attempted. But to seek to devise new codes for the operation of that instinct would be once more to intellectualise a faculty which is not in my opinion primarily an intellectual one at all.

If therefore an architect should reply: 'The first part of my problem is one of means and ends—mechanical means to a practical end—and is purely one of reasoning; but the further problem is one of taste, and here I can see for myself, and no mere argument can upset my felt preferences': if he says this I have little to object. Such an attitude was precisely that of the great architects of the past. Only there is this difference: that they were really saturated in their manner of building, and had not been schooled in the fallacies, while our modern teaching is haunted by the ghosts of all these errors, and in default of a live and veritable tradition we are expected to form an encyclopaedic, and therefore necessarily superficial familiarity with 'styles.'

What we feel as 'beauty' in architecture is not a matter for logical demonstration. It is experienced, consciously, as a direct and simple intuition, which has its ground in that subconscious region where our physical memories are stored, and depends partly on these, and partly on the greater ease imparted to certain visual and motor impulses.

But, just as this process goes on (in most cases) below the field of consciousness, and rises to consciousness simply as 'pleasure,' so the training of the creative faculty will not lie so much in the analysis of that process, but rather in rendering it more sensitive. And we can only do this by *habituation* to such works as actually embody the values of Mass, Space, Line and Coherence—the 'humanised' values spoken of in Chapter Eight.

But if so, it matters greatly what our eyes habitually see, apart from our moments of concentrated research. Here, again, our masters had the advantage over us. They lived and moved among buildings where the values of Mass, Space and Line were often coherently displayed: their eyes were habituated to architecture of a relatively uniform intent. Our eyes, even if we know clearly what serious architecture is, have to search for it questioningly in a welter of commercial and municipal monstrosities. It is as though one had to tune a violin in the midst of a railway accident.

Meanwhile the heritage of humanist architecture on which we must depend for any education of architectural sensitiveness is being allowed to disappear.

Epilogue, 1924

有多少。

　　如果说,在这片清理干净的空地上,我并没有打算建立起一个全面开花结果的庞大理论体系,这与我当初写这本书的观念是一致的。我认为,一个人所能够做的就是把目前的混乱状态清理干净,然后指明创造力能够存在于什么地方,看清楚创造力都包含了些什么内容。这些恰恰是我在努力做的。但是,如果试图发明出一套具体实用的操作规则,幻想凭借着这套规则便可以充分发挥出每个人的创造才能,这又是在重复着把人的才能进行理性化的妄想。在我看来,人的才能是根本不可能通过理性系统分析来获得的东西。

　　如果有一位建筑师因此对我说:"我的问题的前一部分就是手段和目的的问题,亦即技术的手段和使用的目的的问题,这部分是单纯的理性分析;后一部分就是审美观念的问题,我个人认为,不可能有什么理性分析会让我改变我已经形成的个人喜好。"他这样讲,我找不出反对的理由。这种观念也正是过去历史上伟大建筑师们的立场观点。我们今天与过去所不同的是:这些伟大的建筑师们完全沉浸在自己时代继承下来的传统建造手段之中,他们没有听说过任何这些所谓理论的谬论,根本不会受到这些东西的影响,而我们今天一直被这些错误理论笼罩着,谬论像魔鬼一样盘旋在我们头上,总是在纠缠着我们。过去的那个活生生、实实在在的传统被一些百科全书式的形式取代了,一切都从五花八门所谓的建筑"样式风格"等肤浅的表面入手。

　　我们在建筑艺术中发现的"美感",不是用逻辑推导出来的东西。它是有意识地感觉到的一种亲身体验的结果,是直接又简单的本能直觉。我们所谓的直觉实际上是以潜意识中储藏着的真实记忆,加上视觉和动作、姿态所传给我们的直接刺激等为基础的。

　　因为这个过程是发生在潜意识层次,是在我们清醒意识所控制的范围以下,因此,我们的意识能够感受到的只有它的"愉快"的结果。所以,创造能力的训练和培养不可能从分析这个感受过程中获得多少帮助,真正有启发、有帮助的工作是培养自己的感受能力。我们能够做的就是*设身处地、身临其境*地体验并熟悉这些艺术作品,看它们是如何体现体量、空间、线条、和谐一致等价值的,亦即前面第八章里论述的那些"人文主义"的价值。

　　即便是这样,最关键的地方还是在于我们的眼睛平时最习惯于注意到什么东西,并不在于在特定的情况下,精神集中在某些特别感兴趣的东西上面。在这一点上,过去传统上的伟大艺术家们要比我们有着更多的优越条件,因为他们生活的环境里,建筑的体量、空间、线条都和谐一致地出现在那里,所到之处,亲眼所见,都是具有这样价值观的建筑。他们的眼睛已经很习惯于这样的环境。而我们的眼睛就不同了,即使我们知道严肃认真的建筑艺术是什么,我们还是不得已要满腹狐疑、小心翼翼地,在数不清的商业和市政丑陋建筑中寻找我们的对象。这就好像是我们必须要在铁路意外事故发生的时刻给一把小提琴调音一样。

　　我们的建筑教育所依赖的人文主义的传统建筑艺术作品却没有受到任何应有的保护,人们可以随意让这些作品消失。我们基督教会的领袖们一觉醒来,突然发现我们的教堂建

Our ecclesiastical authorities have awakened to the discovery that there are still some Wren churches which may, at a financial profit be pulled down; and our Government, which lavishes money on futile academies and finds doles for the maintenance of museums and other cemeteries, is proceeding merrily to reimburse itself by the destruction of Regent Street; the only point of hope lying in the fact that the indignation aroused by this latter undertaking is at any rate markedly greater than when Mr. Norman Shaw's tumid edifice in the Quadrant first revealed how far we had outgrown the humanism of Nash.

Where then, practically speaking, are we led? Simply to the necessity of a more habitual, a more saturated familiarity with the tradition of humanist architecture. There is nothing novel, certainly, in this conclusion. But we shall perhaps have added this: that the tradition no longer will seem to hang on mere prestige and authority, but will be reinforced by a coherent view of the nature of architectural aesthetics. More than this, we shall look at the tradition itself from a slightly different angle. We shall no longer be concerned to square the works of Wren or Vanbrugh or Nash with our nineteenth-century code of fallacies, and they will appear in rather a different light.

Perhaps this is already happening. I noticed in the literature that appeared on the occasion of the recent Wren Bicentenary, that a number of writers were at pains to present Wren in the character of a great baroque architect. This is, I think, the 'new angle': it implies a better understanding both of the nature of baroque, and of the true inspiration of Wren. It is to attend to Wren's language where previous critics had looked at his parts of speech.

Here I would like to add a word on the subject of baroque. I find this book frequently referred to as though its main purpose were the defence of that style. But that is to take the part for the whole, and to confuse my thesis with my illustration. The baroque is in the highest degree interesting, because of its purely psychological approach to the problem of design, its freedom from mechanical and academic 'taboos,' for its use of scale, its search for Movement, its preoccupation with Mass composition and Spatial values. The mastery of these elements shown by the great baroque architects from Vignola to Bernini entitles the style they evolved to a very different estimate from that which was accorded to it by English writers. But my argument goes essentially beyond that particular question, and if I have come back so frequently to the subject of baroque, it is because it furnished a kind of acid test to the views I was considering. Further, if I defend the 'theatricality' of certain baroque buildings, it is in order to prove that there is no *a priori* architectural law to preclude such devices as they displayed. But 'theatricality' may be in place in one setting and out of place in another: on this point the enthusiasm of some

Epilogue, 1924

筑中还有一些伦恩当初所设计的教堂,这些是可以拆除掉的,而且从财务的角度来看也很划算;我们的政府部门总是大把大把地把钞票花在一些无关紧要的所谓学术研究上,花在博物馆和墓地的维护上,却大肆地拆除摄政大街(Regent Street)沿街的古建筑来弥补自己财政上的亏损。拆除沿街古建筑的举动遭到了巨大的反对声,这一点足以让人们感觉到,有那么一线希望还是存在的。实际上,这种反对政府拆毁古建筑的声音要远远超过人们第一次看到诺尔曼·萧(Norman Shaw)先生拿出的华丽夸张的大厦设计方案时所发出的惊叹声。萧先生的设计让我们第一次看到自从纳什的人文主义之后我们已经走出去多远了。

如此说来,那么从可以实际操作的角度来讲,我们目前到底被引导向何处呢?简单地概括来说,就是走向更加认同那些习以为常的东西,更加彻底地熟悉人文主义建筑艺术的传统。可以肯定地讲,这个结论中不会有什么新奇的东西。但是,我们必须要在这里强调一点:这个人文主义建筑艺术的传统再也不像从前那样具有威信和权威,再也不像从前那样令人信服,但是,它还是可以从建筑艺术美学和谐一致的观点中得到强化。不仅如此,我们也应该略微变换一下我们自己看问题的角度来看待这个传统。我们不应该再用十九世纪那些谬误理论来往伦恩、范堡卢、纳什等大师的作品上面生搬硬套;如果做到这一点,那么这些大师的作品就会呈现出不同面貌。

实际上,这种情形已经在发生了。在最近纪念伦恩大师诞辰二百周年所举办的活动中,我注意到有些文章的作者在绞尽脑汁地证明伦恩是一位具有巴洛克建筑艺术特点的伟大建筑师。我认为,这就是一种看问题的全新角度,它说明人们对于巴洛克建筑艺术开始有了新的认识,对于伦恩建筑艺术的根源也有了新的认识。现在人们开始注重伦恩建筑语言本身了,而从前的理论家们只是关注他语言中所使用的词汇的词性。

关于巴洛克艺术我想在这里多讲几句。我发现在我的这本书里我常常提到这个风格的艺术,好像我的这本书的主要目的是要为它辩护一样。但是,这个看法是以偏概全的,是把我用来说明问题所举的例子当作是我的论点了。巴洛克艺术是最有趣味性的一种艺术风格,因为它在对待艺术设计问题的时候充分考虑到人们的心理因素,采用的设计手段是纯粹的心理学方法;在面对设计视觉尺度的问题时,它不受单纯建造技术的约束,也不受学院式教条清规戒律的约束,它追求强烈的动感,一门心思地关注着体量的组合关系和空间的价值。对于这些技巧的掌握已达到炉火纯青地步的那些巴洛克时期的伟大建筑艺术家们,从维尼奥拉到伯尔尼尼,他们的成就绝对应该让我们重新检讨我们英国艺术评论家对巴洛克艺术所作出的各种评价,我们应该重新认识巴洛克艺术风格。但是,我在这本书中讨论的重点显然不在这里,而在于更广泛的话题。我之所以在论述中常常回到巴洛克艺术这个话题上来,那一定是因为它如同检验黄金时所使用的那一滴硝酸一样,可以最简单、直接地检验出黄金的真假,巴洛克最能帮助我说明自己的观点。不仅仅是这样,如果我在讨论巴洛克建筑艺术的时候,捎带着也为它的"戏剧布景式"的建筑效果辩护过几句,那也是因为我想借此说明一点——在建筑艺术领域里,根本没有任何*前提*条件规定我们必须排除巴洛克的这些内容。但是,"戏剧布景式"的建筑效果在一个地方或许合适,绝不能说明它在另外一个

writers who generously declare themselves my disciples has occasionally outrun my own. In any case, too much, it appears to me, has been made of the theatricality of baroque. If rather more attention had been paid to Roman palaces and rather less to Neapolitan churches, the essential gravity of the style would not have been so widely overlooked. There is a considerable affinity between baroque architecture and seventeenth century prose: the conceits, the flourishes, of Donne, for example, may or may not be to our taste—it is an open question; but if they cause us to overlook the underlying rhythms, the spacious period and the weight, we have missed the very essentials of prose. Some critics of baroque lack, it may be, an ear for architecture.

While the contentions of my earlier chapters seem to have met with fairly wide assent—at any rate among the younger generation of architects, the view of architectural design which I have outlined in the eighth chapter has, I think, been sometimes misinterpreted. Thus I do not for a moment contend, as one writer has asserted, that physical memory supplies '*the whole* explanation of the nature and appeal of architecture.' Still less do I suggest that, in practice, 'an architect when faced with the problem of designing a house... proceeds to imagine physical states, and to take them as his theme.' It is obvious that in the sequence of considerations which confront an architect there are a hundred questions of fitness and common sense, which supply the first block and shape of his problem and compose nine-tenths of his difficulties, and there is an undoubted satisfaction in seeing these difficulties lucidly and reasonably resolved. But I have sought to prove that there is in architecture the possibility of a beauty which lies beyond this and cannot be reduced to these terms. Sooner or later the architect in considering his alternative solutions, falls back quite simply upon an aesthetic preference. In attempting to show on what that perference is psychologically founded, I am not suggesting (1) that it forms the starting point of his problem, nor (2) that in practice he must necessarily be conscious of the psychological process which determines his choice. Most often he will experience it as a purely intuitive judgment of beauty into the sources of which he need not explore. Nevertheless I believe it useful to explore them, because the more conscious these processes become the more clearly will our creative effort be guided, and the less likely shall we be, at each juncture of hesitation, to be misled by some false theory into accepting a solution on quite irrelevant grounds.

If my critics have sometimes seemed to have overlooked essential passages, this, I recognise, is the natural penalty I pay for a condensed argument. Yet I was unwilling to expand it, thinking the wood might not be seen for the trees. The

Epilogue, 1924

地方也一定合适；它很可能就变成是格格不入的东西。我说这句话的目的是试图说明，有些文章的作者很热情地自称在追随我的思想观点，但是他们的实际论述已经超出了我的认识范围。不管怎样讲，我个人觉得我们对关于巴洛克建筑艺术中的"戏剧布景式"的建筑效果投入过多的精力了。如果我们能够把精力和注意力更多地投入到罗马地区的宫殿建筑的研究上，减少一些对于那不勒斯地区的教堂建筑的关注，那么，巴洛克建筑艺术的本质或许就不会像现在这样被人们普遍地忽略了。巴洛克建筑艺术与十七世纪散文艺术之间有着非比寻常的血缘姻亲关系：都是那样自以为是的狂妄，都是那样华丽地堆砌辞藻，如约翰·多恩（Donne）的文章。这些东西或许不对我们今天的口味；这也是一个见仁见智的问题，不会有一个统一的答案。但是，如果我们不能从这些东西身上注意到其中所包含的韵律节奏、空间与重量的关系的话，那么我们已经错过了散文中最基本的形式内容。或许可以说，有些批评巴洛克建筑艺术的理论家们就是缺乏一种倾听其中韵律节奏的能力。

  本书中前面的那些章节的一些观点和主张似乎受到不少人的认同，而且从任何角度看，年轻建筑师居多。但是在第八章中所讨论的关于建筑设计的一些观点，我认为有不少人误解了我的意思。有一位理论家认定我的观点是"身体力行的亲身经历构成了我们对建筑艺术的喜好和建筑性质的*全部*解释"。关于这一点我可以毫不犹豫地表示我的不满。我更不会对于建筑实践提出下面这样的建议："当一个建筑师面对着设计一座住宅的任务时……他需要从假想自己身体的行为状态入手，并且把自己的身体状态当作自己的设计主题。"很显然，建筑师在设计的时候，他会同时思考很多问题，验证自己的常识性判断是否正确，这些内容构成了他的设计需要解决的主要问题，帮助他确定了主要的体块和大致的形状，而这些问题一个接一个地得到彻底的解决，这时无疑会让建筑师产生一种满足感。但是，我想证明的是，在建筑艺术中，还有一些东西不是仅仅依赖于这些具体问题的解决就可以获得的，那是一种美，它不能把问题简化为一、二、三，然后只要是满足了这些条件就可以从中获得美感。不是这样的。这种美观，是建筑师在考虑过所有的解决问题的方法之后，完全出自于对美观的考虑而作的最后选择。我想证明的是，这种出于美学考虑的选择是一种心理学过程，而我在证明影响他做出自己决定的心理过程都包括哪些东西的时候，我绝对没有认为这些东西代表了：第一，这些东西构成了他需要面对解决的问题的起点，第二，从实际工作角度来看，他必须要清醒地了解决定自己选择的心理过程。在绝大多数情况下，建筑师在对美观作出判断的时候，自己所认识到的过程仅仅是他们的直觉而已，对于这种直觉，他们根本不需要去研究发生的过程。但是无论如何，我个人认为，对于这个心理过程进行一番了解还是有帮助的。因为我们越是清醒地了解这个心理过程，我们就会越是有意识地引导自己的创造力走向；而且在遇到岔路口的时候，或者在思想犹豫的时候，我们就越是不会被那些错误的谬论以毫不相关的理由把我们引导向歧途上去。

  对我的观点提出批评的理论家们，有时给人的感觉是他们忽略了我文章中很多的关键内容。我清楚这一点。我想这是我把自己的论证过程尽可能进行压缩所必须付出的代价。然而，我还是不想对这些文字进行扩展，因为我担心扩展的结果会让我们只看到树木而不见

argument is close; but the book is short. And I myself would prefer, if need be, to read some things twice than to have read everything twice repeated.

The other day I was looking through the report of a discussion on this volume at the Royal Institute of British Architects. Among several too generous speakers was one, of whom I cannot think without gratitude and remorse. 'I have read (he averred) Mr. Scott's book fourteen times; and it is rather a tiresome book.'

Fourteen is too much. But to those who have kindly urged me to add a second volume, while at the same time raising objections which are explicitly answered in the first, I can only offer the example of my patient friend.

VILLA MEDICI,
FLORENCE, March 9, 1924.

Epilogue, 1924

了森林。这里的论述是被压缩了,这本书也因此变薄了。如果有必要,我倒是情愿把其中的某些章节读它两遍,也不愿意看到一本书里的内容被作者自己重复叙述了两遍。

有一天我在阅读英国皇家建筑师协会的一份报告时发现,其中有一篇关于我这本书的讨论。在众多慷慨热情的读者发言中,有一位的发言让我觉得既感动又惭愧。"斯科特先生的这本书我前后总共读了十四遍。这本书读起来很辛苦。"

十四遍的确是有些太多了。我对于这本书里所认真讨论过的问题都力争给出详细的解答。但是,对于某些固执地坚持自己的不同意见,同时却又热心地建议我尽快出版本书续集的读者,我只能提议他们,请向这位耐心阅读十四遍的朋友学习一下吧。

于佛罗伦萨的美第奇别墅
一九二四年三月九日

译者的话

斯科特的《人文主义建筑艺术》一书是100年前完成的。自从出版发行以来就一直受到人们的关注,直到今天从未间断过。几位在建筑界有着巨大影响力的人物,例如菲利普·约翰逊、文森·斯卡利、柯林·圣约翰·威尔逊、查尔斯王储等,在自己的文章和讲演中常常引用这本书的内容和主要论点。到了20世纪80年代末,此书开始有了中文译本,并于2012年又重新更新再版。那么,我为什么还要着手推出这一个英汉对照版本呢?

在回答这个问题之前,我们先来了解这本书本身,然后再了解一下目前的中文版。在了解了这两个方面的情况之后,读者也就自然会得出自己的答案了。

斯科特其人与他的《人文主义建筑艺术》一书

斯科特的出生日期有两个说法,一个说法是他出生于1883年,另一个说法是他生于1884年6月11日。这个简单的事实说明,斯科特的生平和职业生涯其实并没有什么杰出的成就而让人们去特别地关注他。斯科特生长在伦敦,高中之后就读于牛津大学。大学期间因为诗歌和作文获得过学校的奖励,这说明他的文学水准在牛津同学中是非常优秀的。这一点可以从他的《人文主义建筑艺术》中略见一斑。除了严谨的论述之外,这本书也是非常优秀的英文散文。他的一篇获奖作文题目是《英国建筑艺术中表现出的国民性格》,可见,他对建筑艺术的关注很早就开始了,但是,他从来都是作为一名建筑艺术的局外人、一名旁观者,来冷静、客观地看待建筑艺术。他父亲自己经营着一家生产地板建筑材料的公司,他叔叔主持编辑着曼彻斯特的一份杂志。如果他有志于建筑艺术理论的写作,他有很好的机会。但是,他显然更感兴趣的是到异国旅行,所以他放弃了在建筑艺术理论和评论方面的发展机会,大学毕业以后,于1907年去了意大利的佛罗伦萨。

斯科特到了佛罗伦萨之后,结识了一位名叫平森特的建筑师,他俩一起做过一些住

译者的话

宅的设计和改造工作,并且前前后后一起分租公寓有四、五年的样子。由于生活在文艺复兴时期的建筑艺术环境之中,斯科特自然地开始了对身边那些对古典建筑艺术风格的关注和思考,并打算从中总结出一些基本原则,作为设计师的参考。也正是在他到达佛罗伦萨的那一年,他结识了著名的艺术史学家、古董鉴定专家贝伦森,这样斯科特随即就不再和平森特做建筑设计,而是给贝伦森当起了秘书和图书资料管理员。在贝伦森夫妇的鼓励下,斯科特开始构思和写作自己的这部《人文主义建筑艺术》,但在完成的时候,斯科特把这本书题献给了平森特,而不是鼓励自己的贝伦森夫妇。

斯科特在这部书出版之后,只写过一些诗歌文学和一部名人传记之类的文章,并没有任何其他的关于建筑艺术理论的文章,唯一不间断的就是到处旅行。他于1929年5月从意大利回到英国,却不幸受寒并发展成为肺炎。8月4日到达美国纽约,然而三天后肺炎发作住进医院,8月14日在纽约病故。

这本《人文主义建筑艺术》可以看作是斯科特对当时以拉斯金为旗手的艺术理论家们所鼓吹的各种说法的一次集中回击,而且是借助于文艺复兴时期的建筑艺术来阐述自己的见解。但是,这样的定位并不能削弱它的价值,因为这本书的内容所阐述的问题,绝不仅仅局限于风格的争论或者个人之间的恩怨,而是从建筑艺术的最基本、最本质的问题入手,以真实生活中人的使用、感受和体验作为依归,而不是从建筑艺术之外的道德、技术、文学、进化等似是而非地进行类比。这使得这本书在百年后的今天,在面对现代建筑发展中出现的崇尚各种奇奇怪怪的建筑风气的时候,仍然具有生命力,激励着我们从这些最基本的问题去思考,重新认识建筑艺术发展过程中出现的各类思想谬误,重新认识并掌握建筑艺术中以人为本的基本原则。

这本书并没有从一般的学术定义开始,对人文主义建筑艺术进行抽象的定义,然后再逐渐展开自己的理论论述。相反,斯科特针对当时流行的建筑艺术理论中的种种荒谬言论和观念展开了揭露和批判。斯科特的论述有理有据,不慌不忙,把各种谬论批驳得体无完肤。同时,他在批判中建立起自己的观点:这些流行的从道德理论、浪漫主义、科学技术、进化论等理论引申出来的建筑理论,其实都是在外围打转,而且全部都是似是而非的谬论。建筑艺术不外乎空间、体量、线条三种视觉元素通过光和影的组合关系呈现在我们面前,其目的就是满足人们的需要:实用、坚固、愉悦。

关于这本书的中文译文

这本书已经有了一个中文版而且在发行20年后重新再版,又是名家推荐、德高望重的前辈执笔完成的。那么,我为什么还要出译本呢?其实,我的想法和初衷很简单,我并不完全认同已有的中文版本。理由来自两个方面：一是在有些地方，我认为我的翻译更接近原文的本意；二是在有些地方，我的译文更兼顾了汉语的语言习惯和表达方式。

因此，我们决定把英文原文附上，因为它一来可以让读者了解到英文原文的优雅，二来也可以让我们的翻译接受读者的检验和监督。关于从英文到中文的翻译，我非常认同赵梦蕤先生的见解："我必须要说的是，尽管我想忠实于原文，但也得考虑中文的流畅。"因为英文中的时态和各种分词、从句等内在结构和逻辑关系是中文中缺少的，因此，很多句子的中文翻译就必须根据中文习惯重新组织句子结构，使用增减副词连词等手段，把原来的涵义表达清楚。正因为如此，我允许自己在忠实于原文的前提下，尽可能用流畅的中文加以陈述，而不拘泥于字面上的对应关系。

吴家琦
写于朝阳常营
2015 年 3 月 18 日